This book is intended for pupils entering their first external examinations in biological science. It is designed to suit the standard of the present Ordinary Level examination; the material included is relevant to the syllabuses in biology of most Examining Boards. In many places the depth and breadth of treatment is greater, so that the book could be considered as a useful aid to the sixth form student, certainly so if study is being made for an examination of somewhat less than Advanced Level standard. Much of the information in the book is also relevant to the Certificate of Secondary Education though no concessions have been made to the lower standard of this examination.

The book is designed primarily for reference. It is not intended to be read page-by-page from beginning to end. Ideally it should act to supplement, either in or out of the classroom, what the student has been taught. A textbook of this sort should not replace the teacher. Nevertheless it is hoped that the book is sufficient in itself for those students, such as many private examination candidates, who are unfortunate enough to lack the guidance of a teacher.

As far as possible the book has been assembled with the minimum of text. Much use has been made of annotated diagrams, either in sequence or by themselves, to convey information in a relatively short space. Such compression, together with a system of cross-references, should enable the reader to find information more easily. Some topics do not lend themselves to this treatment and have necessarily been considered at greater length (e.g. parts of Section III).

The book is divided into three sections. The order of these sections has no special significance since there is no single way in which biology unfolds in a logical sequence. Section I attempts to do no more or no less than its title suggests—to give the student some appreciation of the range of plants and animals. Many of the larger groups of living organisms are included. It is not the purpose of this section to show relationships between one organism and another: nor is there any attempt, with such a restricted range of examples, to demonstrate Evolution. Section II deals with the functioning of living organisms, particularly flowering plants and mammals. Anatomy has been described only as far as it is necessary for the understanding of function. Section III attempts to describe and discuss the environment of living organisms. Throughout the book, emphasis has been placed on the relevance of biology to everyday life. Little reference has been made to molecular biology. The authors feel that it is not crucial to an understanding of biology at this level. Also it is remote from experience and not easily treated experimentally.

Accounts of experiments in this book are included wherever they are relevant and wherever the experiments are straightforward enough to be performed by the students themselves. All these experiments should be attempted, although normally expected results have been given. The aim is to show the experimenter that positive results can easily be obtained and also to assist the examination candidate working by himself.

Biology, however, is not only the study of an experimental science. It is equally important as a source of information. Principles and informed opinion about the biological world cannot be arrived at without a sound factual basis. A major aim of this book is to provide this basis.

NOTE TO THE TEACHER

The success of *Life: Form and Function* has prompted us to produce a pack of twenty-four full-colour slides, specially selected to illuminate further the text. These slides are available in a transparent plastic wallet and accompanied by teaching notes written by Dr Burrow and Mr Brewer.

The pack contains photographs of:

Amoeba	Penicillium	Blood vessels
Cysticercus	Moss	Kidney
Trachea	Fern	Root tip
Larva	Bacteria	Buttercup root
Pupa	Bone	Sunflower stem
Adult gnat	Muscle	Leaf V.S.
Spirogyra	Skin	Stoma
Volvox	Blood	Pollen

Full details available from
Sales Dept. Macmillan Education,
Houndmills, Basingstoke, Hampshire

CONTENTS

INTRODUCTION

The term *biology* is derived from the Greek—*bios*, life; *logos*, learning or knowledge. It means the study of life. As a field of study, biology has a long history. Originally it was mainly concerned with animal and plant husbandry, with medicine (a knowledge of herbs and of human anatomy being of great importance), and with natural history reflected in a love of nature for its own sake.

The most eminent early biologist was probably **Aristotle,** who lived in the fourth century BC. The works of **Galen** (c. 129-200 AD) served as the basis for medical practice until the Renaissance. Modern biology began to develop, with other sciences, in the seventeenth century; **William Harvey** (see p. 152) who demonstrated the circulation of blood, is perhaps the first important modern biologist. Many of the fundamental theories about life were formulated in the nineteenth century—for example, **Darwin** (evolution—see p. 246), **Mendel** (inheritance—see p. 236), and **Pasteur** (micro-organisms and disease).

With increase in depth and scope of biological knowledge, the science has separated into a number of clearly-defined but related branches. Some of these are:

Anatomy—the study of structure

Physiology—the study of function

Embryology—the study of development

Ecology—the study of the interactions of living organisms in their environments

Genetics—the study of inheritance

Each of these can be studied at different levels. Thus genetics can be a study of inheritance in individuals, or in whole populations.

Biology serves as a basis for a whole range of separate specialized studies. **Psychology**, the study of behaviour in animals and in Man, has its roots in biology. Similarly, **sociology** (the study of human beings living in communities), **agriculture, medicine, pathology** (the study of diseases), **public health** and **hygiene, forestry, food technology** and **nutritional science** are all forms of biological science.

Students often ask why biologists use so many long words. This question has a simple answer. Everyday language is appropriate for everyday life; it refers to everyday experience. It is not adequate for a subject which extends beyond everyday experience, nor can it meet the needs of a field of study which, like all science, is international.

Biology has always been an important bridge subject between subjects such as economics and geography and the exact sciences of physics and chemistry. With the development of sophisticated experimental methods, biology is becoming increasingly more involved with the exact sciences. An example of this is the influence of control engineers on biologists: living organisms, either as individuals or as populations, are now regarded as self-regulating automatically-controlled systems (see pp. 90, 129, 162, 168 and 173). Biologists are becoming aware that the environment itself is a self-regulating system whose balance can easily be disturbed (see p. 261).

Section I

THE VARIETY OF LIFE

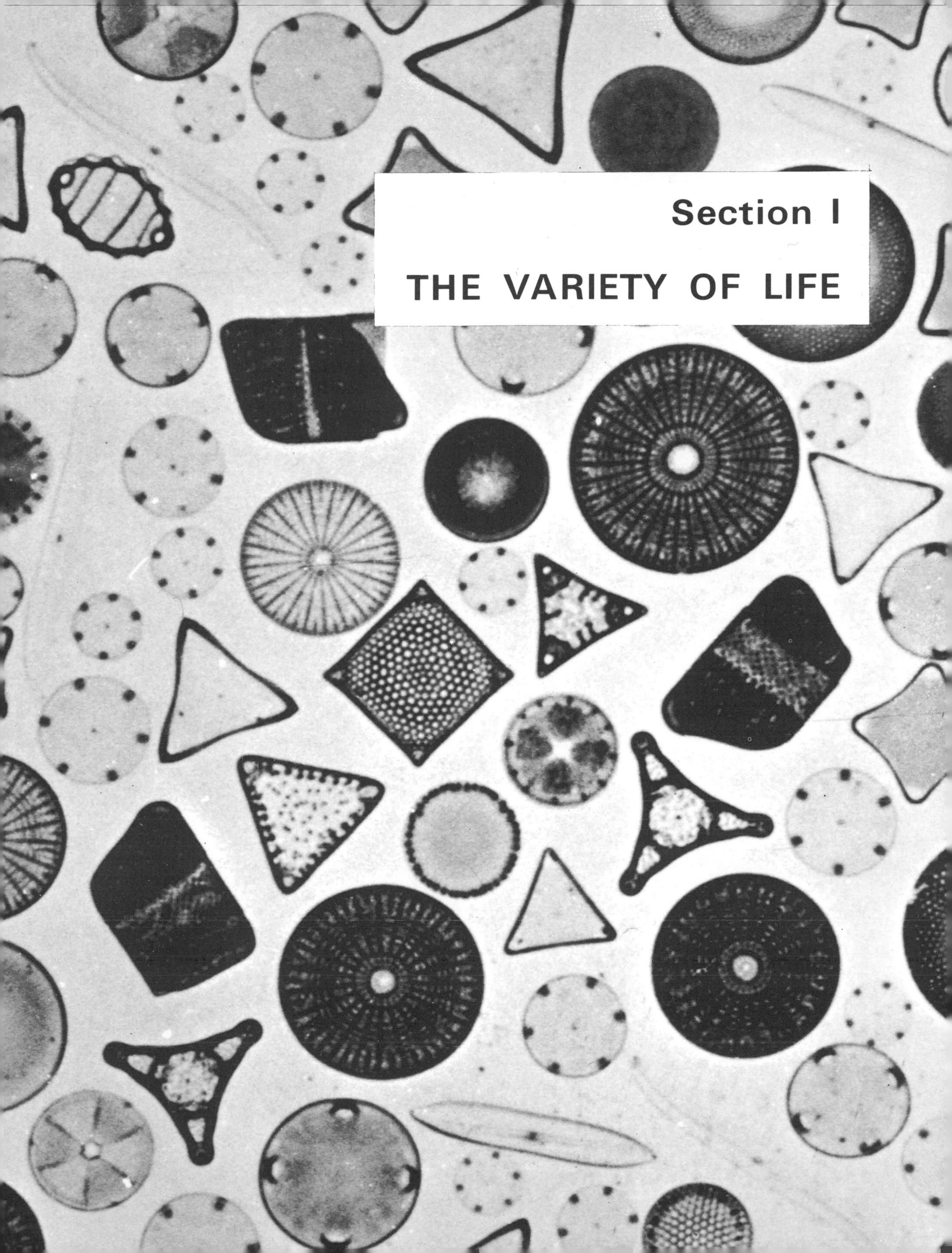

1 CLASSIFICATION

Since classical times biologists have been interested in the task of sorting an ever increasing variety of living things—increasing, that is, in terms of the number of known and described species. There are two main reasons for this. The first is that when we are dealing with collections of different things they become much easier to handle, or refer to, if they are arranged in a rational system. This is the case whatever we happen to be dealing with, whether it is words in a dictionary, books in a library or stamps in a collection. It is also true of living things. There is another good reason for classifying animals and plants. If our system of organizing and sorting is a suitable one it may well shed light on the origins and connections of its subjects.

All classification is based on a knowledge of structure supplemented by information about development and life history. Structure alone does not always provide enough information. Barnacles, for example, were at one time thought to be shell-fish, like oysters and scallops. It was only as a result of following their life history through that they were seen to be rather specialized members of the group to which crabs and shrimps belong.

There are no clear-cut rules to be followed in the classification of living things. This is because the differences between organisms which are useful in classifying them are not measurable. This does not mean that there are no measurable differences but simply that they are not important in classification. No purpose would be served, for example, by classifying animals according to size. Such rules as a biologist applies in sorting species will depend on what it is that he is sorting. Obviously when he is dealing with mammals he is concerned with entirely different properties from those which are important when he is dealing with flowering plants. Furthermore, since a system of order is a man-made thing, there will sometimes be organisms which refuse to fit neatly into any category. There are, for example, some organisms which are regarded as animals by some biologists and as plants by others. There is clearly no reason why an organism must fit into a particular group just because the biologist likes it that way. Also it is not possible to give exact meanings to the terms he uses. Nevertheless we

need at least a rough guide as to what a few of his terms mean.

Species

This term is used to refer to a group of relatively few organisms which share a great many characteristics. It is a group in which the different individuals are, in principle, able to interbreed and produce fertile offspring, which, in their turn, are also able to interbreed with other members of the group. That this is a loose definition the following two examples will show. All dogs are members of the same species. St Bernards and Pekinese do not interbreed because they are so different in size. Runner bean plants do not interbreed because the flowers are normally self-pollinated. There are many other instances which do not fit this definition of a species but in general we may regard the statement above as a useful working definition.

Phylum

This is a term used by animal biologists and now becoming fashionable among botanists. At this level we cannot use the ability to interbreed as a yardstick and we simply define a phylum as a group made up of a very large number of animals or plants sharing a few basic characteristics. The Arthropoda may be taken as an example. It contains all animals with an external skeleton and jointed limbs.

There are of course many other terms. Similar species are collected together into **genera** (sing. genus). Similar genera form **families**. Families are organized into **orders**, orders into **classes** and classes into **phyla**. Species, in their turn, may be divided into subspecies, races or varieties. These various categories are to some extent provisional and to some extent artificial.

The naming of animals and plants

The modern system of naming living organisms is a **binomial (two name) system**. It was introduced by the Swedish botanist **Linnaeus** (1707-1778). The first name is the name of the genus to which the organism belongs, and has a capital letter. The second name has no capital and is the name of the species. The common frog, for example, is called *Rana temporaria*. The edible frog, which is placed in the same genus, is *Rana esculens*. The scientific names used to refer to species are internationally agreed. *Musca domestica* means the common housefly. Many names are derived from the common names used by the Greeks and Romans, for example *Canis* (dog). Many more species have been described and named than were known to the ancients and so names have had to be invented, and Latinized by adding bits or by changing the spelling slightly. The names of famous scientists are sometimes used. The magnolia tree belongs to the genus *Magnolia*, named after a French botanist, Pierre Magnol. Place names are often used. *Trypanosoma gambiense,* the organism which causes sleeping sickness, takes its specific name from Gambia in West Africa.

The meaning of variety

It was not until the nineteenth century that enough was known about animal and plant species to provide a sound basis for speculation about their origins and the meaning of the differences, and resemblances, they show. In 1858 a sound, reasoned explanation was put forward in a famous paper by **Charles Darwin** and **A. R. Wallace** on the origin of species (see p. 246). Much of the evidence they produced was based on classification. The basic assumptions they made can be illustrated by reference to man.

If we begin by considering a pair of identical twins, that is, twins derived from a single fertilized egg, we know that it is very difficult to tell them apart, unless we know them extremely well. Two ordinary brothers or sisters, close together in age, are easier to distinguish, although they may be so alike that the family relationships would be assumed at once by someone meeting them for the first time. We cannot *measure* the degree of resemblance, or difference, although we can recognize it in practice.

When we consider first cousins, general family resemblances may be evident, but less so than with brothers and sisters. The further back the common ancestor, the more different from each other any two related individuals are likely to be.

This idea of relationship can be extended to include all men, wherever they come from, and the concept of an extended family tree can be built up, the relationships between its branches being measured in terms of the degree of difference, or resemblance. We know that all human beings are members of a single species because they can, and sometimes do interbreed. That different races have not done so in the past is a consequence of their being separated by great distances. That they do not interbreed much today is probably due to differences of habit and culture. It is important to stress that there are no sharp differences between human varieties. All racial groups gradually shade into one another at their geographical boundaries. They are in fact continuously varying members of the same family tree.

The idea of family relationships as applied to members of a single species can be extended to include all living things and this is the basic unifying idea behind the classification of living organisms. Man resembles an ape more closely than he does a horse, but all three have more in common with each other than any one of them has with a reptile. With such a much more extensive family tree the time scale is very much greater, extending over hundreds of millions of years. Nevertheless, the only meaning we can attach to the patterns of variety among living things is to assume that all living plants and animals are descended from common ancestors. Although early attempts at sorting organisms were based on structure alone, all modern work is based on the idea that it is the expression of very real relationship by descent. It necessarily follows that if organisms differ, although related, they must have undergone change through descent, and this is the basic biological concept that we call Evolution.

2 PROTOZOA

These are microscopic animals, varying in size from less than ·005 mm to those which are just visible to the unaided eye. Each animal consists of one cell only; all the processes of life take place in this single cell.

Protozoa are found in the soil and in freshwater; many are marine, and some live as parasites in the cells or tissue fluids of other animals.

Amoeba

The large species *Amoeba proteus* is usually studied. It may be found in Britain, in slow-moving freshwater, especially on mud at the bottom of ponds.

Appearance

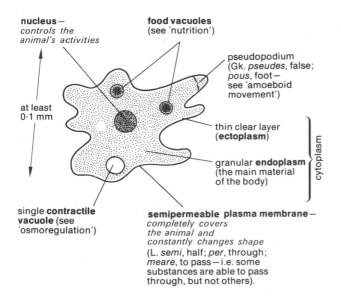

nucleus —
controls the animal's activities

food vacuoles
(see 'nutrition')

pseudopodium
(Gk. *pseudes*, false; *pous*, foot — see 'amoeboid movement')

at least 0·1 mm

thin clear layer (**ectoplasm**)

granular **endoplasm** (the main material of the body)

cytoplasm

single **contractile vacuole** (see 'osmoregulation')

semipermeable plasma membrane —
completely covers the animal and constantly changes shape
(L. *semi*, half; *per*, through; *meare*, to pass — i.e. some substances are able to pass through, but not others).

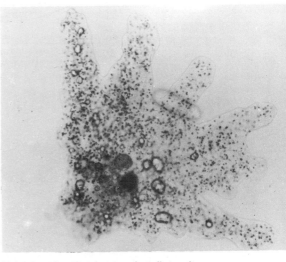

Living Amoeba, *showing pseudopodia, nucleus, cytoplasm and various vacuoles*

SECTIONAL VIEW

nucleus

ectoplasm

endoplasm

plasma membrane

food vacuole

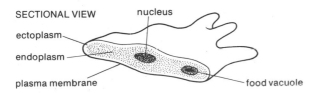

Living Amoeba, *with contractile vacuole enlarged and several pseudopodia extended*

Amoeboid movement

The protoplasm of the endoplasm exists in two different states. There is a region of quite firm jelly (**plasmagel**) just inside the ectoplasm, and internal to this an extensive more fluid region (**plasmasol**). The two states can be changed into each other and this provides the means of movement:

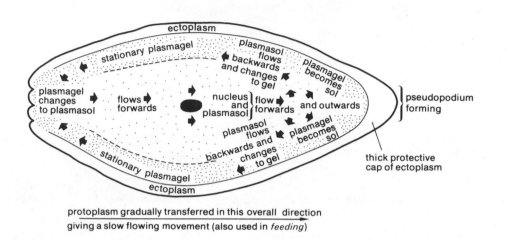

protoplasm gradually transferred in this overall direction
giving a slow flowing movement (also used in *feeding*)

Nutrition

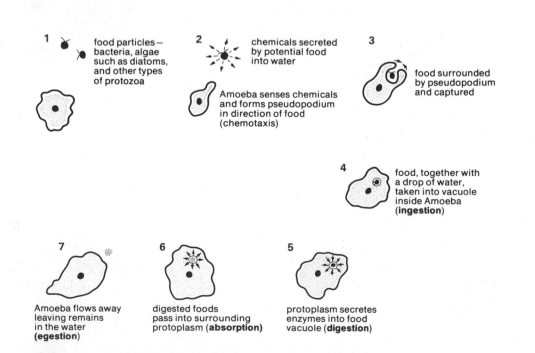

1 food particles — bacteria, algae such as diatoms, and other types of protozoa

2 chemicals secreted by potential food into water
Amoeba senses chemicals and forms pseudopodium in direction of food (chemotaxis)

3 food surrounded by pseudopodium and captured

4 food, together with a drop of water, taken into vacuole inside Amoeba (**ingestion**)

7 Amoeba flows away leaving remains in the water (**egestion**)

6 digested foods pass into surrounding protoplasm (**absorption**)

5 protoplasm secretes enzymes into food vacuole (**digestion**)

Respiration

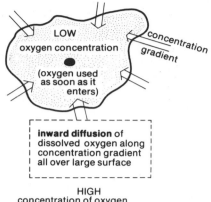

LOW
oxygen concentration

(oxygen used
as soon as it
enters)

concentration gradient

┌ ─ ─ ─ ─ ─ ─ ─ ─ ─ ─ ─ ┐
inward diffusion of
dissolved oxygen along
concentration gradient
all over large surface
└ ─ ─ ─ ─ ─ ─ ─ ─ ─ ─ ─ ┘

HIGH
concentration of oxygen
dissolved in surrounding water

Excretion

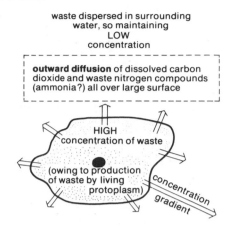

waste dispersed in surrounding
water, so maintaining
LOW
concentration

┌ ─ ─ ─ ─ ─ ─ ─ ─ ─ ─ ─ ─ ─ ─ ─ ─ ┐
outward diffusion of dissolved carbon
dioxide and waste nitrogen compounds
(ammonia?) all over large surface
└ ─ ─ ─ ─ ─ ─ ─ ─ ─ ─ ─ ─ ─ ─ ─ ─ ┘

HIGH
concentration of waste

(owing to production
of waste by living
protoplasm)

concentration gradient

Growth and reproduction

Absorbed food is assimilated to increase the amount of protoplasm up to a **limiting size**, i.e. when the body volume becomes too large in relation to the surface area. Then **asexual repro-** **duction** (meaning that only one individual is involved) takes place; this can occur every three to four days in favourable conditions. No method of sexual reproduction is known.

1

Amoeba stops
moving and
rounds off

2

nucleus becomes
shaped like a
dumb-bell

3

nucleus divides
into two portions;
cytoplasm begins
division

4 final separation of cytoplasm
gives two half-sized cells
(called **daughters**) which resume
moving and feeding

this splitting of one cell into
two cells is **binary fission**

Dormancy and dispersal

Some species of *Amoeba* may **encyst**:

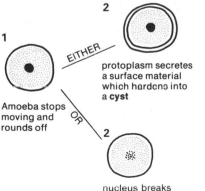

1

Amoeba stops
moving and
rounds off

EITHER

2
protoplasm secretes
a surface material
which hardens into
a **cyst**

OR

2
nucleus breaks
down

3 cyst resistant to
drought and **extreme**
temperatures; may
be dispersed by
wind and in mud
on feet of animals

3
cytoplasm and nucleus
form separate portions,
each covered by
a resistant cyst

4
if carried to a
more favourable
situation, cyst
breaks open and
Amoeba emerges

4
parent Amoeba
disintegrates and
cysts dispersed;
each cyst breaks
open in favourable
situations

Osmoregulation

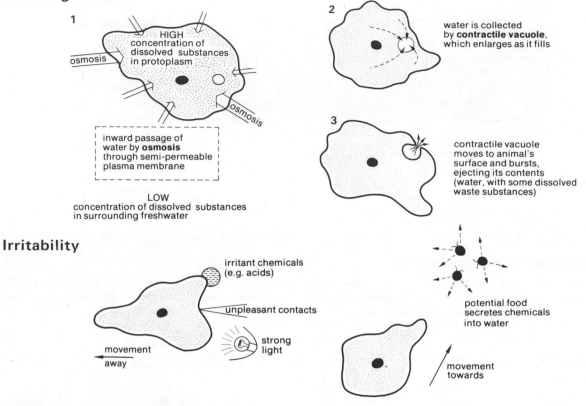

1

osmosis

HIGH
concentration of
dissolved substances
in protoplasm

osmosis

inward passage of
water by **osmosis**
through semi-permeable
plasma membrane

LOW
concentration of dissolved substances
in surrounding freshwater

2

water is collected
by **contractile vacuole**,
which enlarges as it fills

3

contractile vacuole
moves to animal's
surface and bursts,
ejecting its contents
(water, with some dissolved
waste substances)

Irritability

irritant chemicals
(e.g. acids)

unpleasant contacts

strong
light

movement
away

potential food
secretes chemicals
into water

movement
towards

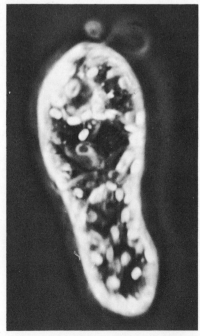

Living Euglena, *showing beating of the
flagellum*

Other protozoa

Some closely related amoebas live in the sea or in
soil. One species lives as a parasite in the intestine
of man and causes dysentery; another is found on
the surface of the teeth and gums, and may be
harmless.

Several protozoa show plant-like characteristics,
e.g. *Euglena*, which lives in ditches and stagnant
water:

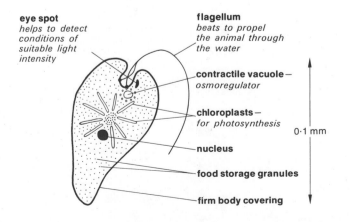

eye spot
*helps to detect
conditions of
suitable light
intensity*

flagellum
*beats to propel
the animal through
the water*

contractile vacuole —
osmoregulator

chloroplasts —
for photosynthesis

nucleus

food storage granules

firm body covering

0.1 mm

Protozoa cause serious diseases in man and his domestic animals; they include the blood parasites which cause **sleeping-sickness** (carried from man to man by the blood-sucking tsetse fly) and **malaria** (carried by the mosquito).

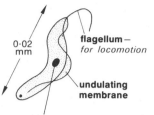

0·02 mm

flagellum —
for locomotion

undulating membrane

nucleus

0·005 mm

nucleus

Trypanosoma
causes sleeping sickness
(found in plasma)

Malarial parasite
(inside red blood cell)

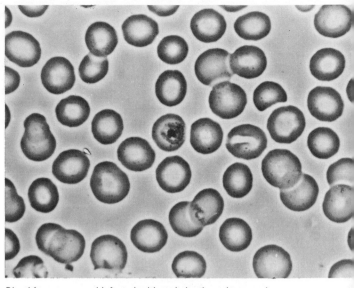

Blood from a mammal infected with malaria; the red corpuscle near the centre of the photograph is almost filled with the parasite which causes the disease

The following two protozoans are called **ciliates**; they can be found in freshwater:

Paramecium
(a free-swimming ciliate)

Vorticella
(a fixed ciliate attached to water plants, etc.)

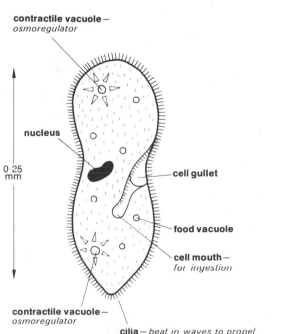

contractile vacuole —
osmoregulator

nucleus

0·25 mm

cell gullet

food vacuole

cell mouth —
for ingestion

contractile vacuole —
osmoregulator

cilia — *beat in waves to propel animal round and carry food such as bacteria into cell gullet*

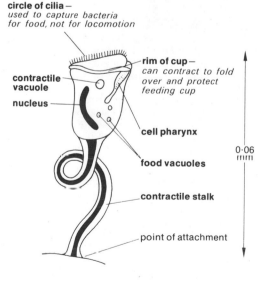

circle of cilia —
used to capture bacteria for food, not for locomotion

rim of cup —
can contract to fold over and protect feeding cup

contractile vacuole

nucleus

cell pharynx

food vacuoles

contractile stalk

point of attachment

0·06 mm

1 Examine fresh pond-water microscopically for different types of protozoans.

2 Investigate the movement of *Amoeba*, and of other types of protozoans under the microscope —the movements of *Paramecium* etc. can be slowed down by adding a drop of 10 per cent methyl cellulose solution.

3 Examine the feeding currents of *Paramecium* in a dilute suspension of yeast.

4 Read further about other protozoa, particularly about those that cause disease.

3 ALGAE

These are plants, usually small, some microscopic. All feed by photosynthesis, though many contain pigments other than chlorophyll, e.g. brown pigments in certain seaweeds. (L. *alga,* seaweed).

Spirogyra

A **filamentous green alga** found in freshwater, floating on the surface of ponds, canals and other stagnant water.

Appearance

The plant is a small slimy green unbranched thread called a **filament**; each filament is composed of similar cylindrical cells joined end-to-end.

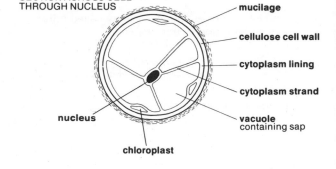

SECTION ACROSS CELL THROUGH NUCLEUS

mucilage

cellulose cell wall

cytoplasm lining

cytoplasm strand

vacuole containing sap

nucleus

chloroplast

DETAIL OF ONE CELL

mucilage covering secreted by cell walls— *slimy, to prevent filaments being eaten by animals and to prevent organisms (epiphytes) fixing themselves to and growing on the cell surface*

cytoplasm lining— *semipermeable, allowing only certain materials to pass through*

cellulose wall-partition between adjacent cells

nucleus embedded in cytoplasm

vacuole— *large space filled with fluid containing dissolved nutrients*

chloroplast— flat green ribbon of chlorophyll spiralling through the cytoplasm lining, *used for photosynthesis*

cellulose cell wall— *rigid, to protect and keep cell in shape*

cytoplasm strand— *holds nucleus in position in vacuole*

pyrenoid— swelling, *centre of starch formation and storage*

Nutrition

The filaments floating at the surface obtain the best possible supply of sunlight for **photosynthesis** (see p. 95).

1   Carbon dioxide gas from the air above the water dissolves in the water surrounding *Spirogyra* and in the mucilage, and diffuses into the cells; some carbon dioxide is obtained from respiration. Water reaches the chloroplast from the surroundings and from the sap in the vacuole.

2   In the presence of sunlight and chlorophyll, carbon dioxide and water are combined to form glucose in the chloroplast; oxygen is also produced, which is either used for respiration, or diffuses out of the cells, eventually to be released as a gas into the atmosphere.

3   The glucose may be changed to starch and stored in the pyrenoids.

4   Mineral salts dissolved in the surrounding water pass into the cells and, together with those salts already present in the sap, may combine with the glucose to form more complex substances, such as proteins.

Respiration and excretion

Nitrogenous waste is not produced under normal conditions.

In sunlight—carbon dioxide produced by respiration is used for photosynthesis; some of the oxygen produced by photosynthesis is used for respiration.

In darkness—oxygen diffuses in from the surroundings and carbon dioxide diffuses out.

Movement and irritability

No independent powers of locomotion; wind blowing over the surface and slight water currents may disturb the filaments.

Spirogyra apparently does not respond to external stimuli such as light, gravity.

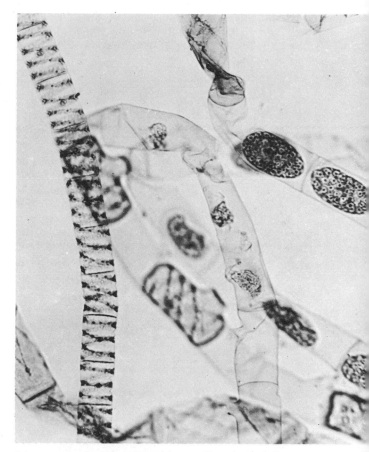

Spirogyra, showing chloroplasts and pyrenoids and reproducing filaments with conjugation tubes, empty cells or zygospores

Growth and reproduction

Food materials produced as a result of photosynthesis are used to enlarge the cells and their contents.

Reproduction is both asexual (involving one filament only) and sexual (involving two different filaments).

(a)   Asexual reproduction

Binary fission—occurs frequently in favourable conditions:

1   nucleus becomes dumb-bell shaped

2   nucleus splits — new cellulose partition grows between the halves, dividing other cell structures, e.g. chloroplasts

3   two half-sized daughter cells formed, which begin feeding and growing

Fragmentation—probably occurs most frequently when the filaments are disturbed; filaments which may have grown long as a result of repeated binary fissions become divided into shorter fragments:

1

cellulose partitions between cells begin to swell

2

swelling increases, forcing adjacent cells apart

3

original filaments separated into two filaments or sometimes into separate single cells

(b) Sexual reproduction

Occurs in Britain usually in late winter when the plants are mature and the filaments often over-crowded. The process is sometimes called **con-jugation**:

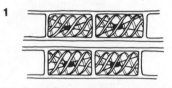

1

two (sometimes more) filaments come together side by side

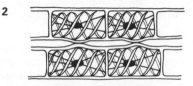

2

cellulose walls of each filament grow outwards where opposite cells are in contact, pushing the filaments apart

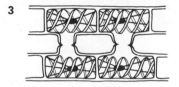

3

outgrowths become larger— cellulose breaks down where outgrowths meet, forming tubes

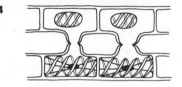

4

opposite filaments completely united by **conjugation tubes**—cell contents of one filament become compact and rounded, producing greenish masses (**gametes**) in which chloroplasts etc. cannot be identified

5

each gamete flows towards and begins to squeeze down its conjugation tube— meanwhile, the contents of the cells in the other filament also round-off to form gametes

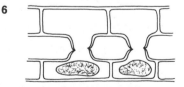

6

each moving gamete *fuses* with (i.e. fertilises) its corresponding gamete, so forming a series of **zygotes** in one filament— the other filament remains empty

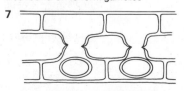

7

each zygote secretes a hard resistant case to become a **zygospore**

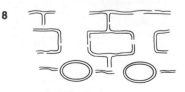

8

parent walls disintegrate, liberating zygospores

9

resistant zygospore dormant in mud etc. — can withstand **drought** and **extreme temperatures**; may be **dispersed** by wind and animals

10

germination in favourable conditions — zygospore wall splits — usually two miniature Spirogyra cells are set free and float to surface — then growth and cell division occurs

Features of particular interest:

(i) no special reproductive structures are formed—any cell may reproduce.

(ii) all filaments look alike—they show no sexual differences.

(iii) both gametes look alike—they differ in that one (which may be regarded as male) moves whereas the other (possibly female) remains stationary.

(iv) sexual reproduction is mainly to protect and preserve the plant during adverse conditions—it does not result in a great increase in the number of individuals.

Other algae

Many algae resemble the protozoa in that all their life processes take place in a single cell.

Chlamydomonas
(found in freshwater)

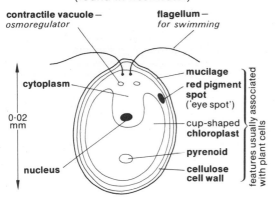

contractile vacuole — *osmoregulator*

flagellum — *for swimming*

cytoplasm

0·02 mm

nucleus

mucilage

red pigment spot ('eye spot')

cup-shaped chloroplast

pyrenoid

cellulose cell wall

features usually associated with plant cells

Pleurococcus
(found as a bright green covering on rocks, etc., and on the moist north side of tree trunks, where it may be used as a makeshift compass)

SINGLE CELL

spherical **cellulose cell wall**

lobed almost spherical **chloroplast**

nucleus

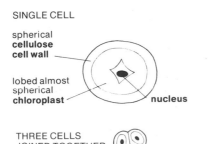

THREE CELLS JOINED TOGETHER

Marine plankton—chain-forming diatoms

Diatoms form a very important constituent of plant plankton. Fish feed on them and they provide the basis of many food chains in the sea (see p. 263).

cell wall — a 'shell' of silica material with sculptured markings (dead diatoms form an ingredient of toothpaste; the shells scour the tooth surface)

position where halves of shell overlap and join

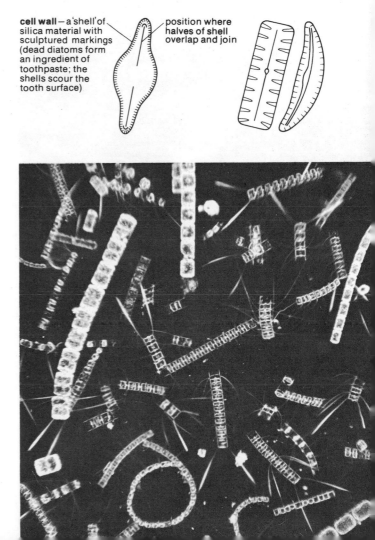

Algae

Many algae are multicellular and attain a much larger size. **Seaweeds** are an example of this:

FUCUS (bladder wrack)

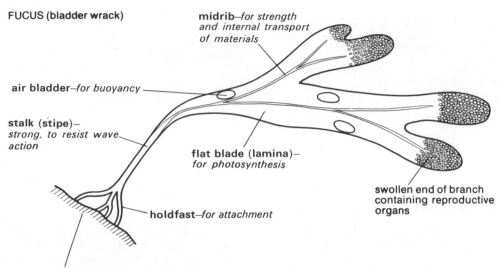

midrib—*for strength and internal transport of materials*

air bladder—*for buoyancy*

stalk (stipe)—*strong, to resist wave action*

flat blade (lamina)—*for photosynthesis*

swollen end of branch containing reproductive organs

holdfast—*for attachment*

rock (on rocky sea shore)

Bladder wrack, a common seaweed of rocky shores, showing holdfast and air bladders

Sunlight gradually loses wavelengths from the red end of the spectrum as it penetrates deeper into water; the light in deep water is dimmer and more blue. This affects the distribution of seaweeds. Green seaweeds (with chlorophyll for photosynthesis) live in shallow waters; seaweeds with brown pigments can live in deeper water. Red seaweeds are able to live at the greatest depths; the red pigment reflects what little red light is present and absorbs the blue-green wavelengths.

1 Collect samples of freshwater algae; examine these microscopically to see the range of different structures.

2 Scrape a sample of *Pleurococcus* from the surface of a tree trunk (note the *aspect* of the surface from which the alga was removed) and examine microscopically.

3 When possible, visit a rocky sea-shore. Note the different colours of seaweeds present; use a guide-book to identify the various species. Note carefully the distribution of the seaweeds between high-water and low-water marks, and also in the rock-pools.

4 Consider the importance of aquatic algae to animal life and the function of algae in an aquarium.

4   A VARIETY OF FUNGI

Fungi do not contain chlorophyll and so cannot feed themselves in the manner of green plants, i.e. they cannot photosynthesize. They are all either **saprophytes** or **parasites** (see p. 110). Fungi are composed basically of a collection of very fine threads called a **mycelium** (Gk. *mykes*, fungus).

There is a vast population of fungi in the world, many in the form of highly resistant spores, in water, on and in the ground, and suspended in the air. Given suitable conditions, these spores show a high rate of germination.

A saprophytic fungus—the pin-mould (Mucor)

Appearance

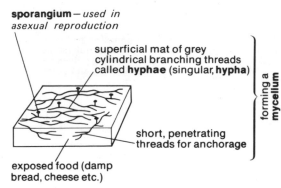

sporangium—*used in asexual reproduction*

superficial mat of grey cylindrical branching threads called **hyphae** (singular, **hypha**)

forming a mycelium

short, penetrating threads for anchorage

exposed food (damp bread, cheese etc.)

DIAGRAM SHOWING MAGNIFIED PORTION OF HORIZONTAL MYCELIUM

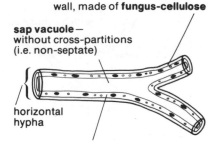

wall, made of **fungus-cellulose**

sap vacuole— without cross-partitions (i.e. non-septate)

horizontal hypha

cytoplasm lining, with *many* nuclei (i.e. multinucleate) and oil droplets

Nutrition

1

hypha secretes enzymes on to food

2

food digested outside plant (**extracellular digestion**) and made soluble

3

soluble foods absorbed back into mould and assimilated - food source is continually eroded away

Respiration

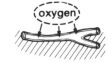

oxygen

diffusion of gas into hypha from air above food

carbon dioxide

diffusion of gas from hypha into air above

Reproduction

(a) Asexual

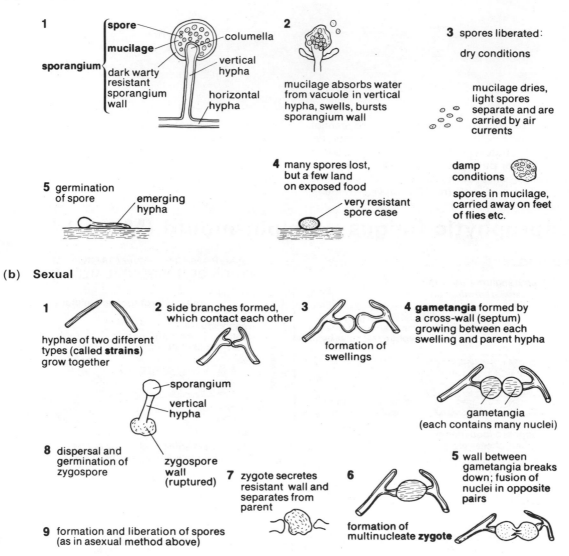

1

sporangium {
spore — columella
mucilage — vertical hypha
dark warty resistant sporangium wall — horizontal hypha
}

2 mucilage absorbs water from vacuole in vertical hypha, swells, bursts sporangium wall

3 spores liberated:

dry conditions

mucilage dries, light spores separate and are carried by air currents

5 germination of spore — emerging hypha

4 many spores lost, but a few land on exposed food — very resistant spore case

damp conditions

spores in mucilage, carried away on feet of flies etc.

(b) Sexual

1 hyphae of two different types (called **strains**) grow together

2 side branches formed, which contact each other

3 formation of swellings

4 **gametangia** formed by a cross-wall (septum) growing between each swelling and parent hypha

gametangia (each contains many nuclei)

8 dispersal and germination of zygospore

sporangium
vertical hypha
zygospore wall (ruptured)

9 formation and liberation of spores (as in asexual method above)

7 zygote secretes resistant wall and separates from parent

6 formation of multinucleate **zygote**

5 wall between gametangia breaks down; fusion of nuclei in opposite pairs

A parasitic fungus

Phytophthora is a parasitic fungus which causes **potato** and **tomato blights**; this disease was at least partly responsible for the Irish potato famine of the mid-nineteenth century. The disease constantly recurs and still causes serious losses to crops in Great Britain, particularly in warm humid seasons.

The fungus develops in the stems and leaves from planted potato tubers (see p. 215) which carry the disease. The first signs of infection are discoloured patches which become progressively darker. A white fungal mycelium can be seen if the leaf is examined closely. Tubers may become infected when sporangia are washed down from the leaves; after being harvested, these tubers rot in storage.

Control of the disease can be achieved by spraying with various fungicides; a solution of copper sulphate and quicklime (Bordeaux mixture) has been effective in the past—copper compounds are more poisonous to the fungus than to the potato.

mesophyll cell of potato host

branching, multinucleate, non-septate **hypha** of parasitic fungus

sucker (haustorium) — *penetrates cell and absorbs food from host*

Reproduction

(a) **Asexual**—begins less than one week after initial infection.

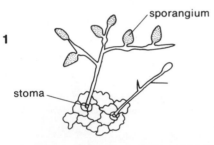

1

sporangium

stoma

hyphae grow out through stomatal pores, branch at leaf surface and develop multinucleate, pear-shaped sporangia at the side and ends of branches.

2 sporangia are held in position only loosely and are easily detached by contact (e.g. with other leaves) or by the impact of raindrops; each sporangium is dispersed intact, by air currents or washed away by rain.

3 further development of sporangium must take place in water within a few hours after dispersal — many potential offspring do not survive.

in *warm* conditions, sporangium produces a hypha.

in *cold* conditions, sporangium contents divide into several uninucleate kidney-shaped **zoospores**. These are liberated through a pore in the wall and swim for several hours in water films; each zoospore eventually ceases swimming and produces a hypha.

4

germinating hypha attaches to surface of host and develops a fine tube which bores through cuticle into epidermal cell.

5

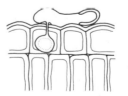

tube inside cell swells to normal size

6 hypha inside epidermal cell develops branching hyphae which grow between host cells and produce haustoria.

(b) Sexual—occurs very rarely.

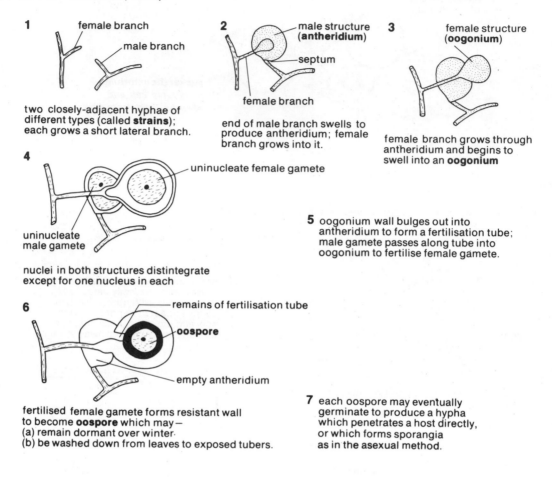

1 female branch

  male branch

two closely-adjacent hyphae of different types (called **strains**); each grows a short lateral branch.

2 male structure (**antheridium**)

  septum

  female branch

end of male branch swells to produce antheridium; female branch grows into it.

3 female structure (**oogonium**)

female branch grows through antheridium and begins to swell into an **oogonium**

4 uninucleate female gamete

  uninucleate male gamete

nuclei in both structures distintegrate except for one nucleus in each

5 oogonium wall bulges out into antheridium to form a fertilisation tube; male gamete passes along tube into oogonium to fertilise female gamete.

6 remains of fertilisation tube

  oospore

  empty antheridium

fertilised female gamete forms resistant wall to become **oospore** which may—
(a) remain dormant over winter·
(b) be washed down from leaves to exposed tubers.

7 each oospore may eventually germinate to produce a hypha which penetrates a host directly, or which forms sporangia as in the asexual method.

Other fungi

Yeast is an unusual plant. Each organism consists of a single cell. The main method of reproduction is asexual, by **budding**; this generally results in the

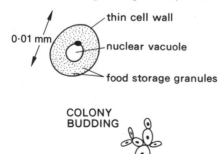

  thin cell wall

0·01 mm

  nuclear vacuole

  food storage granules

COLONY BUDDING

daughter cells remaining attached together, forming characteristic colonies.

Yeast can respire carbohydrates in order to obtain energy for its metabolism, but when it respires without oxygen gas (i.e. **anaerobically**—see p. 121) fermentation occurs; glucose can be respired in this way to produce alcohol and carbon dioxide, and relatively small amounts of energy are released :

$$C_6H_{12}O_6 \rightarrow 2C_2H_5OH + 2CO_2 + 105 \text{ kilojoules}$$
(approximately)

glucose	ethyl alcohol	carbon dioxide which escapes as a gas

(see p. 123)

Commercial uses of yeast

(i) **Wine-making**—the sugar is obtained from grapes and the yeast used to ferment this sugar occurs naturally on the grape skins. In order to make fortified wines (e.g. sherry, port) and spirits, both of which have a higher alcohol content, the original 'natural' wine is distilled and blended.

(ii) **Beer-making**—the sugar is obtained from germinating barley (malt) and yeast is added; hops give the characteristic bitter flavour.

(iii) **Bread-making**—yeast is added to the flour paste and liberates carbon dioxide which makes the dough 'rise'.

(iv) **Vitamin supplements**—yeast cells contain rich stores of vitamins, especially of the vitamin B group (see p. 114).

The fungus *Penicillium* is interesting because it secretes **penicillin**, the first anti-biotic substance to be discovered. A large number of different substances of this type are now used to treat disease. The word **anti-biotic** comes from the Greek meaning, literally, 'against life', but the anti-biotics most useful to man have a selective action effective against plant cells but not affecting animal cells.

In addition to moulds, **toadstools** and **mushrooms** feed saprophytically. The mycelium is below ground. The above-ground portion (commonly called 'the toadstool') is specialized for reproduction; it grows upwards from the underground mycelium and is composed of a compact mass of hyphae. Related closely to the toadstools are the **bracket fungi**, often found growing on dead tree trunks.

Field mushroom

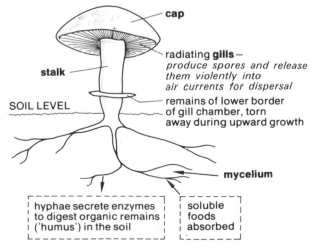

- cap
- radiating **gills**— produce spores and release them violently into air currents for dispersal
- stalk
- SOIL LEVEL
- remains of lower border of gill chamber, torn away during upward growth
- **mycelium**

| hyphae secrete enzymes to digest organic remains ('humus') in the soil | soluble foods absorbed |

Mushrooms at different stages of development; the mature mushroom has gills

Many of the parasitic fungi cause a wide range of serious plant diseases, including tomato rot, soft rot of apples and pears, and 'damping-off' of young seedlings. Members of the cabbage family are commonly attacked by fungi and this gives rise to 'club root' and 'downy mildew'. Fungi are also responsible for diseases in animals, from a fatal disease of house-flies to fin-rot (fishes) and the skin diseases athlete's foot and ringworm (in man).

Fungi form interesting associations with other organisms. **Lichens** are associations between fungi and algae; they are most frequently the first plants to colonize bare ground (see also p. 250).

Mycorrhiza are fungal associations with the roots of certain trees (e.g. beech) or with the roots of heathers.

These associations may be examples of **symbiosis** (see p. 110), with each partner in the pair gaining some nutritional advantage.

---

1 Thoroughly moisten a slice of bread with water; expose the bread to the atmosphere for 24 hours, then cover and leave in a warm place. After about one week, examine small portions of the bread microscopically for fungal mycelia and reproductive bodies.

2 Collect and examine (but do not eat!) various types of toadstools and bracket fungi.

3 Consider the importance of fungi, particularly of soil fungi, in nature.

4 Read further about yeast and its economic importance, including the processes involved in wine- and beer-making commercially. If possible arrange a conducted tour around a local brewery. The essential features of fermentation can easily be demonstrated at home, e.g. with the making of elderberry wine.

5 Read further about the discovery of anti-biotics and their use in medicine.

6 Search for lichens on walls and trees etc.

5 BACTERIA

Bacteria are probably more widely distributed than any other type of living organism. They occur under very varied conditions; nearly every environment has its population of bacteria, e.g. some are found even in the waters of hot springs where the temperature is above 60°C. They almost certainly occur in greater numbers than do other organisms, e.g. one gram of fertile soil is said to contain 1000 million, and 1 cm³ of fresh milk may contain more than 30 000 million bacteria.

Structure

Bacteria are very simple plant-like organisms showing certain similarities with fungi and algae. All are microscopic, and some are so small that they are difficult to see even with the highest magnification provided by a light microscope.

Each bacterium consists of a single cell. The smallest bacteria (cocci) measure less than 0.01 mm diameter; the largest are rod-shaped forms with dimensions of 0.02 mm × 0.002 mm approximately.

Bacteria can be cultured in the laboratory in various solid or fluid media. A common method is to dissolve agar jelly by boiling it with a suitable nutritive solution (such as meat broth). The mixture is then poured into a petri dish and is inoculated with bacteria when it has cooled and set.

The study of bacterial cells with electron microscopy reveals a variety of structures; the diagram opposite shows a generalized rod-shaped bacterium with some of the more commonly found structures.

Note that although the protoplasm contains materials (such as nucleic acids) normally found in the nucleus of plant cells, there is no nucleus in the strict sense. Most bacteria are colourless, though a few contain chlorophyll or other pigments; however, these are always spread throughout the cytoplasm and are not confined to definite bodies such as chloroplasts.

Bacteria can be classified into four main groups, according to the shape of the cell:

(i)   **spherical (coccus)**

(ii) **straight cylindrical rod (bacillus)**

(iii) **curved cylindrical rod (vibrio)**

(iv) **spirally coiled (spirillum)**

In addition, there are several associated groups of bacteria-like organisms such as **spirochaetes**, where the cell is long and spirally twisted and

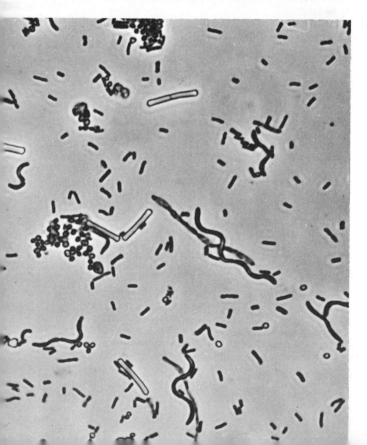

Different types of solitary and colonial bacteria; large rod-shaped and spirally-coiled forms are clearly distinguishable

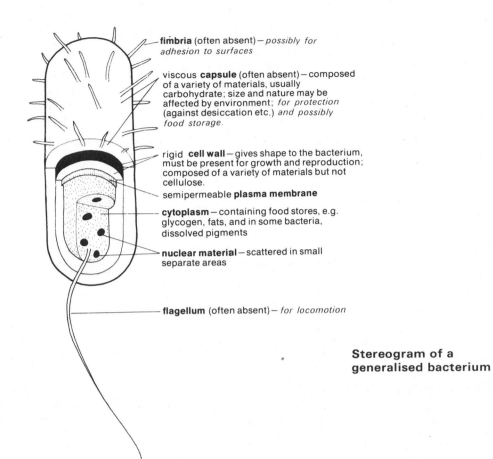

fimbria (often absent) — *possibly for adhesion to surfaces*

viscous **capsule** (often absent) — composed of a variety of materials, usually carbohydrate; size and nature may be affected by environment; *for protection* (against desiccation etc.) *and possibly food storage.*

rigid **cell wall** — gives shape to the bacterium, must be present for growth and reproduction; composed of a variety of materials but not cellulose.

semipermeable **plasma membrane**

cytoplasm — containing food stores, e.g. glycogen, fats, and in some bacteria, dissolved pigments

nuclear material — scattered in small separate areas

flagellum (often absent) — *for locomotion*

Stereogram of a generalised bacterium

Rod-shaped tuberculosis bacilli in lung

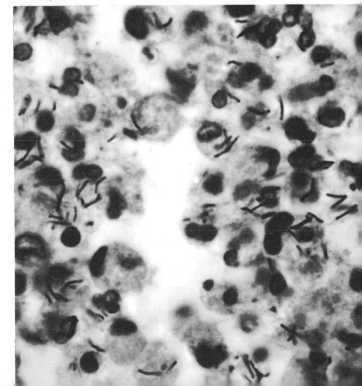

capable of flexings which bring about locomotion. One spirochaete causes the important human disease **syphilis**.

In many species of bacteria, colonies of characteristic shape are formed by the failure of the daughter bacteria to separate completely after cell division. The gummy nature of the capsule may assist in holding the individual members together. This can be commonly seen in the spherical forms. Single cells are called Micrococcus and associations of

two cells together ⬯ Diplococcus; others

may form a linear colony resembling a string of

beads ⬯⬯⬯⬯ (Streptococcus), or irregular

collections ⬯⬯⬯ (Staphylococcus). Similarly,

bacilli may be joined together end-to-end in groups of two or more.

Different types of bacteria are often so similar structurally that other ways of distinguishing between them have had to be found. Important. methods of bacterial classification are based on the fact that bacteria will stain with various dyes. One such technique was developed by the bacteriologist Gram. Bacteria which react in a certain way in this technique are said to be **Gram-positive**; those that react in another way are **Gram-negative**. Furthermore, different features of their behaviour and metabolism can be used to distinguish between bacteria.

Locomotion

Many bacteria show Brownian movement only, but some can move freely in the surrounding medium. Apart from those few bacteria that make slow gliding movements, bacteria can move themselves by their flagella when these are present. There may be just one flagellum per cell, or flagella may be evenly distributed over the surface. Many bacteria shed their flagella as the cells become old. Flagella are rarely found in cocci.

Respiration

Many bacteria respire anaerobically (see p. 121) whereas others can respire *only* in the total absence of oxygen. Some bacteria can live equally well with or without atmospheric oxygen. (It is said that bacteria produce more carbon dioxide than do all the world's plants and animals.)

A stab-culture technique can be used for determining the respiratory requirements of bacteria; here, bacteria are inoculated deeply into a test-tube containing nutrient agar jelly, with the following results:

Nutrition

Bacteria feed in one of two main ways:

By synthesis of food from inorganic sources

This method is found in comparatively few bacteria only. An external energy supply is required for this synthesis. Bacteria obtain their energy from:

(a) **sunlight.** The small number of bacteria that contain chlorophyll or related pigments are able to photosynthesize in a similar manner to higher plants (see p. 95).

(b) **certain chemical reactions in which they take part.** These bacteria are called chemosynthesisers. Some of them are essential in maintaining soil fertility; an example is *Nitrobacter*—this plays an important part in the nitrogen cycle by oxidizing nitrites to nitrates. The bacteria obtain energy for their own metabolism from this oxidation.

By use of organic ('ready-made') food

This method is common to the majority of bacteria, which are either saprophytes or parasites (see p. 110).

(a) **Saprophytes.** These include the putrefying bacteria of soil (see p. 258); they produce humus from plants and animal remains. They may also cause the decay of materials useful to man, including his food.

(b) **Parasites.** These are widespread and responsible for diseases of plants and animals, including many serious diseases of man and his domestic animals. Bacterial diseases of plants are less common, though soft rot of carrots and turnips, for example, is of economic importance.

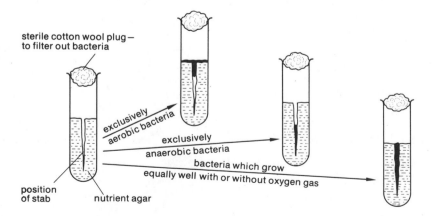

sterile cotton wool plug—
to filter out bacteria

exclusively
aerobic bacteria

exclusively
anaerobic bacteria

bacteria which grow
equally well with or without oxygen gas

position
of stab

nutrient agar

Reproduction and spore-formation

No convincing example of sexual reproduction has been demonstrated in bacteria. Reproduction is asexual by binary fission. When the limiting size is reached the cytoplasm divides transversely; a new cell wall is formed separating the two portions, which may remain together, so beginning colony-formation. In favourable conditions, the limiting size can be reached every 20 minutes. Theoretically, at this rate of division, *each* bacterium would produce more than one million offspring in less than 7 hours; in practice, many factors resulting from overcrowding (see below) tend to prevent this.

Bacteria, particularly rod-forms, are capable of forming hard resistant spores, called **endospores** because they are formed within the original cell wall. The protoplasm shrinks away from the wall and a tough resistant material is secreted around the protoplasm. The number and position of spores in relation to the 'parent' bacterium can be used for identification—for example:

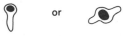

Factors affecting the growth of bacteria

Apart from the obvious considerations of **suitable food supply** and presence or absence of **oxygen** (see above), the growth of bacteria is affected by external conditions. These include **temperature** and **acidity**; most bacteria grow best between 25–40°C and in neutral surroundings. Although bacteria may tolerate very low temperatures (below freezing point), most bacteria, though not their spores, are killed above 60°C. Exposure to **ultra-violet rays** of the sun kills many bacteria, perhaps because they lack protective pigment. Overcrowding, besides depleting the available food, produces accumulated excretions which prevent further growth. Also, bacteria are attacked and destroyed by virus-like organisms called **bacteriophages**.

BACTERIA USEFUL TO MAN

Besides bacteria which are harmful to man there are many that have no effect and others that are positively useful. Those in the soil are dealt with on p. 257. There are also symbiotic bacteria (see p. 110—those that digest cellulose in the alimentary canal of ruminants (p. 109), those that synthesize vitamins in the body of man, and those that live in root nodules and 'fix' nitrogen from the atmosphere (p. 258).

The disposal of human sewage and the decomposition of garden compost also depends on the activities of bacteria.

Bacteria are used commercially in several food-manufacturing processes, e.g. butter and cheese from milk, and vinegar from alcohol. Industrially bacteria are used to make chemicals difficult or impossible to synthesize in other ways.

Food preservation

Food is spoilt by the saprophytic activities of putrefying bacteria. Sometimes, the food becomes contaminated with highly poisonous bacterial excretions, as in 'botulism'. Methods of preserving food attempt to reduce its bacterial content by removing at least one of the conditions essential for the survival of most living organisms—a water supply, oxygen and a suitable temperature. The bacteria either die or are unable to reproduce.

In **refrigeration**, food is kept at temperatures too low for bacterial reproduction. Fruit can be preserved as **jam**. Here, a sugar solution of high concentration draws water out of the bacteria by osmosis: the bacteria die by dehydrating. The same process occurs when meat is preserved in **brine**, a concentrated solution of salt. **Dried fruit** is preserved similarly, because the sugar in the fruit becomes very concentrated when the fruit dries. **Canning**, suitable for a wide variety of foods, attacks the bacteria in two ways. The high temperatures used (up to 120°C) kill both bacteria and their spores; the cans are quickly cooled and vacuum-sealed, so that there is very little air and no more bacteria can enter. Bacteria cannot survive in acid conditions: **pickling** is used to preserve certain foods whose flavour, though altered, is not spoilt by the acetic acid (in vinegar). **Pasteurization** delays the souring of milk. Milk is heated to about 70°C and quickly cooled. Some bacteria are killed: the growth of others is retarded. Boiling milk (**sterilization**) kills more bacteria, but alters the flavour.

1 Bacterial cultures can be grown safely in the laboratory if certain procedures are followed carefully. Many books on the subject of bacteriology give precise details of apparatus and techniques which the student should follow under supervision.
2 Read further about the work of Pasteur and other early bacteriologists, and also about Lister and the development of antiseptic surgery.
3 Many hospitals allow educational visits to their laboratories; the student should take the opportunity to see work performed there. Similarly food-manufacturers may allow visits to their factories. The student should himself attempt simple forms of food preservation (e.g. bottling of fruit).

6 VIRUSES

Apparently all viruses are parasitic, living inside the cells of their hosts. Viruses combine characteristics of the living and non-living. Outside their hosts viruses seem inert and can survive indefinitely in a desiccated crystalline state. Once inside a suitable host cell the virus becomes able to reproduce itself; the virus takes over the cell's metabolism for its own uses, i.e. to reproduce more virus. Eventually the cell dies and bursts, liberating the reproduced viruses.

Viruses are ultra-microscopic; all are able to pass through filters which retain bacteria. Although their presence was demonstrated in 1892, viruses could not be seen until the development of the electron microscope. The largest viruses measure 0.00035 mm, the smallest 0.00001 mm. Each virus essentially consists of the large chemical molecules that are elsewhere associated with the nuclei of plant and animal cells, i.e. nucleo-proteins.

The shapes of viruses vary from spheres, cubes and rods to flat plates and spheres with a tail. In many, the protein components are arranged in complex geometric patterns.

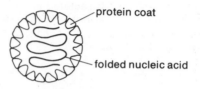

Simple virus (of poliomyelitis)

In some viruses the protein coat is surrounded by a further external covering.

Viruses are generally named after the diseases they cause, though in some cases the symptoms of the virus attack are insufficient to warrant the term disease; for example, some plant viruses cause the production of variegated foliage only. Economically important virus diseases of plants include the leaf mosaics of tobacco, potato and tomato. They are transmitted by insects such as greenfly. In animals, viruses cause poliomyelitis, influenza, the common cold, chickenpox, measles, rabies (in man), distemper (in dogs), myxomatosis (in rabbits), foot-and-mouth (in cattle), fowl pest and swine fever. They also seem to cause certain types of cancer.

Mammals protect themselves against viruses with the same methods that are used against bacteria. However viruses are not affected by anti-biotics (see p. 19).

Electron micrograph of viruses with 'heads' and 'tails'

Hydra is found in ponds, water-filled ditches or in slow-moving streams. It may be found attached to water weeds or suspended from the surface film. There are three species in Great Britain:

Hydra viridissima, (the green hydra—the colour is due to green algae living in the endoderm cells); *H. vulgaris* (greyish colour); *H. oligactis* (brown colour). It is carnivorous, feeding on small animals, e.g., water fleas. When a water flea comes in contact with a tentacle, the sting cells discharge, and paralyze the water flea, which is then drawn

HYDRA

a simple multi-celled animal

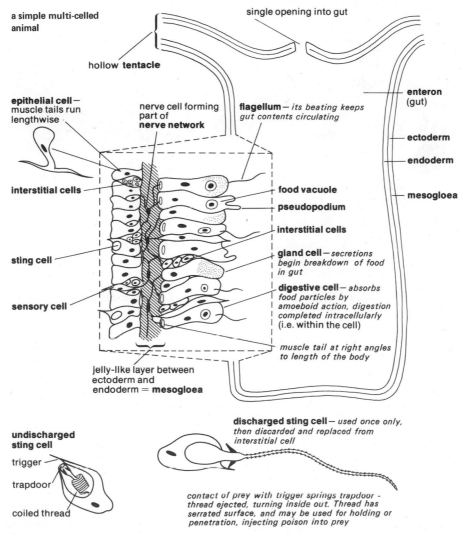

single opening into gut

hollow **tentacle**

epithelial cell—muscle tails run lengthwise

nerve cell forming part of **nerve network**

flagellum—*its beating keeps gut contents circulating*

enteron (gut)

ectoderm

endoderm

interstitial cells

food vacuole

pseudopodium

interstitial cells

mesogloea

sting cell

gland cell—*secretions begin breakdown of food in gut*

sensory cell

digestive cell—*absorbs food particles by amoeboid action, digestion completed intracellularly (i.e. within the cell)*

muscle tail at right angles to length of the body

jelly-like layer between ectoderm and endoderm = **mesogloea**

undischarged sting cell

trigger

trapdoor

coiled thread

discharged sting cell—*used once only, then discarded and replaced from interstitial cell*

contact of prey with trigger springs trapdoor - thread ejected, turning inside out. Thread has serrated surface, and may be used for holding or penetration, injecting poison into prey

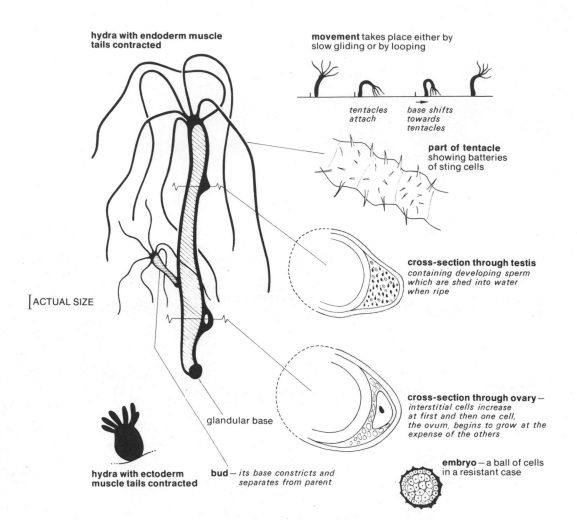

hydra with endoderm muscle tails contracted

movement takes place either by slow gliding or by looping

tentacles attach

base shifts towards tentacles

part of tentacle showing batteries of sting cells

cross-section through testis *containing developing sperm which are shed into water when ripe*

⌐ACTUAL SIZE

glandular base

cross-section through ovary — *interstitial cells increase at first and then one cell, the ovum, begins to grow at the expense of the others*

hydra with ectoderm muscle tails contracted

bud — *its base constricts and separates from parent*

embryo — a ball of cells in a resistant case

towards the mouth by the tentacles and swallowed. Undigestible remains are removed through the same opening, which acts as both mouth and anus. Hydra has no special organs of respiration or excretion, all cells being in close contact with the outside world.

Hydra was first described by Abraham Trembley, a Swiss naturalist, in the eighteenth century. He called it hydra after the many headed monster of Greek mythology. He found that after cutting it into pieces each part would grow into a new hydra. He is credited with having turned a hydra inside out. After a few days the animal re-organized itself, the cell layers actively migrating through the jelly layer to take up their normal positions. This experiment has been successfully repeated in recent years.

Sexual Reproduction. Cross fertilization occurs. In hermaphrodite species (see p. 214) ovary and testis ripen at different times. The egg is fertilized in the ovary and begins its development there.

When shed it drops to the bottom and, under favourable conditions, gives rise to a young hydra.

Asexual Reproduction. Under favourable conditions hydra produces buds which eventually detach. It can also regenerate parts lost, and, if cut in half, each part can **regenerate** to form a complete organism.

Hydra is a member of a group of animals called **Coelenterata** (Gk. *coel*, hollow; *enteron*, gut). With a few exceptions, including hydra, they are marine animals. They all show **radial symmetry** about the vertical axis. This means that a cut which passes vertically down through the centre of the animal, in any direction, will divide it into symmetrical parts. All possess sting cells and all have a structure consisting of two cell layers separated by a **jelly layer (mesogloea)**. They are found close to the shore or drifting in the surface layers of the oceans. Some species form colonies, that is, their bodies are continuous with each other.

The structure of coelenterates

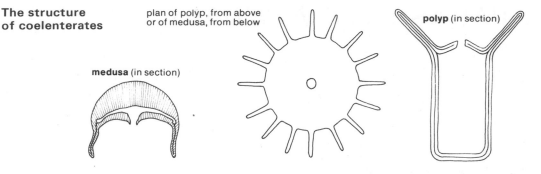

medusa (in section)

plan of polyp, from above or of medusa, from below

polyp (in section)

Coelenterates show two basic forms:

Medusa (named after mythical woman whose tresses were vipers). This is a free floating form found in open waters.

Polyp (L. *polypus*, many feet). Found near the low water mark attached to weeds or rocks.

Examples of coelenterates:

1 **Hydroids.** These are mostly colonial forms. The colony grows by budding. Both polyps and medusae are found. The polyps are fixed and can only feed. The medusae are set free and carry out reproduction and dispersal.

2 **Jellyfish.** These are always medusoid and are found in all oceans.

3 **Anemones.** Found only as polyps. Never colonial.

4 **Corals.** Form massive colonies and secrete skeleton formed of calcium carbonate. They are found in the tropics in the Pacific and Indian Oceans and in the Caribbean Sea.

Hydra with buds and extended tentacles

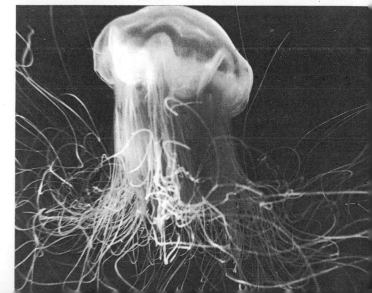

Medusoid shape illustrated by jellyfish

1 Collect water weeds from a pond or water-filled ditch and leave in a dish with some of the water. Hydra, if present, will appear after a few days attached to the sides of the dish. Remove a specimen, place in water in a watch glass and examine under the low power of a microscope. Try feeding with water fleas and observe what happens.

2 When opportunity arises, look for colonies of hydroids on seaweed at the low water mark. Look for anemones in rock pools at low tide.

3 Visit your local natural history museum if you have one and study specimens of coral formation.

8 A PARASITIC FLATWORM

Tapeworms are parasites (see p. 110); they belong to a group of animals called **flatworms**. Characteristically, flatworms have fairly broad bodies but in cross-section they appear very thin.

Tapeworms are **obligate parasites**, i.e. they die if deprived of a host. Tapeworms are also called **endoparasites**, i.e. they live inside the bodies of their hosts. The parasites are remarkably adapted to life in their peculiar habitat.

Structure

The adult **pork tapeworm** (*Taenia solium*) lives in the small intestine of man. The body consists of a creamy-white ribbon or tape which may be nearly 5 metres long when fully mature. The tape is attached at one end to the internal intestine lining and becomes progressively broader towards the opposite, unattached end. Attachment is achieved by a small knob (**scolex**) which has four prominent suckers and a circle of hooks on a crown-like portion (**rostellum**) at the extreme end. The tape itself is composed of a series or chain of as many as 1000 divisions called **proglottides** (singular, **proglottis**) which resemble segments; these become larger towards the free end. A middle-region proglottis may be several millimetres wide.

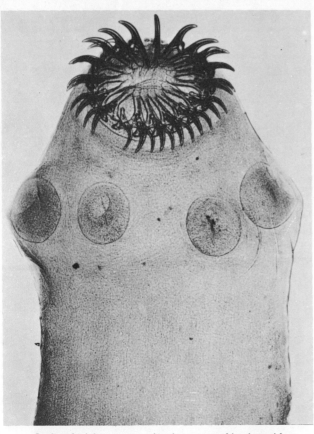

Scolex of adult tapeworm, showing crown of hooks and four suckers

Movement

Although the tapeworm contains muscle tissue the established parasite remains fixed and displays some sinuous wriggling movements only.

Growth

Proglottides continually drop away from the free end of a mature tapeworm (see 'reproduction'). The length of the animal is maintained by a growth process called **strobilation** in the 'neck' region immediately following the scolex. Here, very small new proglottides are constantly produced. Rather as on a factory assembly line these proglottides are steadily pushed down the body by the strobilation of more proglottides. As well as becoming larger, they also become more fully developed and eventually reach the end, where they are shed.

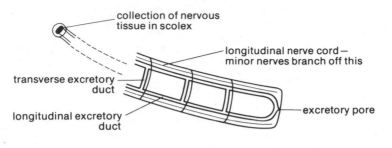

**Diagram showing excretory
and nervous structures**

Nutrition

Tapeworms have no gut and apparently produce no digestive enzymes. In the small intestine, most of the food ingested by the mammal host has been fully digested and is ready for absorption. The tapeworm absorbs a proportion of this food all over its surface; the flatness of the body provides the largest possible surface area for absorption in relation to the body volume.

Respiration and excretion

The same large surface area would provide an excellent area with which to absorb oxygen for respiration. However the small intestine is unlikely to contain much oxygen gas, so that tapeworms presumably can respire anaerobically (see p. 121). Carbon dioxide can escape by diffusion all over the surface.

Other excretory materials are transported by fine tubes into two longitudinal ducts; these open into the host's intestine by an excretory pore at the end of the last proglottis. New pores are formed when proglottides are shed.

Irritability

There are no sense organs. The nervous system is extremely simple and consists essentially of two longitudinal nerve cords. It is possible that tapeworms respond to their hosts' enzymes by secreting anti-enzymes to counter them.

Reproduction

Tapeworms lead a remarkably passive life but, in common with many parasites, much of the tapeworm's activities are concerned with reproduction. Tapeworms are hermaphrodite with a complex reproduction and life-cycle.

Each proglottis undergoes a progressive series of reproductive developments. A newly-formed proglottis is immature. As it becomes pushed further away from the scolex it develops male organs and then female organs. About half-way down the tape there is a region of truly hermaphrodite (Gk. *hermaphroditos*, combining both sexes) proglottides; fertilization occurs here. Beyond this region the male organs and somewhat later most of the female organs degenerate and disappear. When the proglottis reaches the terminal region it is ripe and consists almost entirely of a uterus, filled with masses of fertilized eggs.

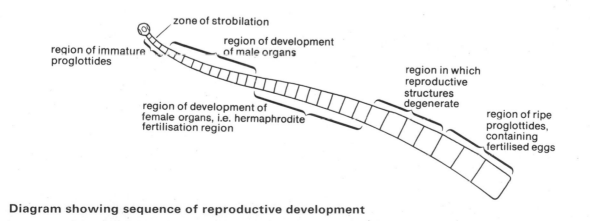

Diagram showing sequence of reproductive development

Flatworms

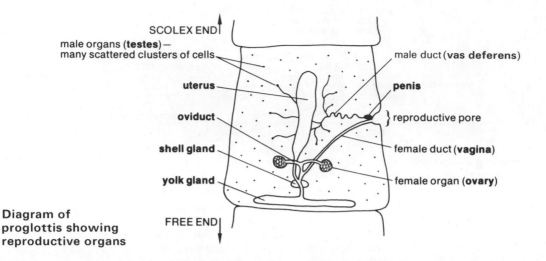

SCOLEX END

male organs (**testes**) —
many scattered clusters of cells

male duct (**vas deferens**)

uterus

penis

oviduct

} reproductive pore

shell gland

female duct (**vagina**)

yolk gland

female organ (**ovary**)

FREE END

**Diagram of
proglottis showing
reproductive organs**

Sperm cells produced in the testes pass along the vas deferens to the reproductive opening. They are then transferred by the penis into the vagina, either of the same proglottis or of an adjacent proglottis which has become coiled or folded into close proximity. Sperm travel up the vagina to fertilize ova shed from the ovaries along the oviducts. Each fertilized egg (zygote) receives a supply of yolk and a resistant shell covering from glands nearby, and passes into the uterus which steadily enlarges to receive more fertilized eggs. Finally, ripe proglottides, either singly or in short chains, drop off into the cavity of the intestine. Each zygote inside its shell and still within the uterus develops into a solid ball of cells with six hooks; this is the **hexacanth embryo** (Gk: *hex*, six; *akantha*, thorn).

**Diagram of
ripe proglottis**

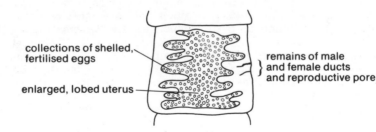

collections of shelled,
fertilised eggs

remains of male
} and female ducts
and reproductive pore

enlarged, lobed uterus

SPECIAL FEATURES OF TAPE-WORMS, ASSOCIATED WITH PARASITIC LIFE

1 **Scolex** with hooks and suckers, for attachment to intestine—prevents removal during peristalsis etc.

2 **Flat shape**: (a) to increase surface area for absorption of pre-digested food.

(b) to lie closely against intestine lining, and so prevent damage by passing food.

3 **Anaerobic respiration**—because of oxygen lack in intestine.

4 **Generally simple body structure:**

(a) no digestive system—food is pre-digested.

(b) poorly developed muscle and sensory structures—tapeworm does not have to move to seek food or escape enemies.

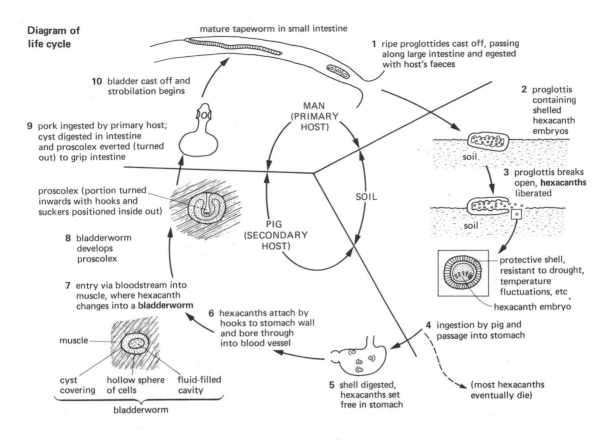

Diagram of life cycle

mature tapeworm in small intestine

1 ripe proglottides cast off, passing along large intestine and egested with host's faeces

10 bladder cast off and strobilation begins

2 proglottis containing shelled hexacanth embryos

MAN (PRIMARY HOST)

9 pork ingested by primary host; cyst digested in intestine and proscolex everted (turned out) to grip intestine

soil

3 proglottis breaks open, **hexacanths** liberated

soil

proscolex (portion turned inwards with hooks and suckers positioned inside out)

SOIL

PIG (SECONDARY HOST)

protective shell, resistant to drought, temperature fluctuations, etc

hexacanth embryo

8 bladderworm develops proscolex

7 entry via bloodstream into muscle, where hexacanth changes into a **bladderworm**

6 hexacanths attach by hooks to stomach wall and bore through into blood vessel

4 ingestion by pig and passage into stomach

muscle

cyst covering

hollow sphere of cells

fluid-filled cavity

bladderworm

5 shell digested, hexacanths set free in stomach

(most hexacanths eventually die)

5 **Creamy-white colour**—camouflage not needed because of internal situation.

6 **Thick cuticle covering body**—resists action of host's enzymes (possibility of secretion of anti-enzymes).

7 **Growth by strobilation** gives continual production of new reproductive segments—increases number of potential offspring.

8 **Hermaphrodite** and **self-fertilizing**—because each host may not be able to support more than one tapeworm simultaneously.

9 **Internal fertilization**—because gametes might be destroyed if liberated into intestine.

10 **Large production of eggs and sperm,** giving vast numbers of zygotes—most offspring will die owing to the difficulties of reaching new primary host.

11 **Embryos protected by shell**—particularly important for hexacanths, which thereby survive longer on soil.

12 **Use of secondary host** in life-cycle—this host is food source of primary host and provides a direct entry into intestine of primary host.

METHODS OF CONTROL

Tapeworm infections are more common in backward countries, where man lives in close contact with domestic animals and the standard of hygiene is low. In Great Britain, the pork tapeworm is almost unknown, largely because of control measures taken against it. A knowledge of the parasite's life-cycle is essential for effective control, as the tapeworm is most vulnerable when it changes hosts—from man to pig, or from pig to man.

The following measures have been used:

1 Good sanitation—so that untreated sewage does not come in contact with soil.

2 Farming pigs in hygienic conditions, with clean uncontaminated foods.

3 Proper inspection of meat for human consumption—pork containing bladderworms has a spotted, 'measly' appearance.

4 Thorough cooking of pork to kill bladderworms.

Prevention of infection is much more satisfactory than cure, even though a drug extracted from ferns can be taken orally and used against the adult parasite. The drug causes violent contractions of the intestine but generally only the tape breaks away, leaving the scolex and neck, which will continue to strobilate.

Adult tapeworms are said to be able to live in man for more than twenty years. They cause abdominal pains, vomiting and nervous disorders, but their effects are usually serious only in children or in adults already weakened by some other condition.

Other flatworms

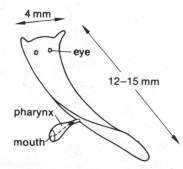

Planarians are **turbellarians**, a group of free-living flatworms always found in damp places; most species live in the sea or under stones in ponds and streams. They crawl over the bottom or in the surface film by the beating of cilia which project from their surfaces; also by muscle contraction throwing the body into a characteristic zig-zag path with the head bending from side to side. The head has a pair of eyes and the beginnings of a brain. In many species the gut can be seen through the thin body; there is a mouth but no anus. Most feed on carrion—the pharynx is pushed out and food sucked in. Separate pieces of turbellarians can regenerate complete new animals. If starved, turbellarians can 'degrow' by losing organs in a definite sequence. Eggs are laid, which hatch into larvae or miniature adults, according to species.

Flukes are parasitic flatworms, covered by a thick spiny cuticle. **Liver flukes** cause fatal 'liver rot' of sheep; they feed mainly on blood and tissue directly from the wall of the bile ducts, to which they attach themselves by suckers. They have a very complex reproduction. Eggs are passed in the host's faeces and these develop into a series of different larvae. A water-snail is needed as secondary host to complete the life-cycle. The larvae reproduce themselves in the snail, devour the host's tissues, leave the dead snail and encyst on grass, where they may be ingested by a sheep.

1 Collect planarians from beneath stones in streams. Investigate their response to light; examine their method of feeding with small pieces of meat. Place a planarian in a dish of distilled water; cut the animal across into halves and keep the dish in a cool dark place—examine daily for signs of regeneration.

2 Examine with a lens whole specimens and microscope slides of parasitic worms including tapeworms. Note carefully any features which might be considered as adaptations to a parasitic life.

3 Read further about the economic importance of parasitic worms and their methods of transference from host to host.

9 MOSSES

Mosses belong to the group of flowerless plants called the **Bryophyta** (Gk. *bryon*, moss; *phyton*, plant). The group also includes the **liverworts**; plants which have no distinct stem and leaves, the body being a flattened, branched structure called a **thallus**. Bryophytes are in some respects the amphibia of the plant kingdom, showing only partial adaptation to life on dry land.

Mosses are usually found in moist, shady places. They are common in woods, both on the ground and on the trunks of trees. Some species have a high tolerance to dry conditions and may be found on porous rocks, in the crevices of stone walls and on roof tiles. Others will thrive in bogs, swamps, or even standing water. In the absence of water, mosses may dry out, losing perhaps most of their body weight, but will revive after a shower of rain. Some have an enormous capacity for taking up water. (Dried sphagnum moss used to be sold by chemists for use in babies' nappies.)

Cluster of mosses, each plant showing thin leaves arranged around a central stem

Life cycle of the moss

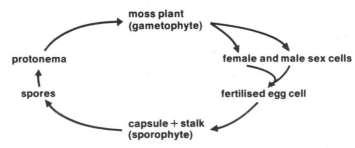

protonema → moss plant (gametophyte) → female and male sex cells → fertilised egg cell → capsule + stalk (sporophyte) → spores → protonema

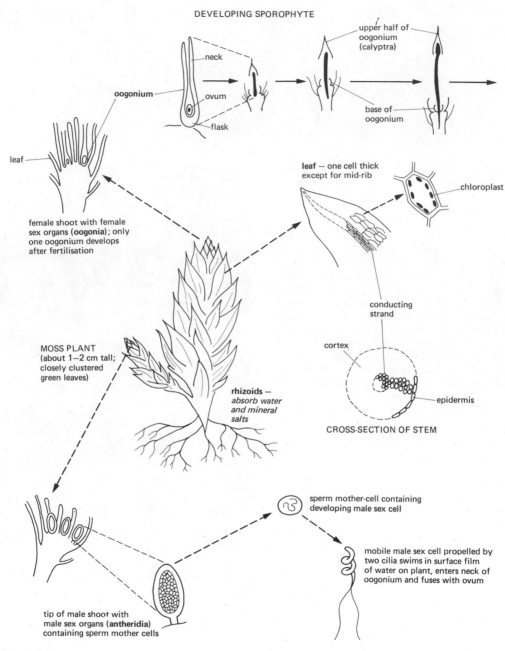

DEVELOPING SPOROPHYTE

upper half of
oogonium
(calyptra)

neck

oogonium

ovum

flask

base of
oogonium

leaf

leaf — one cell thick
except for mid-rib

chloroplast

female shoot with female
sex organs (**oogonia**); only
one oogonium develops
after fertilisation

conducting
strand

cortex

MOSS PLANT
(about 1—2 cm tall;
closely clustered
green leaves)

rhizoids —
*absorb water
and mineral
salts*

epidermis

CROSS-SECTION OF STEM

sperm mother-cell containing
developing male sex cell

mobile male sex cell propelled by
two cilia swims in surface film
of water on plant, enters neck of
oogonium and fuses with ovum

tip of male shoot with
male sex organs (**antheridia**)
containing sperm mother cells

Funaria hygrometica

THE DEVELOPMENT OF THE CAPSULE

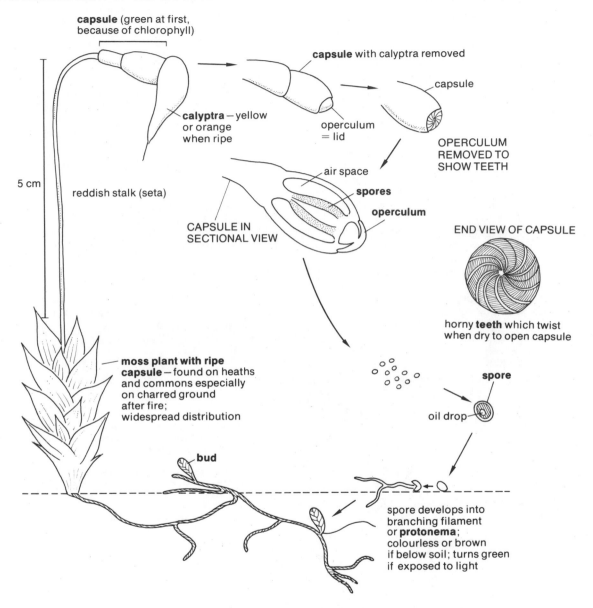

capsule (green at first, because of chlorophyll)

capsule with calyptra removed

capsule

calyptra—yellow or orange when ripe

operculum = lid

OPERCULUM REMOVED TO SHOW TEETH

5 cm

reddish stalk (seta)

air space

spores

operculum

CAPSULE IN SECTIONAL VIEW

END VIEW OF CAPSULE

horny **teeth** which twist when dry to open capsule

moss plant with ripe capsule—found on heaths and commons especially on charred ground after fire; widespread distribution

spore

oil drop

bud

spore develops into branching filament or **protonema**; colourless or brown if below soil; turns green if exposed to light

Moss plants reproduce sexually. The fertilized egg develops on the parent, never becoming completely independent. It develops into a capsule, containing spores, carried on the end of a long stalk, which absorbs materials from the parent plant. The spores are produced asexually within the capsule. They are set free on to the ground and develop into branching threads, **filaments**, from which tiny buds grow. Each **bud** can give rise to a new moss plant.

10 FERNS

Ferns belong to a group of plants called the **Pteridophyta,** (Gk. *pteron,* feather; *phyton,* leaf), so named because of the feathery appearance of the leaves of many ferns. They are much more complex than mosses and have well developed vascular tissue. The leaves are green, and, in addition to carrying out photosynthesis, produce spores. Under favourable conditions each spore grows into a small, inconspicuous, leaf-like structure called a **prothallus**. This will grow only in damp shady conditions. It possesses both male and female sex organs and, provided that a film of moisture is present on the surface of the prothallus, the mobile male gamete swims to the female sex organ where fertilization of the ovum

takes place. The fertilized ovum begins its development in the tissues of the prothallus, eventually giving rise to the familiar fern plant. As in mosses, the life cycle follows a pattern of alternating, dissimilar generations. Unlike the mosses, the spore-producing stage is the dominant one, well adapted to life on dry land. The sexual stage, however, must depend on the presence of external water. This has not prevented ferns from exploiting 'difficult' conditions. Bracken, for instance, often grows profusely in regions where there is little water. In these circumstances it spreads by the growth of an underground stem. It is often a serious problem to farmers on hillside pastures, particular in Scotland and on Dartmoor.

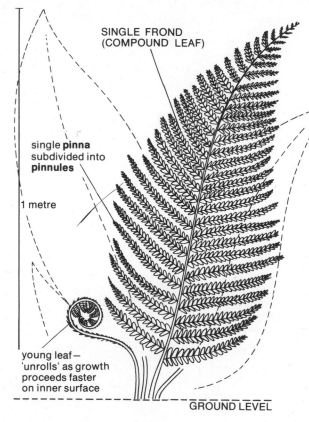

SINGLE FROND
(COMPOUND LEAF)

single **pinna**
subdivided into
pinnules

1 metre

young leaf —
'unrolls' as growth
proceeds faster
on inner surface

GROUND LEVEL

Widespread in temperate regions in hedgerows, woods, mountainsides and some suburban gardens; dark green on upper side, paler green underneath; the stem is underground and vertical. The spore cases are lens-shaped, and possess rows of unequally thickened cells along one edge, which straighten as they dry, splitting open the cases.
The spine also ruptures eventually: acting as a catapult, it flicks the spores clear of the plant.

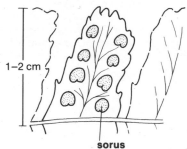

UNDERSIDE OF PINNULE
TO SHOW SORI

1–2 cm

sorus

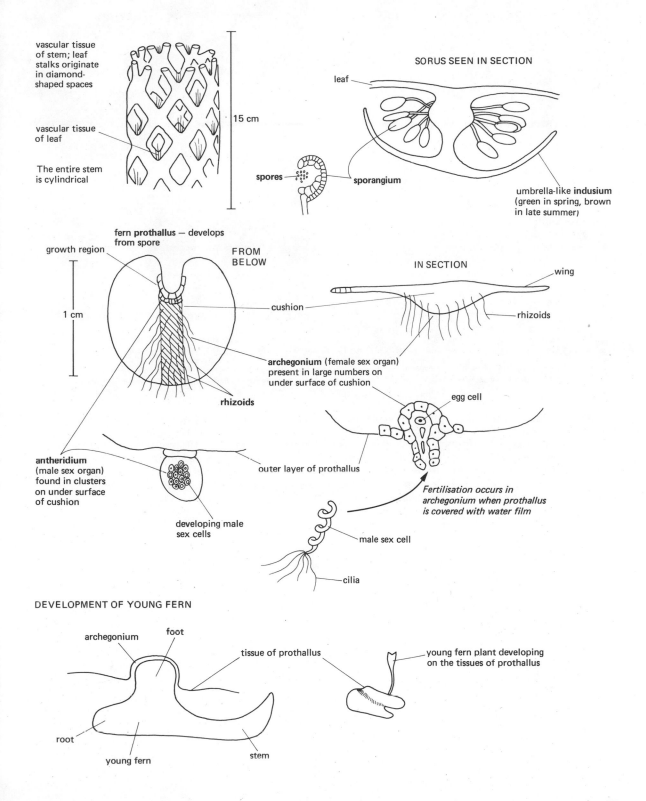

vascular tissue of stem; leaf stalks originate in diamond-shaped spaces

15 cm

vascular tissue of leaf

The entire stem is cylindrical

SORUS SEEN IN SECTION

leaf

spores

sporangium

umbrella-like **indusium** (green in spring, brown in late summer)

fern **prothallus** — develops from spore

growth region

FROM BELOW

IN SECTION

wing

1 cm

cushion

rhizoids

archegonium (female sex organ) present in large numbers on under surface of cushion

egg cell

rhizoids

outer layer of prothallus

antheridium (male sex organ) found in clusters on under surface of cushion

Fertilisation occurs in archegonium when prothallus is covered with water film

developing male sex cells

male sex cell

cilia

DEVELOPMENT OF YOUNG FERN

archegonium

foot

tissue of prothallus

young fern plant developing on the tissues of prothallus

root

young fern

stem

above—*Club-moss, showing trailing stem with small stiff leaves and rootlets*

left—*Horsetails, survivors of a major plant group, showing fertile cones and characteristic rings (whorls) of branches*

Pteridophytes were the dominant vegetation during the Carboniferous epoch, when the Coal Measures were being laid down. The largest ferns were the tree ferns, with a woody stem and growing to a height of sixty feet. All the leaves in ferns produce spores. Other pteridophytes are the **clubmosses** and **horsetails**. Clubmosses produce spores on the terminal leaves only, which are collected together to form cones. They sometimes produce two kinds of spore, developing into single-sex prothalli. Clubmosses are rarely found except on wet hillsides in the north and west of the British Isles, but horsetails are widespread, being particularly common on railway embankments.

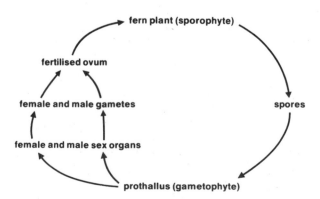

Life cycle of the fern

1 Set up a moss garden. An aquarium or large shallow dish is suitable. Put a layer of soil in the bottom and arrange your collection of mosses to cover the earth completely. Wet thoroughly to begin with and then water occasionally. Mosses are quite common, in woods and marshy areas, on stone walls or roofs, and sometimes in gutters. Watch for the growth of capsules.

2 Examine a moss plant under a microscope. Since the leaf is one cell layer thick it is easy to observe cell structure.

3 Examine prepared slides showing the spore capsules of the male fern. Notice particularly the spine of thickened cells which act as the catapult.

11 EARTHWORMS AND THEIR RELATIVES

The earthworm belongs to a group of animals called **annelids** (L. *annulus*, ring ; Gk. *eidos*, form). Most annelids have their bodies so distinctly divided into segments that the segment markings can be clearly seen externally. They also have stiff bristles which project out of the body wall and which can be moved by muscles. Internally, adjacent segments contain a similar structure, giving a repeating pattern along the length of the body.

Essentially the body consists of two tubes, one inside the other :

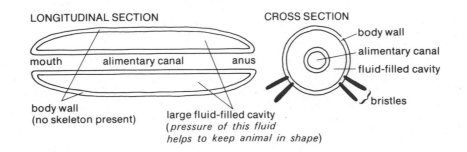

LONGITUDINAL SECTION

mouth   alimentary canal   anus

body wall
(no skeleton present)

large fluid-filled cavity
(*pressure of this fluid
helps to keep animal in shape*)

CROSS SECTION

body wall

alimentary canal

fluid-filled cavity

bristles

The Earthworm

Earthworms live in loose or compact soil. They come to the surface at night or after the burrow has been flooded by heavy rain.

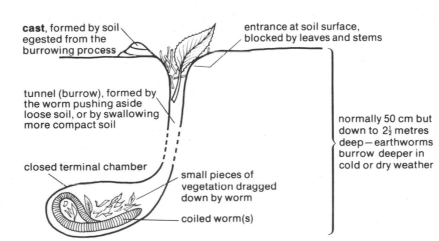

cast, formed by soil egested from the burrowing process

entrance at soil surface, blocked by leaves and stems

tunnel (burrow), formed by the worm pushing aside loose soil, or by swallowing more compact soil

normally 50 cm but down to 2½ metres deep — earthworms burrow deeper in cold or dry weather

closed terminal chamber

small pieces of vegetation dragged down by worm

coiled worm(s)

Appearance

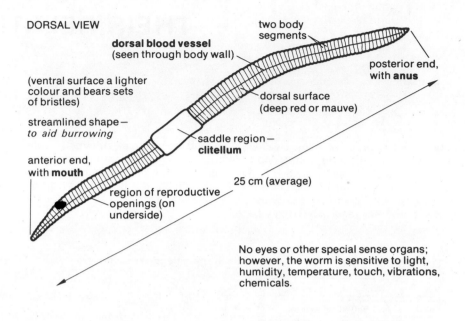

DORSAL VIEW

dorsal blood vessel
(seen through body wall)

two body
segments

posterior end,
with **anus**

(ventral surface a lighter
colour and bears sets
of bristles)

dorsal surface
(deep red or mauve)

streamlined shape—
to aid burrowing

saddle region—
clitellum

anterior end,
with **mouth**

25 cm (average)

region of reproductive
openings (on
underside)

No eyes or other special sense organs;
however, the worm is sensitive to light,
humidity, temperature, touch, vibrations,
chemicals.

Movement

Brought about by the contraction and relaxation of muscles in the body wall, which act on the fluid in the cavity beneath. A series of contraction waves passes down the body from the front to the back end.

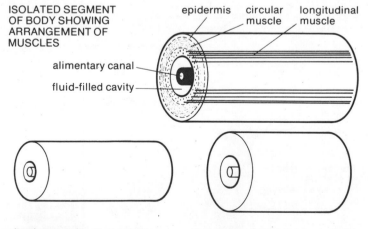

ISOLATED SEGMENT
OF BODY SHOWING
ARRANGEMENT OF
MUSCLES

epidermis   circular   longitudinal
muscle   muscle

alimentary canal

fluid-filled cavity

circular muscles contracted,
longitudinal muscles relaxed;
segment long and thin;
this provides the propulsion

longitudinal muscles contracted,
circular muscles relaxed;
segment short and fat;
this provides the anchorage

If the soil is very tightly compacted, soil is steadily ingested as the worm moves forwards.

Nutrition

Worms feed on organic material which is present in the soil, and on vegetation.

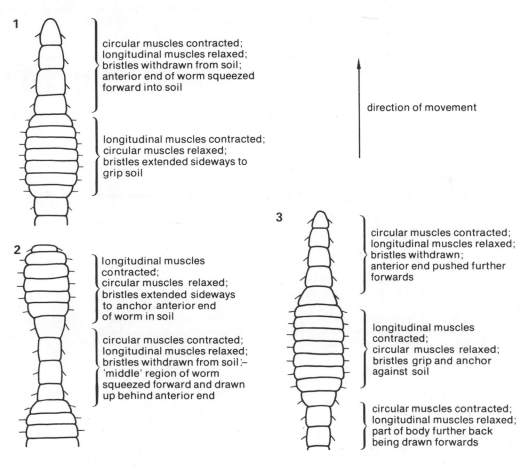

1 circular muscles contracted;
longitudinal muscles relaxed;
bristles withdrawn from soil;
anterior end of worm squeezed
forward into soil

longitudinal muscles contracted;
circular muscles relaxed;
bristles extended sideways to
grip soil

direction of movement

2 longitudinal muscles
contracted;
circular muscles relaxed;
bristles extended sideways
to anchor anterior end
of worm in soil

circular muscles contracted;
longitudinal muscles relaxed;
bristles withdrawn from soil:—
'middle' region of worm
squeezed forward and drawn
up behind anterior end

3 circular muscles contracted;
longitudinal muscles relaxed;
bristles withdrawn;
anterior end pushed further
forwards

longitudinal muscles
contracted;
circular muscles relaxed;
bristles grip and anchor
against soil

circular muscles contracted;
longitudinal muscles relaxed;
part of body further back
being drawn forwards

Respiration

Oxygen is obtained from the atmosphere or from the 'soil air' through the skin:

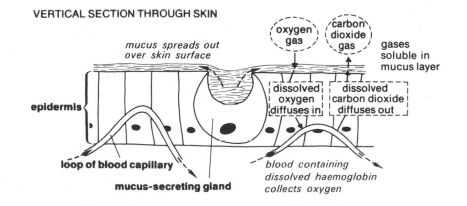

VERTICAL SECTION THROUGH SKIN

mucus spreads out
over skin surface

oxygen
gas

carbon
dioxide
gas

gases
soluble in
mucus layer

dissolved
oxygen
diffuses in

dissolved
carbon dioxide
diffuses out

epidermis

loop of blood capillary

mucus-secreting gland

blood containing
dissolved haemoglobin
collects oxygen

Reproduction

Earthworms are **hermaphrodite**, i.e. each animal has both male and female organs (Gk. *Hermes*, the male winged messenger of the gods; *Aphrodite*, the goddess of love—see also p. 29). There is, however, **cross-fertilization**. Worms come together in pairs on warm damp nights and mate on the soil surface; the whole mating process is completed in about three hours:

1

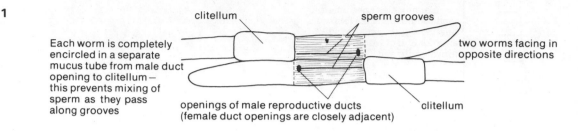

Each worm is completely encircled in a separate mucus tube from male duct opening to clitellum — this prevents mixing of sperm as they pass along grooves

openings of male reproductive ducts (female duct openings are closely adjacent)

clitellum

sperm grooves

two worms facing in opposite directions

clitellum

2 Ventral surfaces of worms placed in contact. Worms held together by a common mucus tube around each clitellum region and by bristles thrust into the opposite member of the pair.

Arrows show movement of sperm along groove from male duct opening into sperm-storing sacs of other worm

3 Worms separate and withdraw into burrows. Clitellum secretes material which hardens into a cocoon, drawn by the worm towards its anterior end.

When cocoon reaches the female opening several eggs are passed into it, and then sperm from the storage sacs.

When anterior end of worm is withdrawn from cocoon the ends are sealed.

Fertilization occurs in the cocoon.

After about twelve weeks' development usually only *one* young worm emerges.

THE ACTIVITIES OF EARTHWORMS IN SOIL

Earthworms are found in all but the most acid soils. In soils rich in humus (see p. 256) as many as two million worms per acre may be present.

The importance of earthworms in maintaining soil fertility was noted as long ago as 1777 by the naturalist the Rev. Gilbert White. Excellent experiments on this aspect of the earthworm's activities were carried out around 1880 by Charles Darwin, who kept worms in pots containing garden soil and described the ways with which they dealt with various types of vegetation given to them. Darwin also collected worm casts from measured areas of land and estimated that about ten tons per acre were produced annually, equivalent to a complete covering of 4mm, though this is probably an overestimate. Darwin spread chalk lumps and cinders on the soil surface and found that in less than thirty years the activity of earthworms had buried them 20cm deep.

Summary of the **beneficial** effects of earthworms in soil:

1 Their burrows

    (i)   provide natural drainage channels

    (ii) bring air down into the soil

    (iii) allow the thicker roots of plants to penetrate more easily.

2 As worms ingest only fine (clay and silt) particles, these are selectively deposited at the surface—sand is left below. The fine particles provide a good tilth (seed bed).

3 The action of the worm's gizzard in breaking-up

the soil and egesting it as fine casts also provides a good tilth.

4 The worm ingests soil at a lower level and egests casts at the surface—this mixes soil from lower layers with those at the surface and increases the depth of topsoil.

5 By dragging organic remains into their burrows, these remains become mixed with soil more quickly—they are decomposed more quickly by soil bacteria to form future plant food.

6 Urine and faeces from the worm add manure to the soil and encourage the activity of soil bacteria.

7 The worm's oesophagus glands neutralize slightly acid soils when these soils are ingested.

The activities of worms are **detrimental** to lawns—the excavations spoil the lawn surface, cause bare patches and encourage the growth of moss.

Other annelids

Ragworm from mud in the Thames Estuary

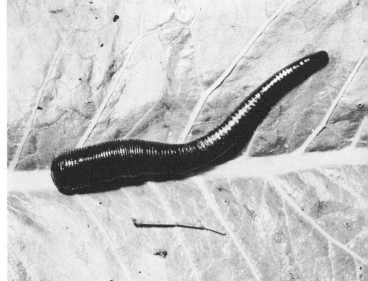

Pond leech on a floating leaf

Ragworms may be found at the lower levels of the seashore or in estuaries, under stones or in burrows in muddy sand. The head is better developed than in earthworms and has sensory tentacles, eyes and jaws for catching and biting food. Each body segment has a pair of flat paddles, used as an oar in swimming and to increase the surface area for respiration.

1 Set up a wormery. Three-quarters fill a glass container with damp soil, place several worms on the soil and cover with dead leaves. Keep the wormery in a dark place. At intervals investigate the effect of the worms on the soil and on the leaves; also examine the activities of the worms seen through the glass sides.

2 Place an earthworm on damp blotting paper in a dish and investigate the animal's movements.

Leeches live in rivers and marshes. The body is rather flat with an anterior and posterior sucker for attachment to skin, which the animal punctures with sharp-toothed jaws. The blood which is sucked can be stored in the leech's body in quite large quantities. Leeches were once used by doctors to 'let' human blood. Like earthworms, leeches are hermaphrodite.

3 Place a leech in a dish of water. Observe its movements; how do they differ from those of an earthworm?

4 Read further about segmented worms, particularly those that live in the sea. Enlarge your knowledge of the animal kingdom—read about other marine invertebrates such as molluscs and echinoderms.

12 ARTHROPODA

(Gk. *arthron*, joint ; *pous*, foot)

Arthropods are mostly small animals, covered by an **exoskeleton**, a horny material secreted by the epidermis. Except at the joints, which must necessarily remain flexible, the exoskeleton is relatively hard and inelastic. For growth to occur the skeleton is shed periodically (**ecdysis**). The body is divided into clearly visible segments, which may possess jointed appendages. These appendages may be used for walking, swimming, feeding, as sense organs, or as structures involved in reproduction. The circulation consists of a dorsal, tubular heart which receives blood from and discharges into a series of blood spaces. Since there are very few blood vessels, this is described as an 'open system' and the organs are, in effect, bathed in the blood, which is usually colourless. It is also a low pressure system. The young arthropod is often markedly different in appearance and habits from the adult.

aquatic larvae which breathe by gills, they breathe through a system of air tubes (**tracheae**) which supply oxygen direct to the cells of the body.

The skeleton

This serves the functions of support and movement. The walking legs, mouthparts and antennae are articulated, jointed, hollow levers moved by sets of opposed muscles.

The outer layer of the cuticle remains flexible at the joints between the segments of the body and between the different segments of the appendages. At various points the exoskeleton is infolded to form an internal framework for the attachment of muscles. The anterior and posterior parts of the digestive system, together with the larger tubes of the respiratory system are also lined with cuticle.

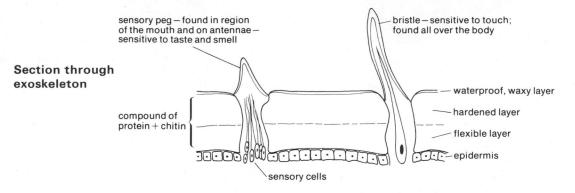

Section through exoskeleton

sensory peg — found in region of the mouth and on antennae — sensitive to taste and smell

bristle — sensitive to touch; found all over the body

compound of protein + chitin

waterproof, waxy layer

hardened layer

flexible layer

epidermis

sensory cells

Insecta

(L. *insectum*, cut into : a reference to the division of the insect body into clearly visible segments)

Insects are a very successful group of animals which has existed for at least five hundred million years, since Devonian times. About three quarters of a million species have been described. They are very small and this is probably mainly due to their method of respiration. They are land arthropods with **six legs** and usually **two pairs of wings**. The body is divided into **head**, **thorax** and **abdomen**. Except for some

The compound eye

The surface of the compound eye in an insect is made up of a number of closely fitting hexagonal facets, each of which is the surface of the lens of a single unit called an **ommatidium**. The number of ommatidia in the eye as a whole varies, according to species, from several hundred to several thousand. Each is surrounded by a sheath of pigmented cells so that each receives light from a limited area of the visual field. The compound eye therefore forms what is called a mosaic image, formed from the slightly overlapping images of the separate omma-

A SINGLE OMMATIDIUM

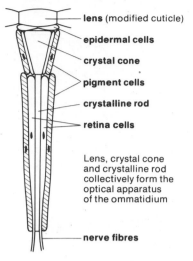

lens (modified cuticle)

epidermal cells

crystal cone

pigment cells

crystalline rod

retina cells

Lens, crystal cone
and crystalline rod
collectively form the
optical apparatus
of the ommatidium

nerve fibres

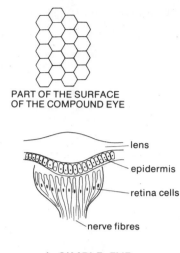

PART OF THE SURFACE
OF THE COMPOUND EYE

lens

epidermis

retina cells

nerve fibres

A SIMPLE EYE

tidia. Such an image lacks definition but the eye is thought to be particularly sensitive to movement of objects in the visual field. Insects also possess **simple eyes** on the dorsal surface of the head. These are probably only sensitive to light and do not form images.

Nutrition

Three pairs of appendages are adapted to form **mouthparts**. For any material that is edible an insect species can be found which will live on it. Mouthparts therefore are extremely variable, according to whether they are used for biting, sucking or puncturing, or whether they are used for hard materials, like wood, or liquids such as nectar or blood.

Respiration

Oxygen and carbon dioxide are carried to and from the tissues by means of a system of air tubes. The blood plays no part in the carriage of respiratory gases. Oxygen enters by way of the **spiracles** found along the sides of the body. These may be controlled by valves. They open into a series of air sacs from which smaller tubes, the **tracheae**, lead off. Ventilation of the air sacs may occur by the bellows-like action of the body wall. Movement of gases along the tracheae is the result of simple physical diffusion, a process which is effective as a means of transportation only over short distances.

Reproduction and growth

Insects reproduce **sexually**, the sexes being separate. **Fertilization** is internal and sperm can sometimes be stored by the female after mating has occurred. In some species, such as the aphids, the females lay eggs which are capable of developing without fertilization. These eggs which are laid during the summer always develop into females. In bees, sex is determined by whether or not the egg is fertilized, male bees developing from unfertilized eggs.

All insects show a pattern of stepwise growth. This is a consequence of the presence of an exoskeleton, which is non-living.

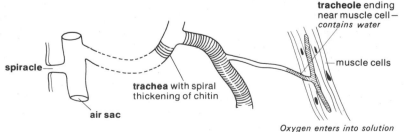

spiracle

trachea with spiral
thickening of chitin

air sac

tracheole ending
near muscle cell—
contains water

muscle cells

*Oxygen enters into solution
before absorption by tissues*

Growth pattern of locust

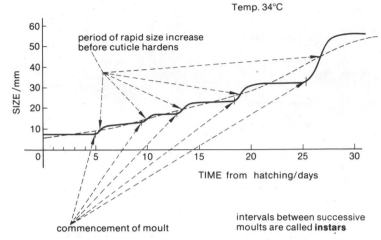

Temp. 34°C

period of rapid size increase
before cuticle hardens

SIZE /mm

TIME from hatching/days

commencement of moult

intervals between successive
moults are called **instars**

Growth rates are dependent upon the temperature. For example, in the house fly (*Musca domestica*), the time taken from hatching to the commencement of the final larval instar may be as little as $5\frac{1}{2}$ days (at 30°C) or as long as 34 days (at 20°C).

The number of moults and the total time taken for the adult to appear varies considerably with species. The maggot of the house fly will moult twice. A mayfly larva on the other hand may moult forty times and take two years to become an adult.

Three basic patterns of development may be recognized among insects:

1 The young may hatch as a miniature copy of the adult. Successive instars therefore look very much alike, only differing in size. This is the case with wingless insects such as silver-fish and springtails.

2 The young form may be fairly similar to the adult, except that it lacks wings, or that they are only partly developed. The locust is an example. Some insects in this group have a young form which is adapted to life in water and which breathes by means of gills. Examples: dragonfly, mayfly, caddis fly. The young insect in either case is called a **nymph**.

3 The young insect is known as a **larva**. It is completely different from the adult in habits and diet as well as appearance. It possesses only simple eyes, or sometimes no eyes at all. A special instar, the **pupa**, occurs during which extensive reorganization of the larval tissues into the adult form takes place. This is a stage in which there is usually little external sign of activity. Examples of larvae are the caterpillar of the butterfly or moth, the grub of bees or wasps, and the maggot of two-winged flies. (See p. 48).

Locomotion

Walking The action of the leg of an insect is essentially similar to that of a mammal, but the fact that there are six legs attached to a relatively rigid thorax means that the basic patterns of movement over the ground are quite different. An insect always has three legs in contact with the ground at any one time. They therefore form a tripod within which the centre of gravity of the insect lies. The members of any particular pair of legs normally alternate in their movements and it is generally the case that a fore- or middle limb is not raised until the one behind on the same side has been lowered. The usual pattern of movement is therefore fore and hind leg on one side together with the middle leg on the other side. Each leg spends the same length of time in the air as on the ground and speed over the ground is determined entirely by how fast the complete cycle of limb movement can be repeated. Since three legs must always be in contact there can be no increase in speed due to increase in stride, as is the case with a galloping mammal. Insects can therefore move only comparatively slowly on the ground. Movement is controlled by reflex action but this is not absolutely fixed. If the insect loses a leg or two the pattern of movement of the remainder is adapted so as to maintain stability. In those insects (e.g. locusts) which hop locomotion is mainly a function of the hind legs, which are very much larger than the others. Their action is comparable to those of vertebrates which move in a similar way. (p. 66).

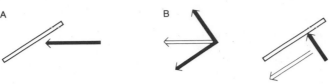

A   B   C

FLIGHT

A body of air (or water) flowing past a flat surface arranged at an angle to the direction of flow exerts a force on that surface (diag. A). As can be seen by reference to any elementary physics textbook, this force can be regarded as equivalent to two smaller forces acting at right angles to each other (diag. B).

One of these forces acts parallel to the surface and can be ignored. The other acts perpendicular to the surface (diag. C). Since for every action there is an equal and opposite reaction we can also regard the surface as exerting a force on the air. The situation is exactly the same if the surface is moving and the air is stationary. There are many examples of this simple principle at work. A kite on the end of a string will rise only on a windy day, not on a calm one. A skier, towed behind a speedboat, can stand supported on a narrow board. When he stops moving the board will not support his weight.

An insect's wing is a similar flat surface. On the downstroke it is held at an angle to its direction of movement through the air. The force in this case provides for both lift and forward thrust. The wing is a simple lever with the flight muscles acting close to the pivot point. Because of the small size of insects and the correspondingly short length of wing, the muscles act through a very short distance. This, coupled with the fact that insect flight muscles have an extremely rapid rate of contraction, provides an extremely efficient flight system.

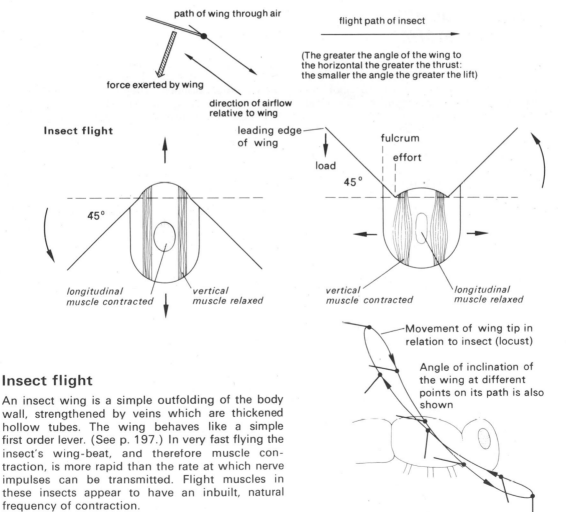

path of wing through air

force exerted by wing

direction of airflow relative to wing

flight path of insect

(The greater the angle of the wing to the horizontal the greater the thrust: the smaller the angle the greater the lift)

Insect flight

leading edge of wing

load

fulcrum

effort

45°

45°

longitudinal muscle contracted

vertical muscle relaxed

vertical muscle contracted

longitudinal muscle relaxed

Movement of wing tip in relation to insect (locust)

Angle of inclination of the wing at different points on its path is also shown

Insect flight

An insect wing is a simple outfolding of the body wall, strengthened by veins which are thickened hollow tubes. The wing behaves like a simple first order lever. (See p. 197.) In very fast flying the insect's wing-beat, and therefore muscle contraction, is more rapid than the rate at which nerve impulses can be transmitted. Flight muscles in these insects appear to have an inbuilt, natural frequency of contraction.

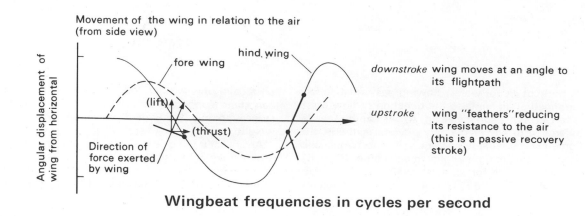

Movement of the wing in relation to the air (from side view)

Angular displacement of wing from horizontal

fore wing

hind wing

(lift)

(thrust)

Direction of force exerted by wing

downstroke wing moves at an angle to its flightpath

upstroke wing "feathers" reducing its resistance to the air (this is a passive recovery stroke)

Wingbeat frequencies in cycles per second

Butterfly 5   Dragonfly 35   Mosquito 580   Locust 18   Honey bee 250   Midge 1000

See also 'bird flight' (p. 74).

The African Migratory Locust
'Locusta migratoria'

The African migratory locust is found in most parts of tropical Africa. It will feed on any green vegetation. Mating occurs almost as soon as the adults emerge from the final moult. It is preceded by courtship behaviour and actual copulation may last for two hours. The female excavates a hole in damp sand with the help of her ovipositors, extending the hole well beyond the normal length of her abdomen, which she is able to extend to almost twice its normal length. The hole is filled with eggs (about 60) which are covered with a frothy substance. This later hardens to form the egg pod.

The adults swarm only after rain. Such a swarm may extend over 500 square kilometres, devastating crops wherever it lands. The young locusts (nymphs) are known as hoppers. They march during the day, climbing up the vegetation at dusk. They, too, cause enormous devastation to crops lying in their path.

Gnat (*Culex*)

The **eggs** are cigar shaped, about 1 mm in length. The eggs are glued together when laid, to form a raft which floats on the surface. They may be laid on any standing water—ponds, puddles, ditches, rainwater butts. The **larva** feeds on particles of plant or animal matter. Some species

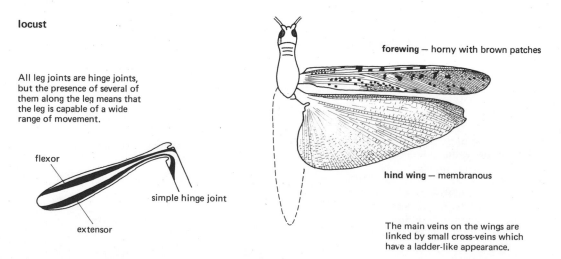

locust

All leg joints are hinge joints, but the presence of several of them along the leg means that the leg is capable of a wide range of movement.

flexor

extensor

simple hinge joint

forewing — horny with brown patches

hind wing — membranous

The main veins on the wings are linked by small cross-veins which have a ladder-like appearance.

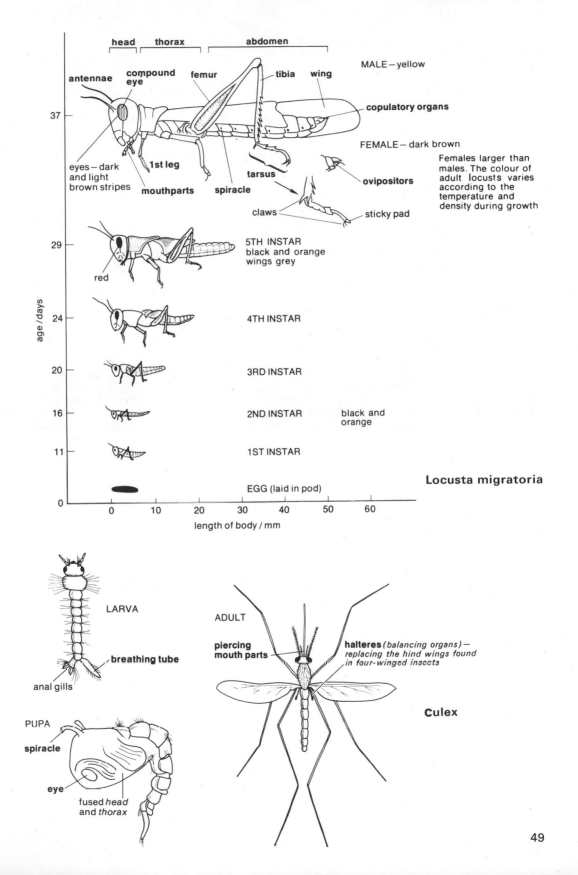

head | thorax | abdomen

MALE — yellow

antennae
compound eye
femur
tibia
wing

copulatory organs

FEMALE — dark brown

eyes — dark and light brown stripes
1st leg
mouthparts
tarsus
spiracle
ovipositors

Females larger than males. The colour of adult locusts varies according to the temperature and density during growth

claws
sticky pad

37

5TH INSTAR
black and orange wings grey

red

29

4TH INSTAR

24

3RD INSTAR

20

2ND INSTAR
black and orange

16

1ST INSTAR

11

EGG (laid in pod)

Locusta migratoria

0

age / days

0   10   20   30   40   50   60

length of body / mm

LARVA

breathing tube

anal gills

ADULT

piercing mouth parts

halteres *(balancing organs)* — *replacing the hind wings found in four-winged insects*

Culex

PUPA

spiracle

eye

fused *head* and *thorax*

49

are predatory. Others may be cannibals if food is scarce. They breathe air, coming at intervals to the surface, where they remain suspended from the surface film by the posterior breathing tube. The larva moults three times. Its development may occupy a few days or several months, according to temperature. At the fourth moult the **pupa** emerges. Although it does not feed it is unusual in that it moves actively about, coming to the surface from time to time in order to breathe through the spiracle on its head. After from one to five days the pupa rests at the surface, the skin splits down its back and the **adult** insect emerges. Adult males feed on nectar but females must feed on blood before they are able to lay eggs. Adult gnats hibernate.

The large white butterfly (*Pieris brassicae*)

First specimens appear in late April or early May. They mate at rest but may fly off, still coupled, if disturbed. The **eggs** are laid on the underside of cabbage or nasturtium leaves.

Eggs hatch in 8–10 days, the young **caterpillar** eating the eggshell. The larvae remain together at first, feeding on the leaf. They separate after the third moult.

The caterpillar has a pair of tough jaws (mandibles) which it uses to bite off pieces of leaf. Its salivary glands are modified to form silk-producing organs which open to the surface behind the mouth by a tube called the spinneret. Its true legs are found on the thorax and each has five sections. The false legs are fleshy and unjointed. They end in a pad with a ring of hooks. The claspers are similar in structure.

The caterpillar moults 4–5 times during a period of about a month. It then leaves the food plant, migrating to a sheltered fence or wall where it spins a silk pad to which it attaches itself by means of the clasper. It also spins a silk rope which suspends the silk pad from its support at the anterior end. The caterpillar then moults, the **pupa** (as it now is) attaching itself to the silk pad by a cluster of hooks at the tail end.

The pupal stage lasts for about three weeks. From August onwards the **adults** appear. They, too, mate, and lay eggs. In this second generation the pupae form the over-wintering stage, emerging as adults in the following April. The adult feeds on nectar and lives for about three weeks. The mouthparts are modified to form a proboscis, which

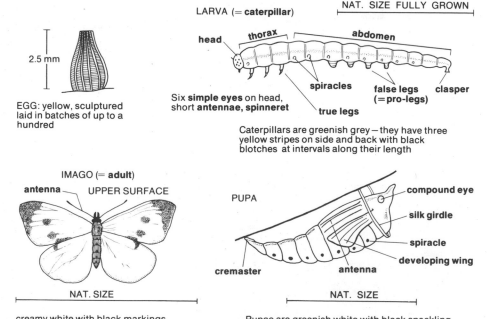

2.5 mm

EGG: yellow, sculptured laid in batches of up to a hundred

LARVA (= **caterpillar**) NAT. SIZE FULLY GROWN

head thorax abdomen

spiracles false legs (= pro-legs) clasper

Six **simple eyes** on head, short **antennae, spinneret** true legs

Caterpillars are greenish grey—they have three yellow stripes on side and back with black blotches at intervals along their length

IMAGO (= **adult**)

antenna UPPER SURFACE

NAT. SIZE

creamy white with black markings (male, slightly smaller, lacks spots on forewings)

PUPA compound eye

silk girdle

spiracle

developing wing

cremaster antenna

NAT. SIZE

Pupae are greenish white with black speckling

is rather like a coiled drinking straw. The numbers of individuals are controlled by the activities of a species of **ichneumonid fly**, which lays its eggs in the caterpillars, large numbers of which never reach the stage of pupation.

The large white belongs to an insect order called **Lepidoptera** (Gk. *lepis*, scale; *pteron*, wing) which also includes the moths. Its life cycle, which is characteristic of the order generally, is summarized below:

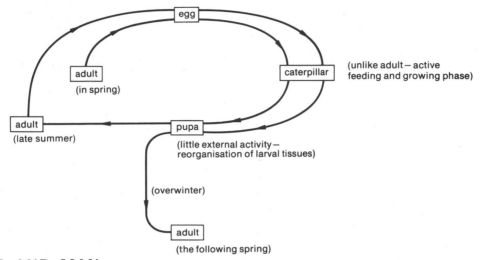

egg

adult
(in spring)

caterpillar
(unlike adult — active
feeding and growing phase)

adult
(late summer)

pupa
(little external activity —
reorganisation of larval tissues)

(overwinter)

adult
(the following spring)

INSECTS AND MAN

Insects may be regarded as Man's most successful rivals. They compete for the use of all sorts of crops and stored foods. The apple tree, for example, provides a source of food or a home for over four hundred species of insect. Whatever the material, provided it is organic in origin, an insect species may be found which will feed on it. Clothes, furniture, books, carpets: these are some of the articles of use to Man which insects can use as food. Insects are involved in the transmission of disease, either accidentally, by the carrying of bacteria on the surface of the body, or because they form a necessary phase, as hosts, in the life cycle of the disease organism. Examples of these are the mosquito (malaria and yellow fever), tsetse fly (sleeping sickness) and the rat flea (plague). Some insects, however, are beneficial to Man. The bees, moths, butterflies and some two-winged flies act as pollinators. Bees also supply Man with honey and beeswax. The silkworm produces silk. There are even some primitive tribes which accept insects as a normal part of the diet, or even as a delicacy.

Dragonfly (adult), showing head with small antennae and large compound eyes, thorax with legs and veined wings, and part of the abdomen

Blood-sucking insect (tsetse fly) with piercing and sucking mouthparts, during feeding

Aquatic bug with hind-limbs forming oars

Cryptic camouflage (leaf insect)

INSECTS AS LAND ANIMALS

The possession of a tough, waterproof exoskeleton has two major advantages. It hinders loss of water, essential to an organism living on dry land, and it provides a support system. It makes possible the development of appendages, useful for a variety of purposes, for example, walking, egg-laying, or the manipulation of food. The development of wings is also clearly important. Insects can get rid of their nitrogenous waste in a semi-solid form which helps in the conservation of water. Their mechanism of respiration, linked with their small size, allows for a high level of metabolism under suitable temperature conditions. Fertilization is internal. They generally have a short life cycle with a high rate of reproduction. This enables them to adapt rapidly to changing environmental conditions. Man has good cause to realize this, because not only do insects quickly exploit food sources grown for man's own benefit, but they also rapidly adapt by becoming resistant to new measures he may take to reduce their numbers. For example, insects are generally able to develop resistance to insecticides more rapidly and effectively than the birds which feed upon them.

Other arthropods

Myriapoda (Gk. *myrias*, many; *pous*, foot): Centipedes and millipedes.

Crustacea (L. *crusta*, rind or shell): Exoskeleton often calcareous and massive. Except for a few species, such as the wood louse, crustaceans are aquatic, both freshwater and marine, and breathe by means of gills. Examples: crab, lobster, barnacle, water flea, freshwater shrimp.

Arachnida (from *Arachne*, a mythical Greek princess, expert at spinning): Spiders and scorpions. All have eight legs.

1   Many insects are fairly easy to keep. Caterpillars, for example, are generally to be found on their food plant and can be observed in the laboratory if supplied with adequate amounts of the food plant. A large cardboard box with a piece of glass for a lid serves as a suitable home.
2   Some insects (stick insects, silk worms, locusts), can be obtained from dealers. Locusts require warmth (light bulb) and jars with several inches of damp sand positioned under holes in the bottom of the cage for egg-laying.
3   A freshwater pond or slow-moving stream yields a rich variety of insects, either as adults or larvae. They may be kept in the laboratory. Identify your specimens, which should be collected separately in specimen tubes. A number of jam-jars will serve as aquaria. Find out as much as possible about your catch. It is important that you keep the fiercer ones separately. The Great Diving Beetle, for example, either as adult or larva, will soon eat everything else in its neighbourhood.

13 FLOWERING PLANTS

Flowering plants are classified into two large groups:

1 **Monocotyledons**—grow from a seed with a single cotyledon (seed leaf) only; their leaves are usually parallel-veined and their flower parts, such as petals and stamens (see p. 218) tend to be arranged in threes or sixes.

2 **Dicotyledons**—grow from a seed with two cotyledons; their leaves are usually net-veined and their flower parts tend to be arranged in fours or fives.

The two groups also differ in several other important respects—method of growth, internal anatomy (e.g. arrangement of vascular tissue) of stem, root and leaf.

Flowering plants can also be classified according to the length of their life-cycles:

A. **Ephemerals**—more than one life-cycle completed in each year.

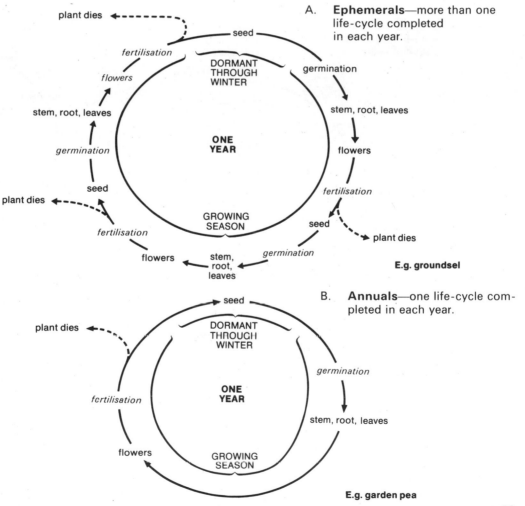

E.g. groundsel

B. **Annuals**—one life-cycle completed in each year.

E.g. garden pea

C. **Biennials**—one life-cycle completed every two years.

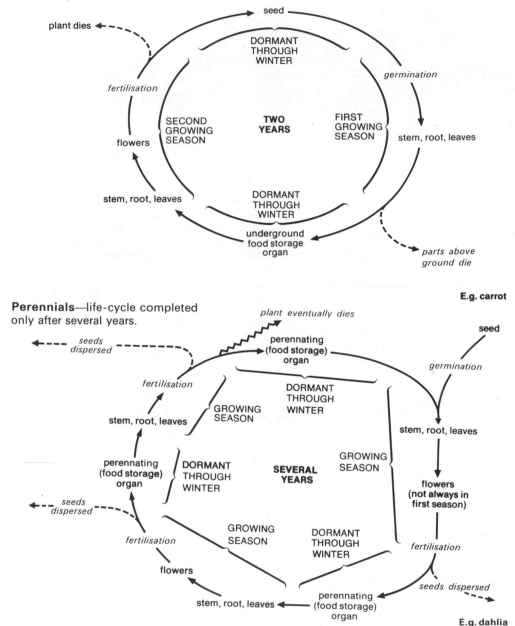

E.g. carrot

D. **Perennials**—life-cycle completed only after several years.

E.g. dahlia

In addition, flowering plants may also be either:

(a) **Herbaceous**—comparatively small plants with unthickened soft green stems, e.g. buttercup, dead nettle, willow-herb.

or—

(b) **Woody**—comparatively large plants with hard brown stems (covered by bark), e.g. privet, hawthorn, oak. Woody plants may be **deciduous**, i.e. shedding their leaves regularly at the end of the growing season, or **evergreen**, i.e. shedding their leaves at irregular intervals. Woody plants are often described as either shrubs or trees; the distinction between the two is fairly imprecise and depends on shape and size differences—

shrubs are usually smaller and more bushy in appearance, i.e. they do not have a single large trunk.

Some plants, e.g. wallflowers, are mainly herbaceous but become woody towards ground level.

Flowering plants are composed of three main organs—stem, root and leaf. At certain times the stem may bear organs of sexual reproduction, flowers.

some are horizontal, and some may be underground (see p. 215); they grow from the plumule (see p. 228) of the seed.

Stems have the following main functions:

1 To hold the leaves to the light (for food manufacture by photosynthesis—see p. 95).

2 To support the flowers in suitable positions for pollination (i.e. to be visible to insects or to hang in air currents—see p. 219).

3 To act as a pipeline between root and leaves.

4 To become modified as organs of food storage and vegetative propagation (see p. 215).

5 To manufacture food by photosynthesis (in green stems).

Stems

Stems are essentially cylindrical structures, though a variety of shapes is found, e.g. four-angled stems in the mint family. Typically, stems are upright, but

Internal structure—herbaceous dicotyledon

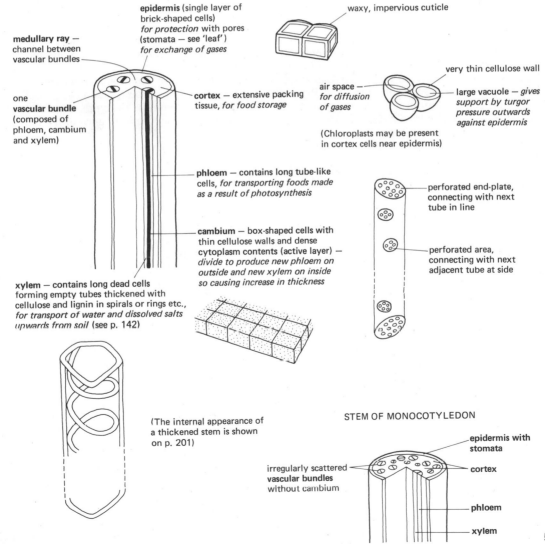

epidermis (single layer of brick-shaped cells) *for protection* with pores (stomata — see 'leaf') *for exchange of gases*

waxy, impervious cuticle

medullary ray — channel between vascular bundles

one **vascular bundle** (composed of phloem, cambium and xylem)

cortex — extensive packing tissue, *for food storage*

air space — *for diffusion of gases*

very thin cellulose wall

large vacuole — *gives support by turgor pressure outwards against epidermis*

(Chloroplasts may be present in cortex cells near epidermis)

phloem — contains long tube-like cells, *for transporting foods made as a result of photosynthesis*

perforated end-plate, connecting with next tube in line

cambium — box-shaped cells with thin cellulose walls and dense cytoplasm contents (active layer) — *divide to produce new phloem on outside and new xylem on inside so causing increase in thickness*

perforated area, connecting with next adjacent tube at side

xylem — contains long dead cells forming empty tubes thickened with cellulose and lignin in spirals or rings etc., *for transport of water and dissolved salts upwards from soil* (see p. 142)

(The internal appearance of a thickened stem is shown on p. 201)

STEM OF MONOCOTYLEDON

irregularly scattered **vascular bundles** without cambium

epidermis with stomata

cortex

phloem

xylem

External appearance

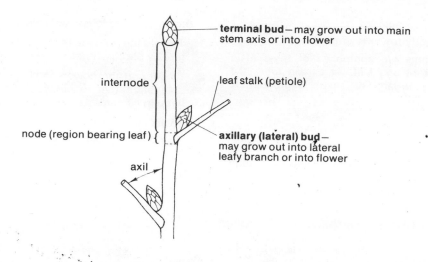

terminal bud—may grow out into main stem axis or into flower

internode

leaf stalk (petiole)

node (region bearing leaf)

axillary (lateral) bud— may grow out into lateral leafy branch or into flower

axil

Buds are compressed embryonic shoots capable of developing into new shoots (which themselves produce buds) or flowers. If the bud produces a flower, new stem growth must take place from another bud elsewhere on the stem. The Brussels sprout is an example of an unprotected green bud:

Commonly, buds contain foliage leaves surrounded by tough scale leaves. This is seen in the buds of plants that are dormant throughout the winter.

The scale leaves may give further protection by a covering of hairs (in willows) or by secreting gummy substances (in horse-chestnut). When the bud begins to sprout in spring the scales are forced open and eventually drop off, leaving a **girdle scar** (see diagram opposite). The foliage leaves inside expand, separate from each other by elongation of internodes, and become green.

Stems sometimes develop buds that are not axillary. Similarly leaves and roots may develop buds in connexion with vegetative propagation (see p. 215). Such buds are called **adventitious**.

L.S. Brussels sprout

inner green foliage leaves

stem apex (growing point)

green outer foliage leaves: cut edge

surface view

axillary bud

compressed (shortened) stem

L.S. Lilac bud

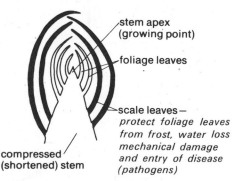

stem apex (growing point)

foliage leaves

scale leaves— *protect foliage leaves from frost, water loss mechanical damage and entry of disease (pathogens)*

compressed (shortened) stem

WINTER TWIGS

Deciduous trees and shrubs shed their leaves at the onset of winter (see p. 141), leaving twigs of characteristic appearance.

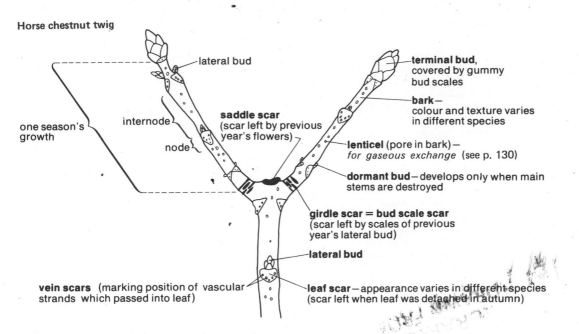

Horse chestnut twig

lateral bud

terminal bud, covered by gummy bud scales

bark— colour and texture varies in different species

internode

one season's growth

node

saddle scar (scar left by previous year's flowers)

lenticel (pore in bark)— *for gaseous exchange* (see p. 130)

dormant bud—develops only when main stems are destroyed

girdle scar = bud scale scar (scar left by scales of previous year's lateral bud)

lateral bud

vein scars (marking position of vascular strands which passed into leaf)

leaf scar—appearance varies in different species (scar left when leaf was detached in autumn)

Roots

These have the following main functions:

1   To anchor the plant in the ground.

2   To absorb water and dissolved mineral salts from the soil (see p. 133).

3   To act as a pipeline between the soil and the stem.

4   To become modified as organs of food storage and vegetative propagation (see p. 215).

Differences between stems and roots

STEMS	ROOTS
1 Develop from plumule	Develop from radicle
2 Bear leaves and buds	Bear root hairs
3 Nodes and internodes present	Nodes and internodes absent
4 Often green	Never green, usually white
5 No cap present at apex	Root-cap present at apex
6 Vascular tissue situated away from central position	Vascular tissue situated as a central core
7 Vascular tissue as separate vascular bundles	Vascular tissue as a single strand

External appearance

mature part of the root

lateral root

main root

zone of root hairs

zone of rapid cell elongation (see p. 200)

root cap

new root hairs being formed here

old root hairs dying here

Flowering plants

There are two main sorts of root system:

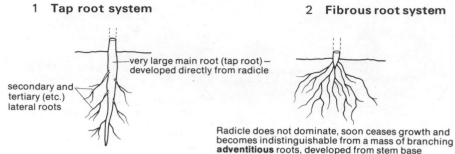

1 Tap root system

very large main root (tap root) — developed directly from radicle

secondary and tertiary (etc.) lateral roots

E.g. carrot, dandelion

2 Fibrous root system

Radicle does not dominate, soon ceases growth and becomes indistinguishable from a mass of branching **adventitious** roots, developed from stem base
E.g. grasses

Many plants develop **adventitious roots**; these are root-like structures that develop from some part other than a true root. Examples are the adventitious roots of rhizomes, bulbs and corms (see p. 215) and of climbing stems, e.g. ivy.

Internal structure (see also p. 134)

(The cells in these root tissues are essentially the same as those in the stem.)

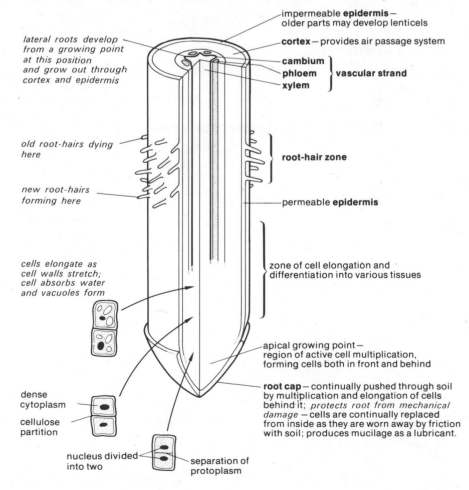

impermeable **epidermis** — older parts may develop lenticels

cortex — provides air passage system

lateral roots develop from a growing point at this position and grow out through cortex and epidermis

cambium
phloem **vascular strand**
xylem

old root-hairs dying here

new root-hairs forming here

root-hair zone

permeable **epidermis**

cells elongate as cell walls stretch; cell absorbs water and vacuoles form

zone of cell elongation and differentiation into various tissues

dense cytoplasm

cellulose partition

apical growing point — region of active cell multiplication, forming cells both in front and behind

root cap — continually pushed through soil by multiplication and elongation of cells behind it; *protects root from mechanical damage* — cells are continually replaced from inside as they are worn away by friction with soil; produces mucilage as a lubricant.

nucleus divided into two

separation of protoplasm

The arrangement of vascular tissue in the stem and root

The xylem of the vascular tissue contains tubes whose walls are thickened by wood (lignin); these provide important strengthening and supporting material.

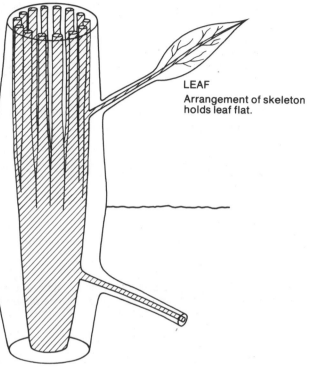

STEM

Stems are subjected mainly to bending forces (e.g. by the wind and passing animals). One side becomes compressed, the other side stretched. The arrangement of veins permits flexibility, but gives greatest strength near the outside.

LEAF

Arrangement of skeleton holds leaf flat.

ROOT

Roots are subjected to pulling forces i.e. to stresses that pass lengthwise. The central core of vascular tissue resists these forces without reducing the flexible, rope-like nature of the root.

Leaves

These have the following main functions:

1 For food manufacture (see p. 98 for the leaf as an organ suited to photosynthesis).

2 For diffusion of gases, for both respiration and photosynthesis (see p. 121 and p. 95).

3 To become modified as organs of food storage and vegetative propagation (see p. 215).

External appearance

Leaves are developed on stems at the nodes; they are arranged variously:

rosette
(leaves in a ring
at ground level)

opposite

alternate
(spiral)

whorl

Flowering plants

Leaves can be simple or compound, net-veined or parallel-veined:

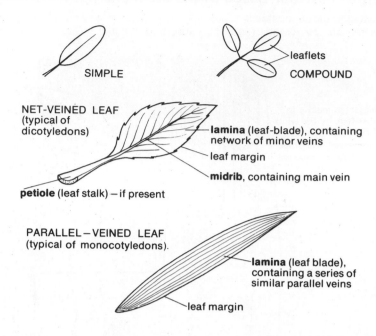

The leaf margin may be entire (as shown in the diagrams), or else 'broken' to form a variety of different wavy or toothed outlines. In many leaves the petiole is absent.

Internal structure

Stereogram of leaf

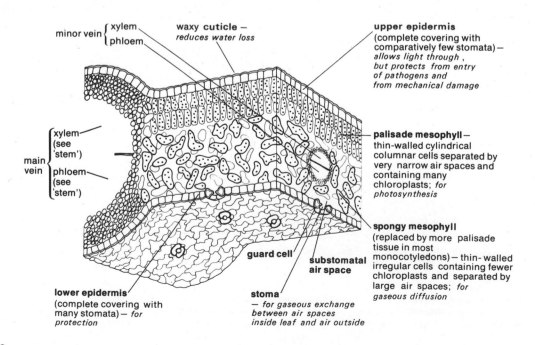

SURFACE VIEWS OF STOMATA

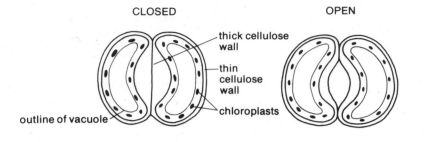

Stomata open and close when the vacuoles of the guard cells gain and lose water respectively. This is because of the thickenings of cellulose where the guard cells are in contact; these two cells behave like balloons which assume sausage-shapes when inflated with water.

1 Classify a range of flowering plants according to the nature of their life-cycles.

2 Either by means of prepared microscope slides or by means of thin sections cut with a sharp razor, compare the layout of tissues in stems and roots. The sections may be examined with a hand lens or in a drop of water on a slide under a microscope.

3 Collect and draw a range of twigs in winter and compare the appearance of these twigs in their summer condition.

4 Cut a vertical section through a Brussels sprout bud and through a bud protected by scale leaves; compare their structure.

14 VERTEBRATES

Main features:

1 The possession of a living, internal, bony skeleton part of which forms a backbone.

2 The presence of gill slits, leading from the throat to the outside of the body, for part or all of the life cycle.

3 Closed system of blood vessels, with a ventral heart, and the respiratory pigment, haemoglobin, found inside the blood cells.

4 Highly organized nervous system, with a much higher level of centralization than in any other group of animals.

Vertebrates are divided into two main groups:

1 Fishes—live in water and breathe by means of gills.

2 Land vertebrates—possess limbs and breathe air. They are divided into four classes: Amphibia, Reptiles, Birds and Mammals.

The herring (*Clupea harengus*)

The herring is a vertebrate animal which spends its entire life in water. It shows the following adaptations to aquatic life.

Shape. It is streamlined. The only projections which break its smooth outline are the fins. These are thin and offer little resistance to forward movement.

Surface. The body is covered by thin bony scales which overlap from front to rear. Being smooth and hard, they minimize the friction caused between the fish and the water flowing past it.

Movement. The backbone is flexible in the horizontal plane, the lateral movements of the backbone being brought about by the coordinated contractions of the segmented, longitudinally arranged muscles along its length.

Control. The dorsal and anal fins make the fish stable about its longitudinal axis. They prevent it from rolling. The paired pectoral and pelvic fins control movement in the pitching plane. The tail fin provides the main propulsive surface in forward movement. The swim bladder is a density-regulating device.

Respiration. The fish breathes by means of **gills**. These are made up of a number of closely packed plate-like folds (lamellae) between which the respiratory water current flows. The lamellae are an effective way of increasing the respiratory surface in an aquatic animal but are ineffective for air breathing. When a fish is removed from water the lamellae stick together by surface tension, like the bristles of a wet paint brush.

Herrings are carnivorous animals, as many animals living in the open sea tend to be. They feed on copepods, which are small arthropods forming part of the **plankton** (minute animals and acellular plants in the surface layers of the ocean).

There is no external difference between the sexes. The eggs are about 1 mm in diameter and the mature ovary may contain up to 30 000 ripe eggs at any one time. They are fertilized externally and spawning takes place in shallow waters where the seabed is stony: the eggs are deposited in clumps in the crevices between the stones. There is no parental care.

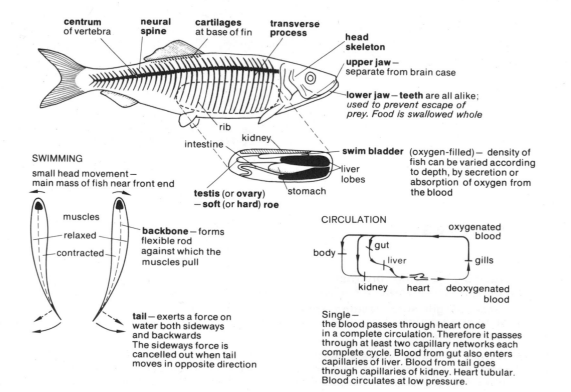

Herring

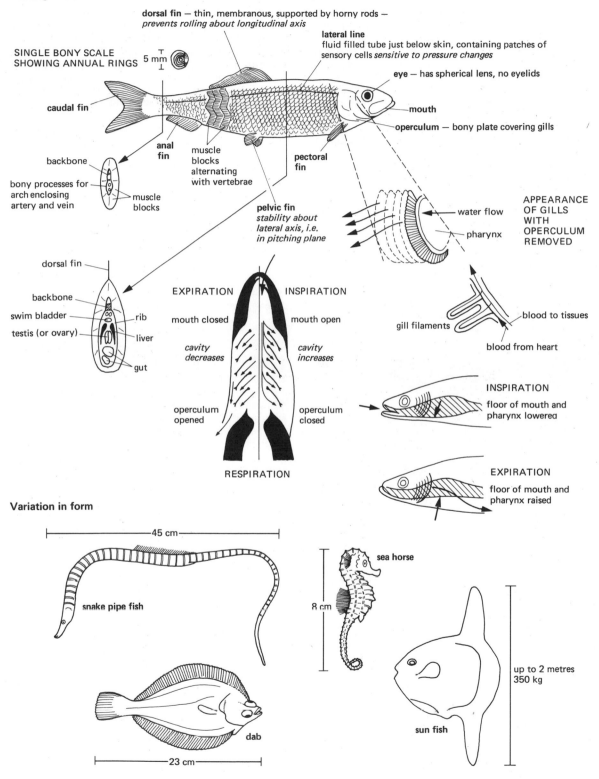

dorsal fin — thin, membranous, supported by horny rods — *prevents rolling about longitudinal axis*

lateral line
fluid filled tube just below skin, containing patches of sensory cells *sensitive to pressure changes*

eye — has spherical lens, no eyelids

SINGLE BONY SCALE
SHOWING ANNUAL RINGS

5 mm

caudal fin

mouth

operculum — bony plate covering gills

anal fin

muscle blocks alternating with vertebrae

backbone

bony processes for arch enclosing artery and vein

muscle blocks

pectoral fin

pelvic fin
stability about lateral axis, i.e. in pitching plane

water flow

pharynx

APPEARANCE OF GILLS WITH OPERCULUM REMOVED

dorsal fin

backbone

swim bladder

testis (or ovary)

rib

liver

gut

EXPIRATION

mouth closed

cavity decreases

operculum opened

INSPIRATION

mouth open

cavity increases

operculum closed

RESPIRATION

gill filaments

blood to tissues

blood from heart

INSPIRATION
floor of mouth and pharynx lowered

EXPIRATION
floor of mouth and pharynx raised

Variation in form

45 cm

snake pipe fish

sea horse

8 cm

dab

23 cm

sun fish

up to 2 metres
350 kg

Economic importance

Herring forms a very large part of the catch of the fishing industry in the British Isles. Spawning time is determined by the temperature and the time at which the fish may be found moving towards shallow coastal waters varies in different regions. The fishing fleets are active off the Hebrides in May. In June the herring are caught off Orkney. They are found off the east coast of England in July and from autumn through to January they are fished off the Cornish coast. They are caught by means of **drift nets**. These form a wall suspended from the surface and the herring are caught as they swim through the mesh. Their heads pass through but not their bodies and they cannot back out because the net catches underneath the operculum. The drifters set their nets at night. The plankton on which the herring feed are found some distance below the surface during the day but migrate upwards as darkness falls. The herring likewise migrate to the surface at night. Herring are generally restricted to regions where the water temperature is below 10 °C.

Sharks

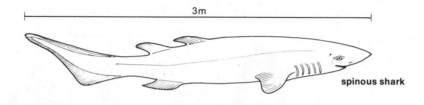

3m

spinous shark

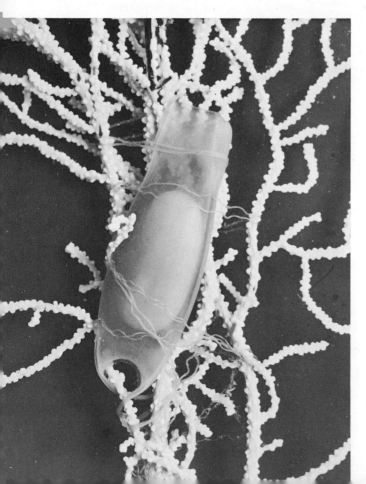

The sharks, together with their relatives, the skates, rays and dogfish, are different from the herring in a number of ways. The skeleton is made of cartilage, not bone. The gill openings are separate, there being no operculum. The skin of a shark is covered with scales that have the same basic structure as teeth. The eggs are large, and only a few are produced—enclosed in elaborate structures called mermaid's purses. Internal fertilization occurs.

Fish can be extremely interesting animals to keep, as anyone who has owned an aquarium will know. Almost any of the smaller freshwater fish found in the rivers, streams and ponds can be kept successfully in the laboratory. It is possible to keep a pair of sticklebacks and to observe their behaviour in the breeding season. Before you start, visit your library and consult a book on keeping fish.

Mermaid's purse (egg-case of dogfish) attached by strings at corners to seaweed. Sometimes the detached egg is found washed-up on the seashore

15 AMPHIBIA

The frog

Vertebrate animals with paired limbs and soft skins; 'cold-blooded' (see p. 72); reproduction usually requires a watery environment with larval stages followed by metamorphosis (see p. 203).

Found in moist shady places, usually where there is abundant vegetation, near ponds or swimming in freshwater. Frogs in Britain hibernate in thick vegetation or buried in mud.

Appearance

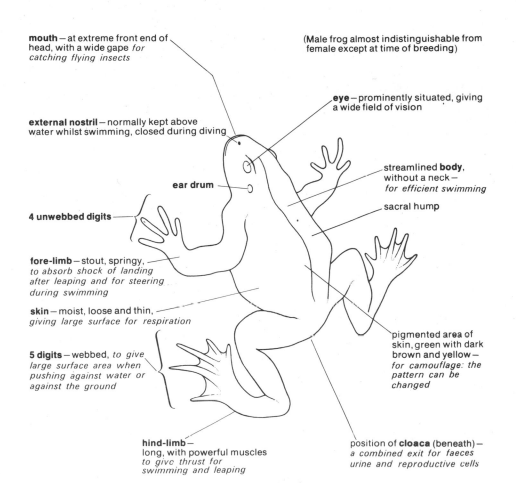

mouth — at extreme front end of head, with a wide gape *for catching flying insects*

(Male frog almost indistinguishable from female except at time of breeding)

eye — prominently situated, giving a wide field of vision

external nostril — normally kept above water whilst swimming, closed during diving

streamlined **body**, without a neck — *for efficient swimming*

ear drum

sacral hump

4 unwebbed digits

fore-limb — stout, springy, *to absorb shock of landing after leaping and for steering during swimming*

skin — moist, loose and thin, *giving large surface for respiration*

5 digits — webbed, *to give large surface area when pushing against water or against the ground*

pigmented area of skin, green with dark brown and yellow — *for camouflage: the pattern can be changed*

hind-limb — long, with powerful muscles *to give thrust for swimming and leaping*

position of **cloaca** (beneath) — *a combined exit for faeces urine and reproductive cells*

Leaping

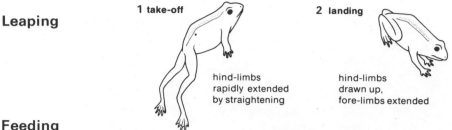

1 take-off

hind-limbs rapidly extended by straightening

2 landing

hind-limbs drawn up, fore-limbs extended

Feeding

Frogs will eat worms and a variety of insects such as beetles and flies, which may be caught in flight.

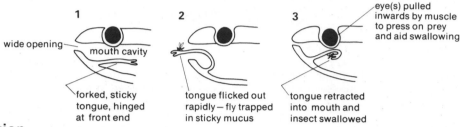

1

wide opening

mouth cavity

forked, sticky tongue, hinged at front end

2

tongue flicked out rapidly — fly trapped in sticky mucus

3

eye(s) pulled inwards by muscle to press on prey and aid swallowing

tongue retracted into mouth and insect swallowed

Respiration

(a) **lining of mouth cavity**—used on land when the frog is relatively inactive—can be seen externally as *gentle* pulsations of the throat:

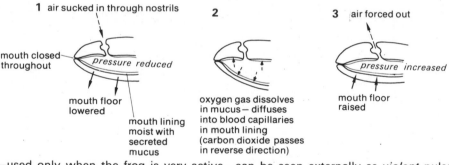

1 air sucked in through nostrils

mouth closed throughout

pressure reduced

mouth floor lowered

mouth lining moist with secreted mucus

2

oxygen gas dissolves in mucus — diffuses into blood capillaries in mouth lining (carbon dioxide passes in reverse direction)

3 air forced out

pressure increased

mouth floor raised

(b) **lungs**—used only when the frog is very active—can be seen externally as *violent* pulsations of the throat:

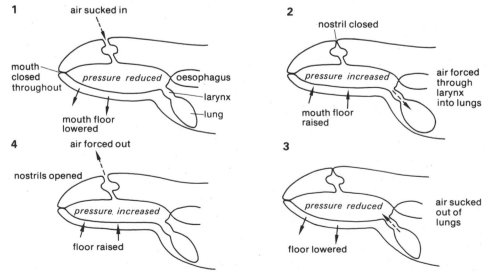

1 air sucked in

mouth closed throughout

pressure reduced

oesophagus

larynx

lung

mouth floor lowered

2 nostril closed

pressure increased

mouth floor raised

air forced through larynx into lungs

4 air forced out

nostrils opened

pressure increased

floor raised

3

pressure reduced

air sucked out of lungs

floor lowered

(c) skin—used all the time but especially in water and during hibernation:

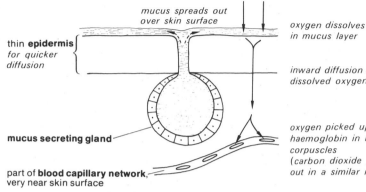

VERTICAL SECTION THROUGH SKIN

mucus spreads out over skin surface

oxygen dissolves in mucus layer

thin **epidermis**
for quicker diffusion

inward diffusion of dissolved oxygen

mucus secreting gland

*oxygen picked up by haemoglobin in red corpuscles
(carbon dioxide passes out in a similar manner)*

part of **blood capillary network**,
very near skin surface

Reproduction

In Britain frogs normally begin breeding in March, when the adults emerge from hibernation stimulated by a rise in temperature. The female is distinguished by her larger size, particularly her lower trunk region swollen with unfertilized eggs. Unlike the female, the male can croak and has a characteristic development of the hand. The males move to a pond and croak to attract the females. The male selects a female and the two frogs swim around together for several days.

male

female

HAND (MALE FROG)

first digit

fourth digit

swollen horny
pad on underside,
for grasping
female

In this position the male frog helps to squeeze eggs out of the female. As the eggs pass down the oviduct they are coated with a secretion and then they pass out of the cloaca into the water. The male immediately sheds fluid containing sperm from his cloaca into the water. Fertilization occurs when one sperm fuses with one egg. As this happens in the water outside the parents it is **external fertilization**. The oviduct secretion rapidly absorbs water and swells to form jelly.

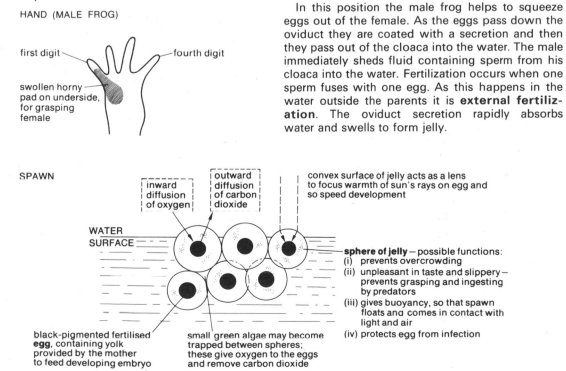

SPAWN

inward
diffusion
of oxygen

outward
diffusion
of carbon
dioxide

convex surface of jelly acts as a lens
to focus warmth of sun's rays on egg and
so speed development

WATER
SURFACE

sphere of jelly — possible functions:
(i)   prevents overcrowding
(ii) unpleasant in taste and slippery —
      prevents grasping and ingesting
      by predators
(iii) gives buoyancy, so that spawn
      floats and comes in contact with
      light and air
(iv) protects egg from infection

black-pigmented fertilised
egg, containing yolk
provided by the mother
to feed developing embryo

small green algae may become
trapped between spheres;
these give oxygen to the eggs
and remove carbon dioxide

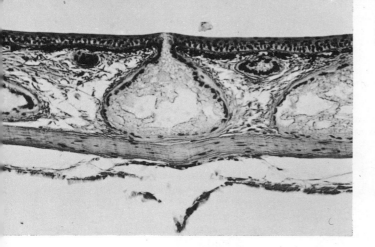

V.S. Frog skin, showing large mucus gland opening to the surface and thin epidermis with a layer of pigment cells beneath

About 10 days after fertilisation

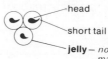

head
short tail } comma-shaped **embryo**— *soon begins wriggling*

jelly— *now begins to liquefy, making emergence of tadpole easier*

About three days later the wriggling movements increase and the embryos break free from the jelly as incompletely-developed **larvae** (**tadpoles**).

Tadpole development

(The times and length measurements given are only approximate and depend on temperature, food supply etc.)

1 Newly-hatched tadpole, 12–14 days after fertilisation

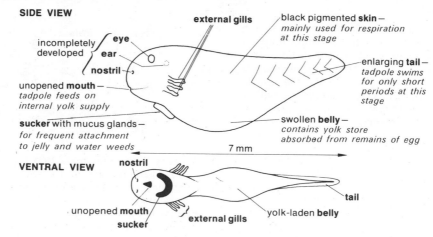

SIDE VIEW

external gills

black pigmented **skin**— *mainly used for respiration at this stage*

incompletely developed { **eye** **ear** **nostril**

enlarging **tail**— *tadpole swims for only short periods at this stage*

unopened **mouth**— *tadpole feeds on internal yolk supply*

sucker with mucus glands— *for frequent attachment to jelly and water weeds*

swollen **belly**— *contains yolk store absorbed from remains of egg*

7 mm

VENTRAL VIEW

nostril

unopened **mouth**

sucker

external gills

yolk-laden **belly**

tail

2 Tadpole about 20 days after fertilisation—1 week after hatching

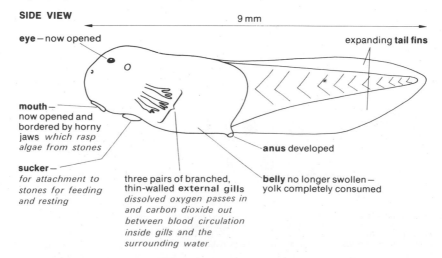

SIDE VIEW

9 mm

eye—now opened

expanding **tail fins**

mouth— now opened and bordered by horny jaws *which rasp algae from stones*

anus developed

sucker— *for attachment to stones for feeding and resting*

three pairs of branched, thin-walled **external gills** *dissolved oxygen passes in and carbon dioxide out between blood circulation inside gills and the surrounding water*

belly no longer swollen— yolk completely consumed

3 Tadpole 2 weeks after hatching

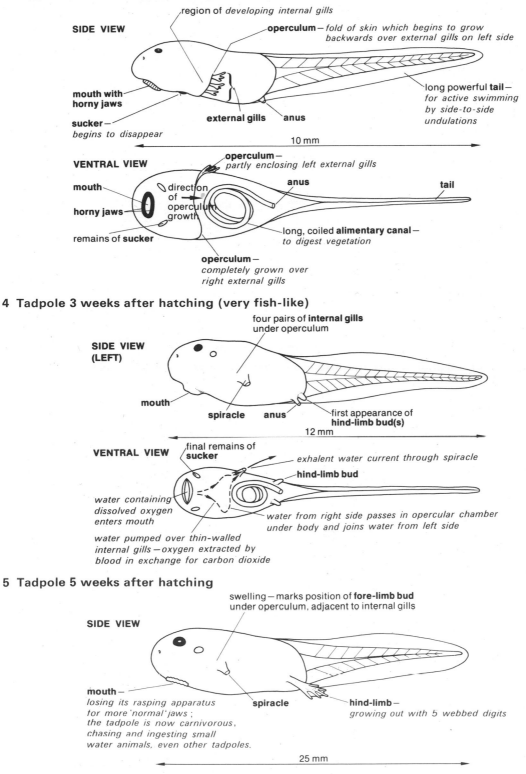

SIDE VIEW

region of *developing internal gills*

operculum — *fold of skin which begins to grow backwards over external gills on left side*

long powerful **tail** — *for active swimming by side-to-side undulations*

mouth with horny jaws

sucker — *begins to disappear*

external gills   **anus**

10 mm

VENTRAL VIEW

operculum — *partly enclosing left external gills*

mouth

anus

tail

direction of operculum growth

horny jaws

remains of **sucker**

long, coiled **alimentary canal** — *to digest vegetation*

operculum — *completely grown over right external gills*

4 Tadpole 3 weeks after hatching (very fish-like)

SIDE VIEW (LEFT)

four pairs of **internal gills** under operculum

mouth

spiracle   **anus**

first appearance of **hind-limb bud(s)**

12 mm

VENTRAL VIEW

final remains of **sucker**

exhalent water current through spiracle

hind-limb bud

water containing dissolved oxygen enters mouth

water from right side passes in opercular chamber under body and joins water from left side

water pumped over thin-walled internal gills — oxygen extracted by blood in exchange for carbon dioxide

5 Tadpole 5 weeks after hatching

swelling — marks position of **fore-limb bud** under operculum, adjacent to internal gills

SIDE VIEW

mouth — *losing its rasping apparatus for more 'normal' jaws ; the tadpole is now carnivorous, chasing and ingesting small water animals, even other tadpoles.*

spiracle

hind-limb — *growing out with 5 webbed digits*

25 mm

6 Tadpole 7 weeks after hatching

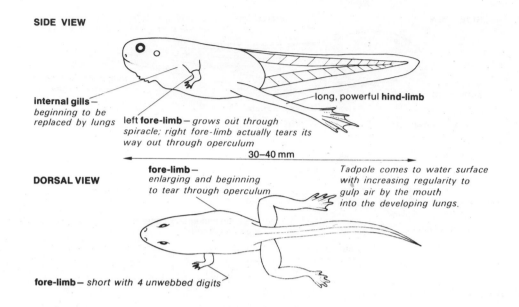

SIDE VIEW

internal gills—
*beginning to be
replaced by lungs*

left **fore-limb**—*grows out through
spiracle; right fore-limb actually tears its
way out through operculum*

long, powerful **hind-limb**

30–40 mm

DORSAL VIEW

fore-limb—
*enlarging and beginning
to tear through operculum*

*Tadpole comes to water surface
with increasing regularity to
gulp air by the mouth
into the developing lungs.*

fore-limb— *short with 4 unwebbed digits*

**7 Tadpole just before final metamorphosis—10 weeks after hatching,
3 months after fertilisation**

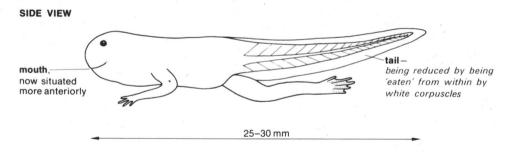

SIDE VIEW

mouth,
now situated
more anteriorly

tail—
*being reduced by being
'eaten' from within by
white corpuscles*

25–30 mm

The tadpole moves to the side of the pond,
ceases swimming and feeding (food is provided
internally by the digesting of the tail by white
corpuscles). **Metamorphosis** (Gk. *meta*, change
of; *morphe*, form) is stimulated by a hormone from
the thyroid gland. The skin is shed, carrying away
with it larval features such as the horny jaws. Many
of the features of the adult frog are shown opposite :

During the several days after casting the skin the
tail is reduced even further by digestion from
within until no stump is visible at all. The young
frog needs *two* more *complete* seasons of growth
and development before it becomes sexually
mature and can breed.

8 Young adult

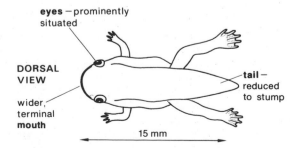

eyes —prominently
situated

DORSAL
VIEW

wider,
terminal
mouth

tail —
reduced
to stump

15 mm

Other amphibia

Only a comparatively small number of different amphibians live in the present world; yet more than 230 million years ago in the Carboniferous period of geological time, amphibians were a major group of animals.

Almost certainly, the earliest vertebrates lived in water and the amphibians of the Carboniferous period were the first vertebrates to leave a watery environment and exploit life on land. Although the frog and its modern relatives are very unusual amphibians and quite different in many ways from their ancestors, they are still partly dependent on water, if only for breeding.

Newts and **salamanders** are amphibians with tails, and are mostly confined to N. America, N. Europe and Asia.

Smooth newt (Triturus vulgaris), *male*

Spotted salamander (Salamandra maculosa), *a common European amphibian*

Three species of **newt** live in Britain, for most of the year in damp places (e.g. under stones). They are voracious nocturnal feeders, eating a wide range of invertebrates (worms, snails, slugs, insects). They travel to freshwater to breed in April. The male develops vivid body colours and a dorsal crest. Fertilization is internal, and eggs are laid singly attached to water weeds, sticks and stones; they hatch into tadpoles, which take from four months to become mature.

Salamanders live in rivers, marshes, damp woods and among rocks. Eggs may be laid in strings on water plants. Some species are eel-like, without hind legs; in others the legs are greatly reduced. All move rather clumsily on land, with fish-like sideways bending of the body. Some species become mature whilst still resembling larvae, and have gills as well as lungs; some are lungless. One species lives in caves, is white and apparently blind.

---

1 Collect spawn in spring and study the development of the tadpole in pondwater in the laboratory. Note in particular the methods of feeding and breathing, and the development of the limbs. Development may be compared with that of the African clawed toad, *Xenopus*.

2 Place a living tadpole in a small drop of water on a slide under a microscope. Examine the thin skin of the tail to see the movement of blood through the skin capillaries.

3 Suggest why tadpoles hatch at such an early stage of development.

4 Read further about different types of amphibians and if possible study the behaviour of adult frogs and newts in vivaria.

16 REPTILES

Reptiles are vertebrate animals adapted for life on land. Over a period lasting for approximately 150 million years reptiles were the dominant land vertebrates. They were superseded by the birds and mammals about 70 million years ago, and today are represented by the following groups:

Crocodiles and **alligators**
Lizards and **snakes**
Tortoises, **turtles** and **terrapins**

Although some of them live for most of the time in water, they return to the land to lay their eggs. Reptiles are cold-blooded animals; that is, their body temperature is only a little above the external temperature, and varies with it. Most reptiles are therefore confined to tropical or sub-tropical regions. Their main adaptations to land life are:

(a) The possession of **dry horny scales** on the skin; these reduce water loss from the surface of the body.

(b) The laying of **large**, **yolky eggs** protected by a **hard shell**. The embryo, like those of the mammal and the bird, is protected within a special membrane, called an amnion, which contains fluid and functions rather like a miniature freshwater pond.

17 BIRDS

Birds, like mammals, are descended from ancient reptiles, although their origins were quite independent from those of the mammals. They are efficiently adapted to terrestrial life and to an aerial existence. Some of the more important points to remember about birds are summarized below.

ADAPTATIONS TO TERRESTRIAL LIFE

Temperature regulation

Birds are able to maintain their body temperature at a constant level (40°–45°C). Their internal chemical processes (metabolism) are carried on at a much higher rate than those of reptiles and are independent of external temperatures. They possess an efficient, high pressure blood system with complete separation of pulmonary and body circulations (see also p. 152). The heart is a double pump with four chambers. Its rate of contraction varies from about 100 per minute to about 500 per minute according to the size of the bird. Heat loss is prevented by the **feathers** which form very efficient insulators, especially when fluffed out. Their insulating property is so high that the temperature on the surface of the skin, under the feathers, remains at about 35°C. Constriction or dilation of the surface blood vessels, except in regions not covered by feathers, is of no value in controlling heat loss. Special mechanisms exist to prevent heat loss from the legs in cold conditions. The flow of blood into the leg may be reduced but in addition many birds, for example, waders, have heat exchange systems in the upper part of the leg. Outgoing arteries are closely associated with the incoming veins. Much of the heat in the arterial blood is therefore transferred to the colder blood in the veins. If this were not the case the situation would be rather like a house with central heating in which some of the radiators were fixed to outside walls.

Most heat loss in birds occurs through the expired air and this may be increased by increasing the breathing rate. Where this may involve an excessive loss of water, as vapour, birds are able to tolerate a rise in body temperature of several degrees. They are also able to tolerate lowered temperatures when heat output is reduced through shortage of food.

Respiration

A high rate of metabolism requires an extremely efficient system for gaseous exchange. In birds the lungs are attached to the ribs, and when not in flight breathing movements are brought about by special respiratory muscles, attached to the ribs, whose action has the effect of enlarging and contracting the body cavity. During flight this mechanism is reinforced by the action of the flight muscles pulling against the sternum, which is alternately raised and lowered. The space inside the lungs is continuous with the cavities of a number of air-sacs. When the bird breathes in, these, as well as the lungs, are filled with air. When the bird breathes out, air from the air-sacs sweeps the used air from the lungs completely. (In mammals, which have no air-sacs, the air in the lungs is only partly changed in each breathing cycle (see p. 129).) Birds have oval, nucleated red blood cells but the oxygen they carry is more readily given up as the oxygen level falls than is the case with mammals.

Water regulation

Birds have dry skins with no sweat glands. The kidney is particularly efficient at the process of water re-absorption. Nitrogenous waste products are formed into the relatively insoluble uric acid. After leaving the kidney still more water is absorbed, the mixed faeces and uric acid crystals forming the semi-solid droppings, or guano. Most of the water loss takes place during the breathing out of expired air saturated with water vapour.

Reproduction

Fertilization in birds is **internal**. Semen is transferred from the male during mating, when the vent of the male is brought into close contact with that of the female. (The vent, or cloaca, is the single common opening from the hinder end of the food canal, the excretory and the reproductive systems.)

The fertilized egg together with its food supply forms the yolk. Albumen, that is the white of the egg, is added, as well as the shell, as the egg passes down the oviduct. The shell is permeable to air, and apart from oxygen which diffuses through the shell, all the needs of the embryo are supplied within the egg. The development of the embryo requires that it shall be maintained at 40°C. The newly hatched birds are virtually helpless. The parent birds therefore display a great deal of **parental care** both before and after hatching. This involves building nests, which are often very complex, incubation of the eggs, and the care and feeding of the young birds. As a result, both male and female parents may play an important role, the male bird assisting with both incubation and feeding. They form pairs which persist throughout the nesting period and in some cases from year to year. Birds often exhibit elaborate courtship patterns, with colourful display by the males.

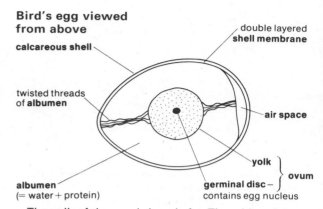

Bird's egg viewed from above

The yolk of the egg is largely fat. The white of the egg acts mainly as a source of water. The yolk is suspended between the twisted threads of albumen and whatever position the egg takes up, the embryo, being at the lighter end, always remains uppermost.

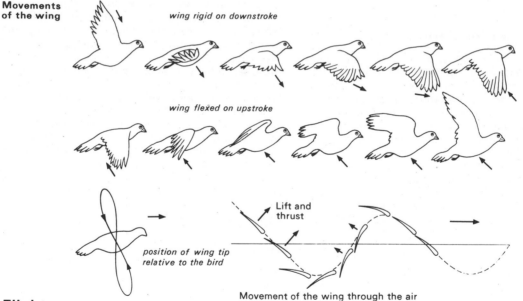

Movements of the wing

wing rigid on downstroke

wing flexed on upstroke

position of wing tip relative to the bird

Lift and thrust

Movement of the wing through the air

Flight

Because of the length of the wing skeleton and the distance through which it must move, it is physically impossible for a bird's wing to beat at the frequencies encountered in insects. (See p. 47). Only humming birds, which are very small, have wingbeat frequencies (60–70 per sec) which compare with those of most insects. A pigeon's wings, for example, beat at 7–8 cycles per second. Since birds' wings operate at frequencies so much lower than those of insects it may be safely assumed that a bird's wing is a much more efficient system for generating lift. Birds generally are able to glide, and this insects are quite unable to do.

When viewed in cross-section a bird's wing resembles what aero-engineers call an aerofoil section.

Air flowing over such a shape generates an upward force acting on the aerofoil. For level flight this force must just balance the force acting downwards, i.e., its weight. The magnitude of the upward force (lift) depends on a number of factors such as the thickness of the aerofoil, the speed of the airflow over it, and the angle it makes with the airflow (angle of attack). The total surface area of the wing is also important. The greater the wing area in relation to the total weight of the bird the lower the wing loading, i.e., the weight per unit area of wing.

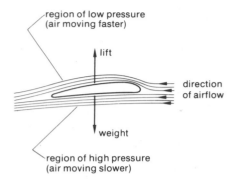

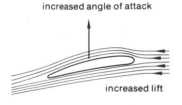

In an aeroplane the wing has only to provide a lift force, the engines providing the thrust. The bird's wing has to supply both lift and thrust. As in the case of the insect wing, the downstroke is the power stroke. As the wing moves down and forward the force acting on it is directed both upwards and forwards, generating both thrust and lift.

The recovery stroke is passive, the feathers parting to allow air to pass between them. Throughout the whole cycle of movement the inner part of the wing moves through a much smaller angle and can be regarded as a fixed aerofoil generating lift only.

Take off and landing present special problems. Since birds have to land on legs and not wheels, their speed relative to the ground must be near zero at the instant of touchdown. They therefore always land into wind. The wings will generate true lift only if air is flowing over them. At take off, and in 'powered' landings, the wings therefore beat faster and through a greater distance. They almost meet at the front of the bird at the end of the power stroke. The angle of attack of the wings is also increased. The tail feathers fan outwards and downwards. This increases lift, and in landing also acts as a brake.

When the airflow over the surface of an aerofoil falls to a certain critical level it ceases to flow smoothly and there is a sudden loss of lift. The wing is said to stall. The stalling speed of a bird's wing is reduced at the moment of landing by the raising of the flight feathers on the leading edge of the wing, allowing air to flow under them. This has the effect of speeding up the flow of air over the wing surface thereby reducing the tendency for it to stall.

Gliding

Being heavier than air, a bird must necessarily lose height, *relative to the air*, in order to maintain the airflow over the surface of its wings to generate lift. If, at the same time, the body of air through which the bird is moving is actually rising the bird may not only maintain its height, *relative to the ground*, but actually increase it. Soaring birds, like the hawk, take advantage of upcurrents of warm air (thermals) caused by differences in surface temperature. Seagulls take advantage of onshore breezes which are deflected upwards at a cliff face. Ocean gliders, such as the albatross, use the differences in windspeed at different levels over the sea, in order to maintain height by gliding.

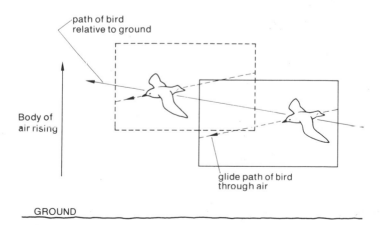

Adaptations to flight

Some important points to remember are given below.

1 Most birds are small. For successful flight, wing loading (i.e. weight supported per unit area of lifting surface) must be low. Doubling all the linear dimensions of a bird will increase its weight by a factor of eight. The strength of a muscle is proportional to its area of cross section and its power will therefore be increased four times (see p. 206). Increase in overall size requires an even bigger proportionate increase in the weight of the flight muscles. The proportion of a bird's weight taken up by its flight muscles must clearly be limited and therefore its size must be limited.

2 Weight is kept to a minimum by the possession of a very light skeleton with hollow bones and air-sacs.

3 The body is streamlined.

4 Wing area is achieved by the use of feathers. Feathers are extremely light, and provide a large area without increase in the area across which body heat may be lost. (In the only true flying mammals, the wing is formed partly from stretched skin which is subject to considerable heat loss.)

5 In very large birds efficient flight is achieved by taking advantage of vertical air movements, i.e. by gliding and soaring.

6 Other important adaptations involve organs not directly connected with flight, for example, the eye. The eye is shaped so that all parts of the visual field are in focus on the retina at the same time. (In mammals only the central part, the fovea, receives a focused image.) The number of cone cells in a day-flying bird is far greater than in a mammal. In a bird such as the eagle the maximum density may reach a level six times as high as that in the fovea of Man. They are also densely packed over the rest of the retina. This means the perception of fine detail over the whole of the visual field. The sense of smell is poorly developed in birds.

7 The brain is well developed, in keeping with the great mobility of birds, and they exhibit many elaborate behaviour patterns.

Birds and mammals are two highly successful groups of land vertebrates. That they are both successful is due probably to the fact that their habits and their diets overlap only to a limited extent. Most birds can fly, whereas few mammals can. **Flightless birds** are found where there are no mammals, for example, in the Antarctic (penguins), or on oceanic islands. Where there are mammals, flightless birds are large and are able to run at high speed (ostrich). There are many examples of competition for the same food (fish-eating birds and mammals; carnivores and carrion-eaters in both groups), but birds are mostly seed- and insect-eaters whereas mammals generally are not.

Birds display great variety and range of adaptation in the form and function of their **beaks** (they possess no teeth) and their **feet**, but relatively little in other respects, since the requirements for flight place strict limitations on what is possible in the way of variation.

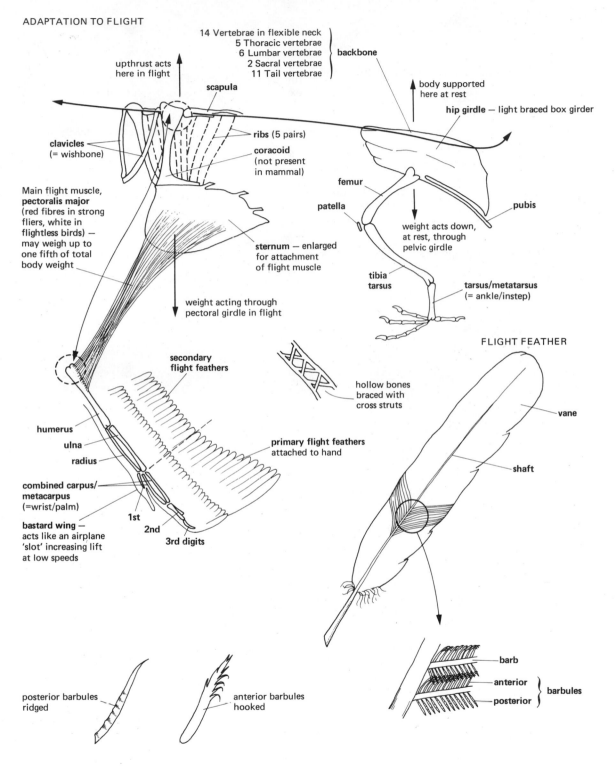

ADAPTATION TO FLIGHT

14 Vertebrae in flexible neck
5 Thoracic vertebrae
6 Lumbar vertebrae
2 Sacral vertebrae
11 Tail vertebrae
} **backbone**

upthrust acts
here in flight

scapula

body supported
here at rest

hip girdle — light braced box girder

clavicles
(= wishbone)

ribs (5 pairs)

coracoid
(not present
in mammal)

femur

pubis

patella

Main flight muscle,
pectoralis major
(red fibres in strong
fliers, white in
flightless birds) —
may weigh up to
one fifth of total
body weight

sternum — enlarged
for attachment
of flight muscle

weight acts down,
at rest, through
pelvic girdle

**tibia
tarsus**

tarsus/metatarsus
(= ankle/instep)

weight acting through
pectoral girdle in flight

FLIGHT FEATHER

**secondary
flight feathers**

hollow bones
braced with
cross struts

vane

humerus

ulna

radius

primary flight feathers
attached to hand

shaft

**combined carpus/
metacarpus**
(=wrist/palm)

1st

2nd

bastard wing —
acts like an airplane
'slot' increasing lift
at low speeds

3rd digits

barb

anterior
} **barbules**
posterior

posterior barbules
ridged

anterior barbules
hooked

hooks engage in ridges so that
adjacent barbs are zip-fastened together

Golden eagle, showing curved pointed beak—flesh eating

Avocet, showing long upward-curving beak for probing

Green woodpecker (Picus viridis)—insect eating and wood-borer

Mallard (Anas platyrhynchos), drake; the Latin name refers to the flat beak—dredger

VARIATION IN BEAK FORM

18 MAMMALS

Mammals are divided into three main groups:

1 Egg-laying mammals

A group containing only two species:

(a) The duck-billed platypus. It is found along the east coast of Australia and in Tasmania.

(b) The spiny anteater. Restricted to eastern Australia and New Guinea.

Both lay hard-shelled eggs which they incubate. The young, after hatching are fed by mammary glands.

2 Marsupial mammals
(L. *marsupium,* a pouch)

Marsupials are found in Australasia (except New Zealand), South America, Central America and the southern part of North America. The young are born at a very early stage in their development. They crawl up the mother's body and finish their development in a pouch on the abdomen, where they become attached to a nipple through which they receive milk.

3 Placental mammals

This group includes the majority of mammals. The young are born in an advanced state. The young receives its food before birth through the placenta (see p. 231).

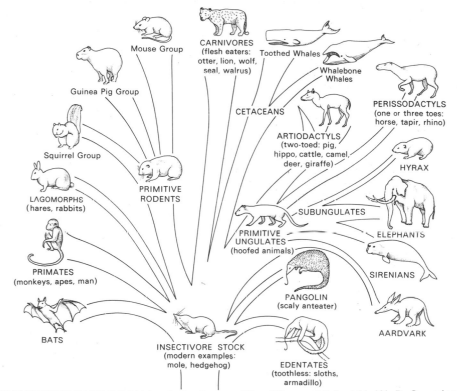

From THE VERTEBRATE BODY by A. S. Romer. 4th edition published by W. B. Saunders Company

Placental mammals

These are regarded as a very successful group of land animals. What does success mean? It means that, taken as a group, mammals are able to occupy and flourish in a great range of physical and climatic conditions. They are found in the rarified atmosphere of high mountain ranges and also, like the whale when diving, several hundreds of feet below the surface of the sea. They can live in the permanent snows of the polar ice cap, or in steamy hot tropical rain-forests. They may live permanently in fresh or salt water, or in regions where there is no free water at all. Together with the birds, they have become independent of external physical conditions to a greater extent than any other group of animals. The main features of their anatomy and physiology which makes this possible are:

1   The ability to conserve water

The cells that make up a mammal's body, except for the dead cells on the surface of the skin, are effectively immersed in a pond whose composition remains remarkably constant. It remains so whether the atmosphere is dry or humid, or even if the body is surrounded by water. The ability for the composition of the body fluid to remain constant depends upon the skin, which is almost waterproof, and the kidney, which maintains the water balance of the body (p. 162).

2   A high rate of internal chemical activity (metabolism)

The rate at which a chemical reaction proceeds doubles, roughly, for every ten degrees (C) rise in temperature. As in all animals, the heat gain in a mammal from the burning of food equals the loss of heat from the surface. The skin, however, acts as a heat barrier, storing up heat behind it (see p. 90). High metabolic rate depends also upon an efficient blood system. This is realized in two ways: (a) by the heart being a double pump, as well as a two stage pump, and (b) by reduction in the number of blood vessels carrying blood from the heart. The circulation is therefore a high pressure system (see p. 145). It also depends on rapid and effective uptake of oxygen. There are two points to note here: (a) the joining of the ribs and their restriction in number to form a cage which, with its associated muscles and the diaphragm, forms an effective bellows (see p. 128); (b) the carriage of oxygen by haemoglobin in cells which have lost their nuclei.

3   Locomotion and support (see p. 189)

The legs are carried underneath the body, which is therefore raised clear of the ground. The backbone is much more complex and is more effectively articulated. It plays an important part in locomotion by its movement in the vertical plane. There is a distinct neck, not present in the lower vertebrates, which increases the mobility of the head. The ends of the long bones of the limbs are separated from their shafts during growth by a layer of gristle. The articulation between bones is therefore not disturbed by the changes in size of the bones during growth.

4   Sensory and nervous systems

Mammalian sense organs (see pp. 177 and 181) have reached a high level of development. The eyes possess a specialized region which is particularly sensitive. This is the yellow spot. Unlike other vertebrates, the mammal has three bones in the middle ear, which not only transmit vibrations but act as a gear to reduce their amplitude. The ear drum lies below the surface, and the external ear (pinna) is important as a direction finder.

Nerve impulses travel much faster, and there is a high degree of centralization of incoming signals from the sense organs; hence the relatively enormous size of the forebrain and the cerebellum. The brain/body weight ratio is higher than in any other animals (see p. 169).

5   Reproduction

The eggs are fertilized internally, and develop in a specialized part of the oviduct (uterus). The embryos are thus protected by the mother. They are supplied with food and oxygen from the mother's blood through a special organ, the placenta, which is actually part of the embryo. They are born at an advanced stage and are fed by milk supplied by the mother. Growth and development is independent of external temperature.

6   The palate

The internal opening of the nose is separated from the cavity of the mouth by a bony shelf, the palate. This means that the presence of food in the mouth does not interrupt breathing. Instead of having to be swallowed at once, food can be chewed. The teeth are specialized for different functions and mammals can exploit a wide range of diets—for example, grasses, flesh, fruits, nuts or mixed diet.

Section II

THE PROCESSES OF LIFE

19 CELLS, TISSUES AND ORGANS

Living organisms are composed of units called **cells**. Generally those organisms that are visible without magnification are **multicellular**; they consist of many interacting cells. Nevertheless there exists a large variety of organisms each of which consists of only a single cell (see pp. 5 and 13).

One of the problems of biology, unlike physics, is the difficulty of formulating precise statements. It is especially difficult to define a cell. There is no definition that includes all those structures which biologists call cells; the following definition is given with this reservation in mind:

A cell is the basic working unit of life.

Isolated cells, maintained in a suitable nutritive solution and supplied with oxygen, will show properties normally associated with life—a muscle cell will contract, a nerve cell conduct impulses, a palisade cell photosynthesize. But the components from within a cell, when isolated from each other, do not show these properties. This is what is meant by describing cells as basic *working* units.

A cell is a small room. The word was given a biological meaning by **Robert Hooke** who in 1665 examined thin sections of cork under the microscope. (Cork cells *are* little rooms; their walls are all that remain of living cells.) All living organisms are now known to be cellular.

CELL STRUCTURE

Cells are usually microscopic. Some bacteria (see p. 20), for example, are scarcely visible with a light microscope. Certain types of cell are large enough to be seen by the unaided eye; egg cells, of birds in particular, may be as much as 3 cm in diameter.

Cells consist of:

(i) **plasma membrane**—an extremely thin outer boundary.

(ii) **cytoplasm** (Gk. *kytos*, hollow; *plasma*, form, mould)—a rather fluid jelly, transparent, often appearing granular; cytoplasm makes flowing movements in some cells.

(iii) **nucleus** (L. *nucleus*, kernel)—bounded by its own nuclear membrane and always surrounded by cytoplasm but composed of a denser material; nuclei show a variety of shapes, commonly between disc and sphere; the nucleus controls cell activities and is essential for cell division (see p. 85).

Cytoplasm and nucleus are together sometimes called **protoplasm** (Gk. *protos*, first). The electron microscope gives a much higher magnification than the light microscope; it reveals that protoplasm has its own very fine structure and contains a variety of minute bodies.

In addition, many cells contain non-living materials such as stored food, pigments (e.g. haemoglobin in red corpuscles, see p. 146) and insoluble waste (see p. 162). In plant cells, non-living material forms a thick boundary outside the plasma membrane. This layer constitutes a clearly visible **wall**, composed of a transparent semi-elastic substance called **cellulose**. As a result,

Flat cells from the lining of the mouth cavity, showing nucleus and cytoplasm

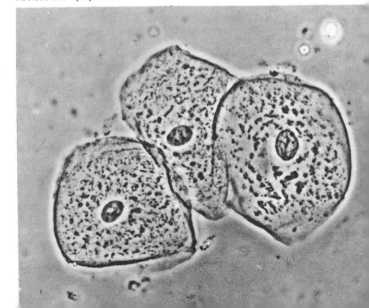

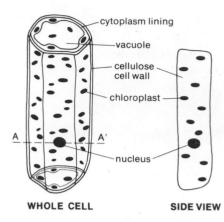
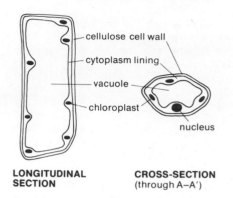

| WHOLE CELL | SIDE VIEW | LONGITUDINAL SECTION | CROSS-SECTION (through A–A') |

Cell from palisade tissue in a leaf, drawn from different aspects

plant cells are more resistant to distortion than are animal cells. Other materials, such as lignin (wood) may be added later on the inner surface of the cellulose wall (see pp. 55 and 162).

The large central space (**vacuole**) which is found in plant cells is filled with cell sap; this is a watery fluid containing various dissolved nutritive substances—sugars and salts. In some cells, e.g. those of beetroot and the epidermis of flower petals,

pigments are dissolved in the sap. The outward pressure of this fluid in the vacuole plays a large part in maintaining the shape of the cell; this in turn helps to support the entire plant (see p. 137).

The structure of the cells varies with their age; this is particularly true of plant cells. The following diagrams illustrate some of the important features common to most cells, and also the differences between cells in plants and animals:

Animal cell
(from central region of flatworm)

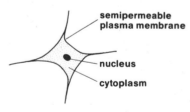

Plant cell
(from outer cortex region of a flowering plant)

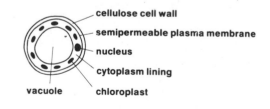

Brief comparison

ANIMAL	PLANT
Plasma membrane as outer boundary	Cellulose wall as outer boundary
Cytoplasm almost fills cell	Cytoplasm as outer lining only
Large vacuoles absent (small vacuoles concerned with secretion may be present temporarily)	Large permanent vacuole filled with cell sap
Smaller size	Larger size
Nucleus variably situated—often near cell centre	Nucleus commonly in cytoplasm lining—occasionally near cell centre suspended by cytoplasm strands
Cells joined by intercellular cement	Cells joined by middle lamella
Chloroplasts absent	Chloroplasts often present
Carbohydrate stored as glycogen	Carbohydrate stored as starch

Tissues and organs

In those organisms where the body consists only of a single cell, all the processes of life are carried out within this single unit and the unit is completely independent.

In multicellular organisms each constituent cell becomes specialized, i.e. capable of performing only more limited functions—biologists refer to this as **division of labour**; it may lead to a cell performing its functions more efficiently. Each cell in such an organism cannot survive independently of the cells which surround it.

Examples of cell specialization include:

1 **Muscle cells** (see p. 195)—cylindrical or spindle-shaped cells performing work by shortening.

2 **Nerve cells** (see p.166)—cells with long thread-like outgrowths carrying signals.

3 **Xylem cells** (see p. 55)—open-ended tubes carrying fluids.

4 **Root-hair cells** (see p. 133)—cells with elongated outgrowths providing a large absorptive surface.

Multicellular organisms are formed by cells organized into larger units of structure. Cells are organized into **tissues**, tissues into **organs**, organs (in the more highly evolved animals) into **systems**.

A tissue is a collection of similar cells performing similar functions, e.g. palisade mesophyll tissue (in the leaf) for photosynthesis; muscle tissue (in the stomach) for moving and churning food during digestion.

An organ is a collection of different tissues, each tissue making its own particular contribution to the functioning of the organ as a whole. A leaf is a plant organ, a stomach an animal organ.

Different, but related organs may then be collected together to constitute a system.

CELL DIVISION

All cells are formed from pre-existing cells by cell division. Two types of division are known:

(a) **Meiosis**—a special type of division involved in sexual reproduction (see pp. 89 and 214).

(b) **Mitosis**—division that results in growth, including the replacement of wounded, diseased and worn-out cells, e.g. red corpuscles (see p. 146) and epidermal cells (see p. 91), and also occurs in asexual reproduction. The phrases of mitosis are shown on pages 86 to 88.

The products of such division are called **daughter cells**. Not all cells are capable of division; highly specialized cells such as nerve and muscle cells cannot reproduce themselves.

Within the nucleus there are minute strands of protein material called **chromosomes** (see pp. 86 to 88).

The chromosomes provide a 'blueprint' for the metabolism of the cell and therefore of the whole organism. When a cell divides by mitosis, each daughter cell possesses the full complement of chromosomes.

The duplication of chromosomes and the equal separation of these duplicated chromosomes into the daughter cells is an important feature of mitosis.

1 Cut thin pieces from a cork using a razor blade or a sharp knife etc. and examine cells in a drop of water on a slide under a microscope.

2 Repeat this with cells scraped by means of a flat wooden stick from the inner lining of your cheek.

3 Examine under the microscope a range of prepared slides showing different types of cells and tissues. Suggest why various stains are used in the preparation of these slides.

4 Using prepared microscope slides examine the organization of leaf and stomach as seen in section. Suggest why it is so much easier to see cells in the leaf than in the stomach.

5 Read further about investigations into the nature of cells since the time of Robert Hooke. What is known at present about the organization of the cell?

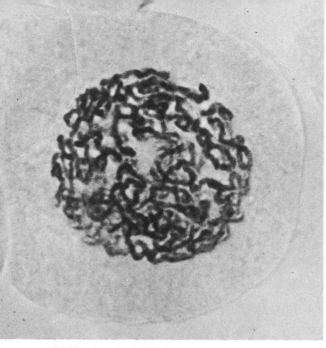

(a) *Mitosis—chromosome strands become visible in the region of the nucleus, at the start of division*

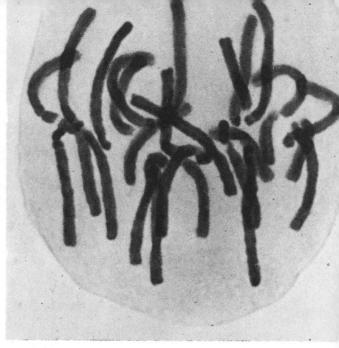

(b) *Mitosis—chromosomes, each consisting of two longitudinal threads, distributed throughout the cell, following disintegration of the nuclear membrane*

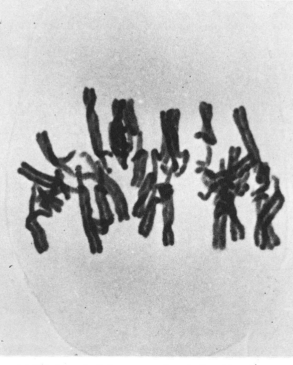

(c) *Mitosis—paired chromosome threads situated at the cell equator, ready for attachment to the nuclear spindle*

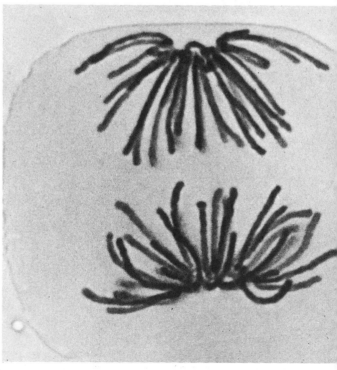

(d) *Mitosis—each pair of chromosome threads separated to opposite poles of the cell*

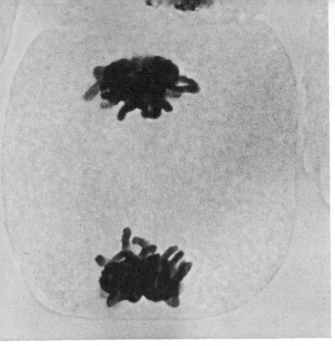

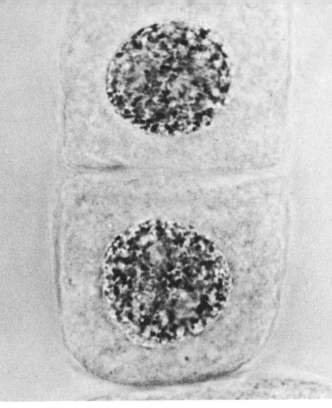

(e) *Mitosis—clusters of equal numbers of chromosome threads forming two nuclei prior to division of the cytoplasm*

(f) *Mitosis—division of the cytoplasm, giving two daughter cells, each with its own nucleus*

Diagram showing those ultra-microscopic structures of an (animal) cell which are important in cell division

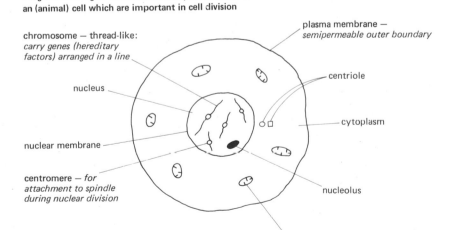

chromosome — thread-like: *carry genes (hereditary factors) arranged in a line*

plasma membrane — *semipermeable outer boundary*

nucleus

centriole

nuclear membrane

cytoplasm

centromere — *for attachment to spindle during nuclear division*

nucleolus

mitochondrion — *rod-like bag, subdivided by shelves: produce energy for all the cell's activities, from respiration*

Cells, tissues and organs

Mitosis - division of the nucleus in which the chromosomes duplicate, so that both daughter cells receive the full set of genes

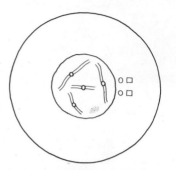

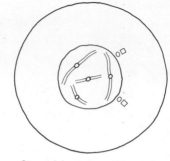

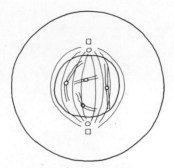

1 *chromosomes (and their genes) duplicate into parallel threads called chromatids: centriole duplicates: nucleolus begins to disappear*

2 *centrioles move apart around nuclear membrane: nucleolus has disappeared*

3 *centrioles reach opposite poles and form a nuclear spindle outside nuclear membrane (spindle made of fine tubes of protein shaped like lines of longitude passing around a globe from pole to pole)*

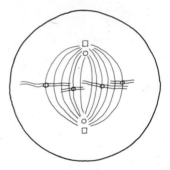

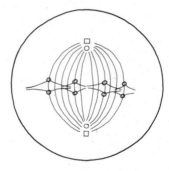

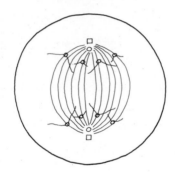

4 *nuclear membrane disappears: chromatid pairs attach to spindle at equator by means of their centromeres*

5 *centromeres of each chromatid pair separate, drawing chromatids apart*

6 *separated chromatids move along spindle towards opposite poles*

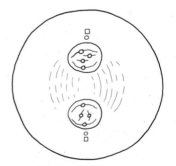

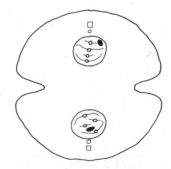

7 *spindle disappears: chromatids become daughter chromosomes surrounded at each pole by a new nuclear membrane: daughter centriole becomes centriole of new cell*

8 *new nucleolus forms in each daughter nucleus: cytoplasm begins to divide*

Meiosis (see also p. 214) — division of the nucleus in which chromosomes are shared equally between the daughter cells, so that the total chromosome number is halved: usually takes place in reproductive organs and results in the production of gametes. Chromosomes in the body cells are paired (called *homologous* chromosomes): for each chromosome given by the male gamete at fertilisation, there is an equivalent (homologous) chromosome given by the female gamete

There are two divisions, producing four daughter cells

First division

1 *homologous chromosomes associate with each other, side-by-side*

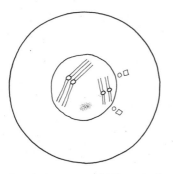

2 *each chromosome duplicates to give 2 longitudinal chromatids: nucleolus disappears: centriole divides and daughter centrioles move apart*

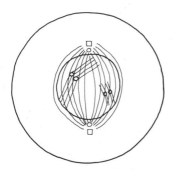

3 *daughter centrioles reach opposite poles and form a nuclear spindle, as in mitosis*

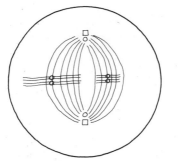

4 *nuclear membrane disappears: pairs of chromosomes arrange themselves on equator of spindle, attached by their centromeres*

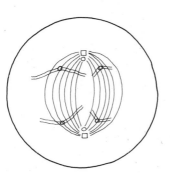

5 *chromosomes separate along spindle towards opposite poles (each separating structure consists of 2 chromatids, not 1, as in mitosis)*

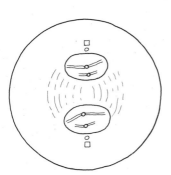

6 *daughter nucleus forms at each pole: nuclear spindle disappears*

Second division

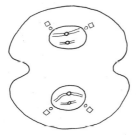

1 *centrioles divide and move apart towards opposite poles*

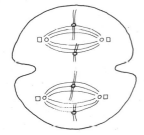

2 *new spindles formed (at right angles to the spindle of the first division): chromosomes attach on to spindle equator*

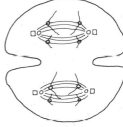

3 *chromatids of each chromosome separate to opposite poles*

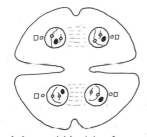

4 *four nuclei (and then four cells) formed, each with half the chromosome number of the parent cell, i.e. the diploid (2n) chromosome number is reduced to the haploid (n) number*

20 SKIN

Skin is an organ, since it is composed of different tissues (see p. 85), but is unusual in that it is widely-spread and covers the whole body surface; it even continues into several of the body openings. Mammalian skin is composed of two main layers—superficial **epidermis** and deeper **dermis**.

SKIN FUNCTIONS

1 Regulation of body temperature (maintenance of the warm-blooded state)

The chemical reactions taking place in all the cells of the body produce small quantities of heat. Heat is lost when any material is passed out of the body, e.g. during egestion and excretion, but the greatest heat loss is under the direct control of the brain and takes place from the skin. Although heat can be lost from a surface by conduction, convection and radiation, the last two of these are most important in the skin.

The skin temperature regulators are:

(a) The blood vessels (See diagram below)

(In the extremities of the body, e.g. ear lobes and fingers, very low temperatures may cause frost-bite; under such conditions, to prevent damage the superficial skin capillaries may dilate so that the blood warms these parts—but in time the animal will have lost too much heat and will die of 'exposure'.)

(b) The sweat glands (diagram p. 93)

(Some mammals, e.g. dogs, have very few sweat glands, and they lose heat by evaporating saliva from their tongues.)

(c) The hairs

The mechanisms of temperature control in mammals (and birds) provide examples of **homeostasis**, i.e. *automatic self-regulation*

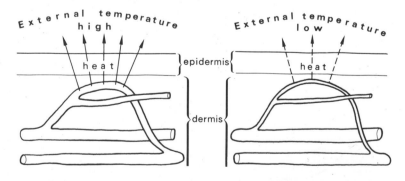

superficial blood vessels dilate—more heat radiated out from blood into surrounding atmosphere—body cooled

superficial blood vessels constrict—less radiation of heat into atmosphere—heat conserved in deeper parts of body

EXTERNAL TEMPERATURE HIGH

erector muscles relaxed:
hair shafts very oblique
in relation to skin surface —
free circulation of air
over hairs (moving air is
a good heat convector)

('X' — point of attachment of
erector muscle to hair follicle)

EXTERNAL TEMPERATURE LOW

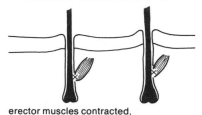

erector muscles contracted,
pull on hair follicles at 'X' —
hair shafts now perpendicular to skin surface.
Air (poor conductor of heat),
trapped between hairs,
acts as heat insulator to warm skin.
(Similar effect achieved in man
by adding layers of clothing).

within the organisms. (In birds (see p. 73), feathers are used in a similar way to hairs to provide insulation).

The adipose tissue of mammals gives additional heat insulation; mammals such as pigs have thick fat deposits, and in whales and seals the adipose tissue is very thick, forming 'blubber'. In other mammals, adipose tissue is built up in autumn to prepare for severe cold.

2 Excretion

The sweat glands extract water containing dissolved salts and a trace of urea from the bloodstream. This 'sweat' flows up the duct, overflows the skin surface, and is periodically washed away (see 1 (b) above).

3 Protection

The dead horny layer is tough, impervious and inert; it provides protection against entry of bacteria (which can enter only through hair follicles and sweat glands or where skin has been damaged), loss (and entry) of water, and against frictional damage. As a consequence, even in a newly-born infant, this layer is thickest where there is greatest friction, e.g. palms and soles. The adipose tissue acts as a cushion and assists in protecting more delicate tissues under the skin from knocks and blows.

Oil from the sebaceous glands maintains the hair shaft in good condition by making it supple and less liable to break; it also water-proofs the skin and destroys certain bacteria.

The skin gives protection from radiation. Cells below the epidermis produce a brown pigment in response to excessive exposure and this acts as a barrier to the sun's rays—hence the skin 'goes brown'. Fair-skinned people produce less of the pigment and are more likely to blister as a result of skin damage; others produce the pigment only in small isolated patches in the skin and become 'freckled' when exposed to the sun.

V.S Mammalian skin, showing follicles with sebaceous glands and erector muscle

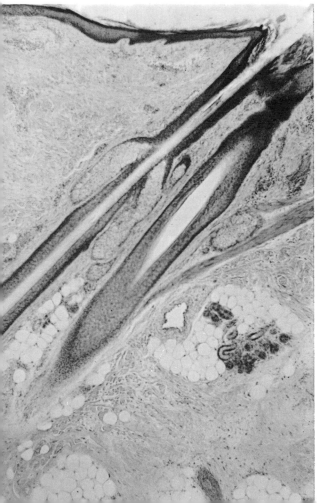

4 Reception of stimuli (see 'Sense organs', p. 175)

5 Vitamin D production

When ultra-violet rays, for example from sunlight, penetrate the skin they react with fatty substances called sterols, in the dermis. These become converted into vitamin D. In this way a proportion of the nutritional requirement of this vitamin (see p. 115) is made naturally by the body except in industrial areas where the ultra-violet light is filtered out by smoke in the atmosphere.

6 Fat storage

In addition to its use as protective cushion and heat insulator, skin adipose tissue provides a site for the storage of fat. When needed, this food reserve is mobilized and distributed around the body in the bloodstream.

7 Camouflage

The pigment contained in the hairs of mammals other than man makes these animals less conspicuous. Camouflage may be:

(a) **cryptic**—making the animal blend with its background, e.g. rabbit.
(b) **disruptive**—breaking-up the outline of the animal's body, e.g. zebra.

When seasonal variations of a climate are so extreme as to produce very different backgrounds the mammal's coat may change colour—the stoat may have brown hair in the summer but a white ermine coat in winter.

Hairs may be useful in other ways. Slight differences in colour may help to distinguish male from female and thus aid recognition prior to mating. Some mammals, e.g. frightened or angered cats, can make themselves appear larger and more threatening by erecting their fur.

In man the hair covering varies in different regions of the body. The palms and soles are hairless, and whereas much of the rest of the body surface is covered with fine hair, the hair of the scalp and armpits is coarse.

OTHER MODIFICATIONS OF SKIN STRUCTURES

1 **whiskers**—very long stout hairs which act as extended touch receptors. The width of the whiskers may give the animal, e.g. cat, an appreciation of the width of its body.

2 **nails**—very thick plates of dead horny material on the upper surface of the terminal digit. Nails continually grow from the Malpighian layer immediately below them. Used in gripping (for feeding, defence and cleaning).

3 **claws**—similar to nails but completely covering the digit-end. The dead horny layer is continually pushed out from base to tip. Claws are kept sharp by the mammal's own activities. Used in climbing, tearing (for feeding, defence) and in cleaning the body.

4 **hooves**—especially in ungulate mammals, e.g. horses, at the end of a single digit only, and cattle, where the hoof is cloven at the end of two digits. Hooves are often associated with fast-running mammals; each hoof has a broad horny outer covering and softer tissue internally acting perhaps as a shock absorber.

5 **horns**—dense horny projections (sometimes composed of compacted hair) often moulded around a bony projection from the head. Used for protection and for establishing the dominance of one male over another.

6 **mammary glands**—probably modified sebaceous glands. Secrete milk (see p. 231) for the nourishment of the very young mammal.

7 **wax-secreting glands**—probably modified sweat glands, found lining the outer ear tube (see p. 182). Used to trap materials which enter the ear and to prevent the ear drum becoming brittle.

8 **scent glands**—usually modified sweat glands. Used for distinguishing different species and male from female, or for marking out a territory. Unpleasant odours, e.g. from skunks, are protective.

---

1 Examine prepared sections of mammalian skin microscopically. Compare the appearance of human scalp skin with skin from elsewhere on the body.

2 Consider the nature of skin in other vertebrates—fishes, amphibians, reptiles and birds. Relate the differences in skin structure to the modes of life of these animals.

3 Compare the skin of vertebrates with the outer body covering of invertebrate animals.

4 Collect information concerning the ways in which their outer coverings protect plants and animals.

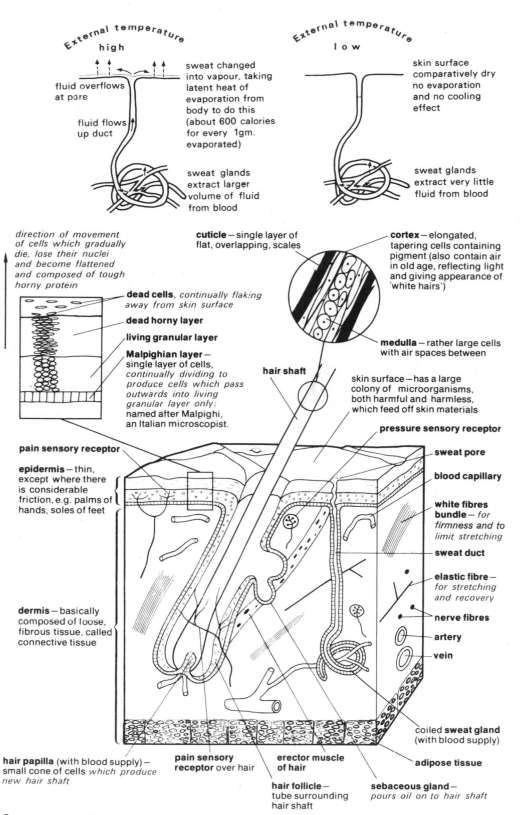

External temperature **high**

fluid overflows at pore

fluid flows up duct

sweat changed into vapour, taking latent heat of evaporation from body to do this (about 600 calories for every 1gm. evaporated)

sweat glands extract larger volume of fluid from blood

External temperature **low**

skin surface comparatively dry no evaporation and no cooling effect

sweat glands extract very little fluid from blood

direction of movement of cells which gradually die, lose their nuclei and become flattened and composed of tough horny protein

cuticle — single layer of flat, overlapping, scales

cortex — elongated, tapering cells containing pigment (also contain air in old age, reflecting light and giving appearance of 'white hairs')

dead cells, *continually flaking away from skin surface*

dead horny layer

living granular layer

Malpighian layer — single layer of cells, *continually dividing to produce cells which pass outwards into living granular layer only;* named after Malpighi, an Italian microscopist.

hair shaft

medulla — rather large cells with air spaces between

skin surface — has a large colony of microorganisms, both harmful and harmless, which feed off skin materials

pressure sensory receptor

pain sensory receptor

epidermis — thin, except where there is considerable friction, e.g. palms of hands, soles of feet

sweat pore

blood capillary

white fibres bundle — *for firmness and to limit stretching*

sweat duct

elastic fibre — *for stretching and recovery*

nerve fibres

artery

vein

dermis — basically composed of loose, fibrous tissue, called connective tissue

coiled **sweat gland** (with blood supply)

adipose tissue

hair papilla (with blood supply) — small cone of cells *which produce new hair shaft*

pain sensory receptor over hair

erector muscle of hair

hair follicle — tube surrounding hair shaft

sebaceous gland — *pours oil on to hair shaft*

Stereogram of vertical section through human skin (diagrammatic)

21 FOOD MATERIALS

All living organisms must be supplied with, or be capable of making, food. The feeding processes of organisms are called **nutrition**.

Substances which can be classed as food for plants and animals are:

Carbohydrates ⎫
Fats ⎬ required in relatively
Proteins ⎭ large quantities.
Inorganic (Mineral) Salts—required in relatively small quantities.
Water—variable amounts required, according to conditions.

In addition, animals are unable to manufacture certain **vitamins** (see p. 114) and they obtain these from plants.

Carbohydrates, fats, proteins and vitamins are *organic* chemical substances, meaning that they are made by living organisms.

In general, food is required by plants and animals for:

1 **A Source of Energy**—energy for mechanical purposes, such as the contraction of muscles and for the functioning of the nervous system; also for the chemical processes which manufacture substances in living organisms.

2 **Growth**—for increasing the size and numbers of cells in organisms and for the repair and replacement of damaged or worn tissues.

3 **Producing Essential Organic Substances**—as raw materials for such secretions as enzymes and hormones.

IMPORTANCE OF WATER

Plants and animals use the water which they take in for very many different purposes; the following are especially worth noting:

(a) as an essential part of protoplasm, which may be more than 75 per cent water.

(b) as a universal solvent—water dissolves or suspends substances for transport into (e.g. gases for respiration), around inside (e.g. foods) and out of (e.g. excretory materials) living organisms. Many chemical reactions such as enzyme actions take place only in solution.

(c) as a replacement for that water constantly being lost, during transpiration, sweating, etc. A man may lose more than four litres of water per day.

(d) as a means of support.

(e) as a lubricating agent.

(f) as a cooling agent.

Mention is frequently made in this book of the importance of water to living organisms.

22 NUTRITION OF GREEN PLANTS

This is often described as **holophytic** (Gk. *holos*, whole; *phyton*, plant). The method is **photosynthesis**, so-called because organic foods (carbohydrates) are manufactured from simple inorganic compounds, using light energy.

Four factors must be present for photosynthesis:

1 **Sunlight**
2 **Chlorophyll** (in *living* cells)
3 **Supply of Carbon Dioxide**
4 **Supply of Water.**

The following equation represents the chemical reactions of photosynthesis in outline and shows the beginning and end products:

$$6CO_2 + 6H_2O \xrightarrow[\substack{\text{in the presence of} \\ \text{chlorophyll}}]{\text{sunlight energy}} C_6H_{12}O_6 + 6O_2$$

carbon water   glucose oxygen
dioxide

Another name for photosynthesis is **carbon assimilation**; the equation shows that the carbon in carbon dioxide is assimilated (i.e. built-up) into the glucose molecule. Energy from sunlight is locked away in the glucose molecule, and this potential energy may be released later and used in the plant's metabolism.

1 SUNLIGHT

Photosynthesis takes place in the **chloroplasts**; these are green, and therefore reflect green light. The graph shows that it is light at the red and blue ends of the spectrum which is absorbed by chlorophyll and used to make sugar.

The light energy is known to be used in the initial stages of photosynthesis to split the water into hydrogen and oxygen. This is **photolysis** of water (Gk. *phos*, light; *lysis*, loosing, in the sense of breaking apart). The oxygen is given off and the hydrogen reduces the carbon dioxide to give sugar.

1 Photolysis—$2H_2O + \underset{\text{ENERGY}}{\text{LIGHT}} \rightarrow 2H_2 + O_2$

2 Reduction—$6CO_2 + 6 \times 2H_2 \rightarrow C_6H_{12}O_6 + 6H_2O$

Most plants convert the **glucose** into **starch** for storage. This fact is used as a basis for experiments to show the conditions necessary for photosynthesis, i.e. if starch is present, then photosynthesis has occurred.

The following experiments form an important part of the study of photosynthesis, and they should be performed by the student.

A: Experiment to show that sunlight is necessary for photosynthesis

1 De-starch experimental plant (e.g. potted Pelargonium) by placing in dark cupboard for 48 hours. Otherwise experiment shows only that starch is *lost* from plant in darkness.

2

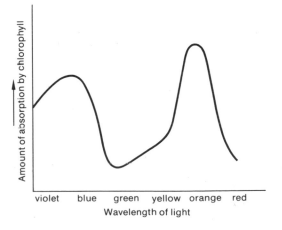

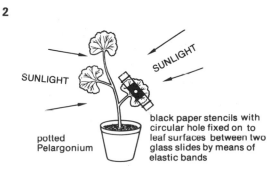

black paper stencils with circular hole fixed on to leaf surfaces between two glass slides by means of elastic bands

potted Pelargonium

3 Leave for minimum period of 5 hours in sunlight.

4 Detach experimental leaf from plant, remove glass slides and stencils.

5 Test leaf for starch as follows:

(a) Kill leaf by dipping briefly in boiling water. This stops any further chemical changes in the leaf cells, makes the leaf less brittle and allows methylated spirits to penetrate it more easily.

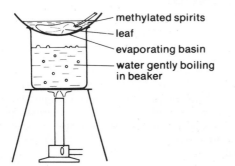

methylated spirits
leaf
evaporating basin
water gently boiling in beaker

(b) Place leaf in methylated spirits in a basin, warmed *indirectly* by a water bath, to remove chlorophyll, which makes the results of the starch test easier to see.

(c) When the leaf is very pale remove from basin, wash in water (preferably boiling to soften leaf) and spread flat on a glazed tile or large watch-glass.

(d) Cover leaf surface with iodine solution. Allow to stand for about 3 minutes.

(e) Wash away excess iodine solution.

Result

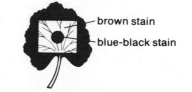

brown stain
blue-black stain

This experiment shows that starch is formed only in those parts of the leaf reached by sunlight. Hence, sunlight is necessary for photosynthesis.

2 CHLOROPHYLL

This is located inside cells only, in structures called chloroplasts. In most plants chloroplasts are small discs present in the cytoplasm which internally lines the cellulose wall.

Cells which contain chloroplasts are found in various situations in the above-ground parts of plants, e.g. just internal to the epidermis in herbaceous stems (see p. 55). Chloroplasts are particularly numerous, however, in leaves, in the palisade mesophyll cells immediately below the upper epidermis (see diagram on page 60). For this reason the upper surfaces of many leaves appear darker green than the lower. In this situation chloroplasts receive the best possible supply of sunlight. This means that the major part of photosynthesis takes place in the palisade tissue, even though chloroplasts are found in the spongy mesophyll cells and in the guard cells of stomata.

Chloroplasts move by streaming in the cytoplasm and take up different positions according to the light intensity.

The function of chlorophyll in photosynthesis seems to be that it absorbs the light (red and blue wavelengths mainly) which would either be reflected from the leaf surface or pass completely through an otherwise colourless leaf. The light energy is then used for the chemical reactions of photosynthesis.

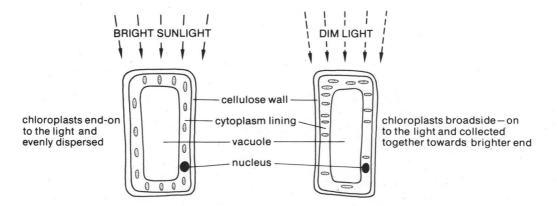

BRIGHT SUNLIGHT   DIM LIGHT

chloroplasts end-on to the light and evenly dispersed

cellulose wall
cytoplasm lining
vacuole
nucleus

chloroplasts broadside-on to the light and collected together towards brighter end

B: Experiment to show that chlorophyll is necessary for photosynthesis.

Chlorophyll cannot be removed from cells without killing them; therefore a plant with **variegated** leaves must be used for this experiment.

1 De-starch potted Pelargonium plant with variegated leaves (see Expt. A 1); otherwise experiment suggests only that the yellow zone *loses* starch.

2

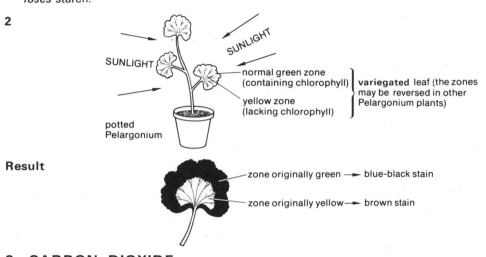

potted Pelargonium

normal green zone (containing chlorophyll)
yellow zone (lacking chlorophyll)
} **variegated** leaf (the zones may be reversed in other Pelargonium plants)

Result

zone originally green ⟶ blue-black stain
zone originally yellow ⟶ brown stain

3 Leave for a minimum period of 5 hours in sunlight.

4 Detach experimental leaf from plant. Carry out 'starch test' (see Expt. A 5).

This experiment shows that starch is formed only in the green parts of the leaf. Hence chlorophyll is necessary for photosynthesis.

3 CARBON DIOXIDE

The atmosphere contains as little as 0.03 per cent carbon dioxide, but this seems sufficient for the photosynthesis of all the world's green plants.

Land plants receive carbon dioxide as a gas through their stomata, minute pores in the epidermis. Carbon dioxide in solution passes into *submerged plants* all over their surfaces from the surrounding water.

In many green stems the cells containing chloroplasts are just internal to the epidermis, so that carbon dioxide does not have far to travel from the stomata. Stomata are generally more abundant in the lower than in the upper epidermis of leaves. After entering the stomatal pore the carbon dioxide diffuses along an intricate system of air passageways between the spongy mesophyll before it diffuses through the cell wall into a chloroplast (see p. 60). This diffusion gradient is maintained because, in sunlight, the carbon dioxide is used up as fast as it reaches the cells.

As shown previously (see p. 95), the carbon dioxide in the chloroplast becomes reduced, using the hydrogen from the water; it is then built up in a series of stages to glucose.

C: Experiment to show that carbon dioxide is necessary for photosynthesis.

1 De-starch potted Pelargonium plant (see Expt. A 1); otherwise experiment suggests only that leaves deprived of carbon dioxide *lose* starch.

2

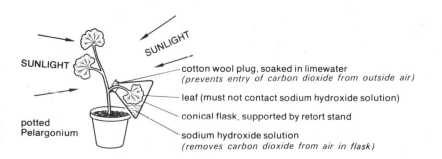

potted Pelargonium

cotton wool plug, soaked in limewater *(prevents entry of carbon dioxide from outside air)*
leaf (must not contact sodium hydroxide solution)
conical flask, supported by retort stand
sodium hydroxide solution *(removes carbon dioxide from air in flask)*

3 Leave for a minimum period of 5 hours in sunlight.

4 Remove experimental leaf from flask and another leaf from elsewhere on the same plant. Carry out 'starch test' (see Expt. A 5).

This experiment shows that starch is formed only where carbon dioxide is available. Hence carbon dioxide is necessary for photosynthesis.

Result

LEAF FROM FLASK   OTHER LEAF

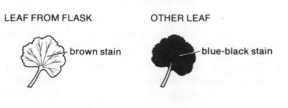

brown stain   blue-black stain

4 WATER

Land plants obtain water for photosynthesis from that which has entered by the roots and been carried upwards in the transpiration stream (see p. 141). Aquatics obtain their water from the surroundings.

Only a small fraction of the water which enters a plant is used for making glucose; because of this no simple experiment can be devised to show that water is necessary for photosynthesis. To deprive a plant of water would kill it for reasons not connected with photosynthesis. More sophisticated experiments using radioactive isotopes, etc., demonstrate that water is essential.

THE LEAF AS AN ORGAN SUITED FOR PHOTOSYNTHESIS

midrib — *carries pipeline (main vein) from leaf base to tip and provides central support*

branching network of **minor veins** — *support, and bring pipelines into lamina*

lamina (leaf blade) — *large flat surface area usually held at right angles to the light*

petiole (leaf stalk) — *carries one or more veins which are pipelines bringing water and mineral salts from the soil, and taking away foods produced as a result of photosynthesis to other parts of the plant; also holds leaf out towards light*

DIAGRAM OF VERTICAL SECTION
THROUGH PORTION OF LAMINA

upper epidermis, with few stomata

palisade mesophyll, containing many chloroplasts

carbon dioxide dissolves in film of water on outer surface of cells

spongy mesophyll, containing some chloroplasts

intercellular air space system

lower epidermis, with many stomata

one stoma

carbon dioxide diffuses in — *used up as fast as it enters, so maintaining diffusion gradient.*

oxygen diffuses out — *continually produced by photosynthesis, so maintaining diffusion gradient*

THE PRODUCTION OF OXYGEN

Oxygen is produced as a waste product of photosynthesis. In land plants it diffuses out along the same pathway (in the reverse direction) that the carbon dioxide uses to enter. So much oxygen may be given off during active photosynthesis that the gas may come out of solution and form bubbles on the leaves of submerged plants. This can be used as the basis of the following experiment:

D: Experiment to show that oxygen is produced during photosynthesis

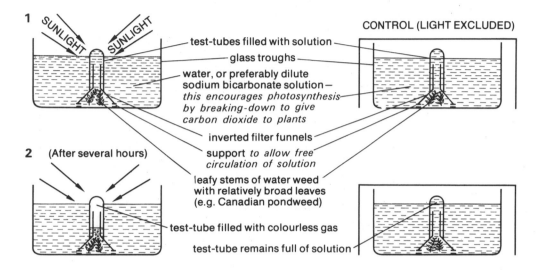

3 Carefully remove test-tube and test gas with glowing splint.

Result → The splint re-ignites, showing the gas to be oxygen. (Alternatively, the gas can be absorbed with alkaline pyrogallol.)

The experiment shows that oxygen is produced during photosynthesis: only the plant in sunlight had all the conditions necessary for photosynthesis.

PROCESSES FOLLOWING PHOTOSYNTHESIS

According to circumstances, in most flowering plants the glucose manufactured by photosynthesis may be:

1 Built up into starch for storage in leaf cells. Starch is stored in preference to glucose because (a) being insoluble, it can have no effect on the osmotic pressure of the cells which store it; (b) being a larger molecule, it cannot pass out through the cell membranes. The starch is later (e.g. during the night, when photosynthesis ceases) reconverted to glucose and used in any of the other three ways.

2 Built up into cellulose for new cell walls, by linking the glucose units together.

3 Built up into proteins for healthy growth and into other organic compounds, using various mineral salts (see Experiment E) which reach the leaf by way of the transpiration stream in the xylem. Energy is needed to synthesize proteins etc. from glucose; this energy is released by the simultaneous oxidation of other glucose molecules (i.e. internal respiration).

4 Transported away from the leaf via the phloem to roots, fruits, storage organs, etc., where it may be changed in any of the above three ways.

Later, the foods synthesized by plants may form the food of animals (see p. 109).

E: Water Culture Experiments

The table on p. 113 indicates the need of mineral salts for healthy growth and development in plants. Experiments involving the culture of seedlings in different watery solutions demonstrate the importance of salts in this capacity.

1   A number of gas jars set up as follows:

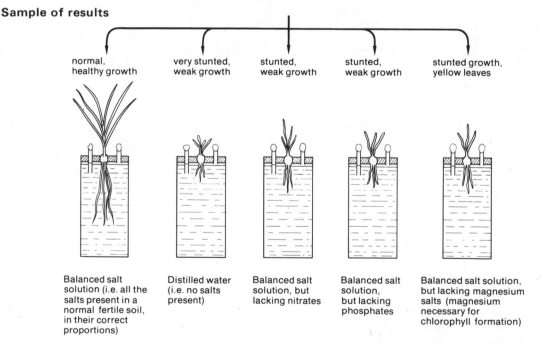

cotton wool plugs

tube for entry of air

watertight cork

clean gas jar

culture liquid

black paper cover — to exclude light and prevent growth of algae

maize seedling

tube for exit of air

Note:
(a) All seedlings used must have reached a similar stage of growth.
(b) Each jar to contain a different culture liquid (examples given below); ideally, each liquid should be repeated in several different jars.

2   Leave experiment for several *weeks*; at regular intervals (three times per week approximately) remove cotton-wool plugs and blow air or oxygen through culture to aid respiration of roots.

Sample of results

normal, healthy growth

very stunted, weak growth

stunted, weak growth

stunted, weak growth

stunted growth, yellow leaves

Balanced salt solution (i.e. all the salts present in a normal fertile soil, in their correct proportions)

Distilled water (i.e. no salts present)

Balanced salt solution, but lacking nitrates

Balanced salt solution, but lacking phosphates

Balanced salt solution, but lacking magnesium salts (magnesium necessary for chlorophyll formation)

The importance of photosynthesis to animal life

1   Photosynthesis produces oxygen gas as a by-product; this gas is essential for the respiration of animals. The waste carbon dioxide produced from the respiration of animals is used by green plants during photosynthesis.

2   Green plants, unlike animals, are able to convert simple inorganic substances into basic organic compounds on which all animals rely for a food supply.

Both these aspects of photosynthesis form an integral part of the Carbon Cycle, i.e. the way the element carbon circulates in nature. As far as living organisms are concerned, the cycle depends on photosynthesis for its completion.

The carbon cycle

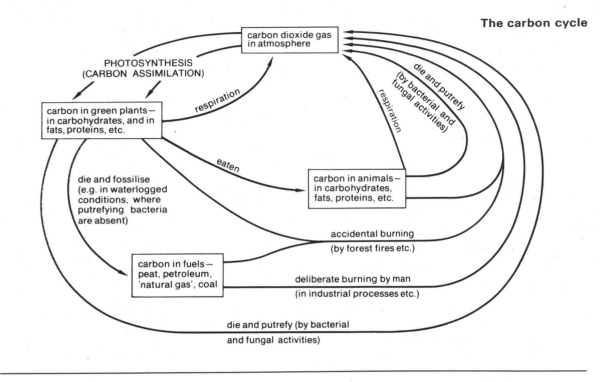

1   Examine microscopically a leaf of a moss plant and/or Canadian pondweed in a drop of water in order to see cells containing chloroplasts and the movement of chloroplasts in the cytoplasm. Add iodine solution to test for the presence of starch.

2   Consider further the importance of photosynthesis to animal life; find a range of different food chains in both water and land environments.

23 NUTRITION OF MAN

This is often described as **holozoic** (Gk. *holos*, whole; *zoon*, animal). It is the typical animal method of nutrition and these stages are present:

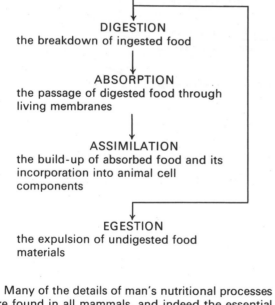

INGESTION
the taking-in of food, which has usually been searched for and/or 'captured'; the food is derived directly or indirectly from green plants

DIGESTION
the breakdown of ingested food

ABSORPTION
the passage of digested food through living membranes

ASSIMILATION
the build-up of absorbed food and its incorporation into animal cell components

EGESTION
the expulsion of undigested food materials

Many of the details of man's nutritional processes are found in all mammals, and indeed the essential features are common to all free-living (i.e. holozoic) animals.

ESSENTIAL DIETARY CONSTITUENTS

1 **Carbohydrates**

2 **Fats**

3 **Proteins**

4 **Inorganic (Mineral) Salts**

5 **Water**

(see p. 111ff)

6 **Vitamins**—sometimes called 'accessory food substances', as they are required only in small quantity, though their effects in the body are out of all proportion to the amount normally ingested. They are not used directly for growth or for providing energy, but are essential in metabolic processes. Absence in the diet leads to **deficiency diseases** (see table, page 114).

7 **Roughage**—material which cannot be digested (e.g. in man, the cellulose of plant cell walls). Roughage provides the alimentary canal muscles with bulk against which to contract. This allows more efficient movement of food along the canal, especially in the large intestine. Absence of roughage (e.g. of vegetables in the diet) leads to constipation.

ADEQUATE AND BALANCED DIETS

It is not enough that all the essential components are present in the diet. The food intake must be adequate, i.e. ideally each meal should provide sufficient

(i)   growth-promoting foods

(ii) energy-providing foods, the energy which they release being measured in joules—see chart A).

The degree to which any particular meal is adequate varies in different individuals, according to (a) sex, (b) age, (c) occupation. For example, manual workers need more than do sedentary 'brain' workers (see chart B). It also varies with external temperature (less food is needed in warm climates), and with body size. As far as small differences are concerned, e.g. between a tall slim man and a short slim man, the taller man usually requires more food; however, with great size differences, e.g. between man and mouse, the mouse requires *proportionately* more because of the continual heat loss from its relatively larger body surface (see p. 206).

For a diet to be **balanced**, the essential dietary components should be present in the *correct pro-*

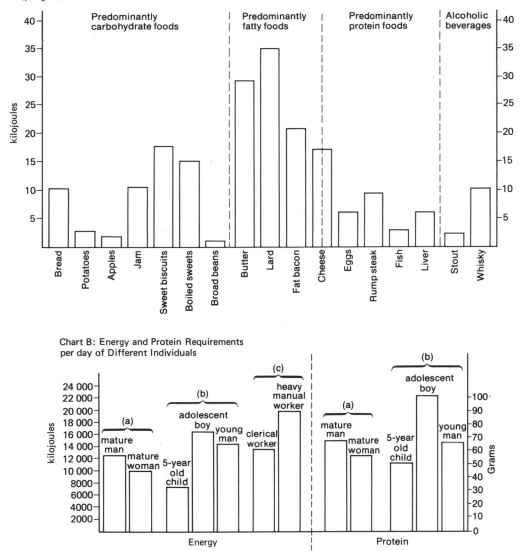

Chart A: Approximate Energy Value (per gram) of some Foods

Chart B: Energy and Protein Requirements per day of Different Individuals

portions, e.g. the energy foods must be evenly distributed between carbohydrates, fats and proteins, on average in a ratio by weight of 5:1:1 respectively (see chart C). Food should also be palatable.

Milk as a food

Milk provides an almost complete balanced food—it contains carbohydrates, fats, proteins, most minerals and vitamins, and a very high proportion of water. Young mammals are fed exclusively on milk for a short time after birth.

However, milk has several disadvantages as the only constituent of a diet:

1 It contains insufficient vitamin C—infants fed exclusively on milk develop scurvy after about a year.

2 It lacks iron—the mother provides the embryo with a store of iron which the young mammal draws upon during the suckling stage, so preventing anaemia.

3 Its protein is not easily digested.

4 The large volume of water in milk causes excessive fluid intake—the diet becomes too bulky.

5 It lacks roughage.

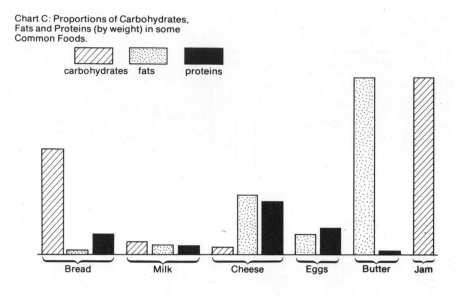

Chart C: Proportions of Carbohydrates, Fats and Proteins (by weight) in some Common Foods.

carbohydrates   fats   proteins

Bread   Milk   Cheese   Eggs   Butter   Jam

The alimentary canal

The organs which together process food are basically tubular and are called the alimentary canal. Associated with the canal are various glands (see diagram on p. 106). Each region of the canal has its own special job in the processing of food, and the foods move along the canal from one region to the next as on a conveyor belt. The main method of propulsion along the belt is **peristalsis**. This is caused by involuntary muscles in the walls of the alimentary canal; once the walls have been stretched by the entry of a **bolus** (ball of food) a wave of muscular contraction begins which is not under conscious control:

Diagrams showing peristalsis

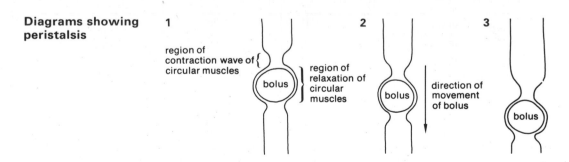

1

region of contraction wave of { circular muscles

bolus

region of relaxation of circular muscles

2

bolus

direction of movement of bolus

3

bolus

Why digestion is necessary

1 Large molecules must be broken down into molecules small enough to permeate living membranes (especially the intestine lining, in order to reach the bloodstream).

2 Insoluble substances must be changed into substances which can be carried in solution around the body (in the blood and lymph).

3 Complex molecules must be converted into molecules simple enough to be used by the cells in assimilation.

There are two basic types of digestion process:

(a) **mechanical digestion**—e.g. the grinding action of teeth

(b) **chemical digestion**—by the use of enzymes and related substances

Enzymes have the following properties:

1 They are organic chemicals, secreted in solution by living cells (though they themselves are non-living).

2 They act as catalysts; in the alimentary canal they speed the rate of digestion which would otherwise occur far too slowly.

3 They are required only in small quantity, though larger quantities of enzyme cause digestion to proceed faster.

4 Each enzyme always causes the production of the same end-product.

5 They are affected by temperature; those in the alimentary canal work most efficiently at body temperature. Below this temperature they work less rapidly, and at higher temperatures enzymes are destroyed.

6 They are affected by the nature of the liquid surrounding them; each enzyme works most efficiently at a certain degree of acidity or alkalinity.

7 They are specific, i.e. each enzyme will act upon only one type of substance. As three types of foods (carbohydrates, fats and proteins) need to be digested, these following three types of enzyme operate in the alimentary canal:

(a) **Carbohydrases**—polysaccharides → disaccharides → monosaccharides

(b) **Lipases**—fats → fatty acids + glycerol

(c) **Proteases**—proteins → peptones → peptides → amino-acids

(The table on p. 110 shows the principal enzymes in the alimentary canal and their activities.) The student can very easily demonstrate enzyme activity using his own saliva.

Experiment to demonstrate the activity of an enzyme
(Using the enzyme **ptyalin** in **saliva**)

1 Prepare some starch solution by boiling a small quantity of starch powder in water. Cool, and test a portion of solution with iodine solution to obtain characteristic blue-black colour.

2 Collect saliva from the mouth into a beaker.

3 Then proceed as described in the diagram at the top right of this page.

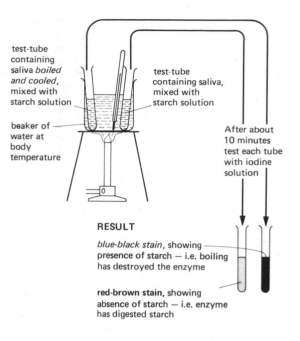

test-tube containing saliva *boiled and cooled*, mixed with starch solution

test-tube containing saliva, mixed with starch solution

beaker of water at body temperature

After about 10 minutes test each tube with iodine solution

RESULT

blue-black stain, showing presence of starch — i.e. boiling has destroyed the enzyme

red-brown stain, showing absence of starch — i.e. enzyme has digested starch

Additional enzyme experiment (to show the effect of **acidity** and **alkalinity** on **ptyalin**)

1 and 2 as in previous experiment.

3

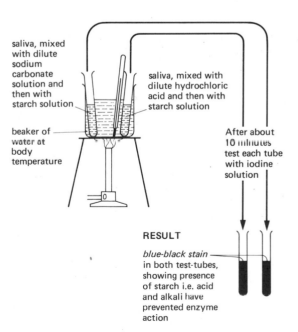

saliva, mixed with dilute sodium carbonate solution and then with starch solution

saliva, mixed with dilute hydrochloric acid and then with starch solution

beaker of water at body temperature

After about 10 minutes test each tube with iodine solution

RESULT

blue-black stain in both test-tubes, showing presence of starch i.e. acid and alkali have prevented enzyme action

Alimentary canal of man (diagrammatic)

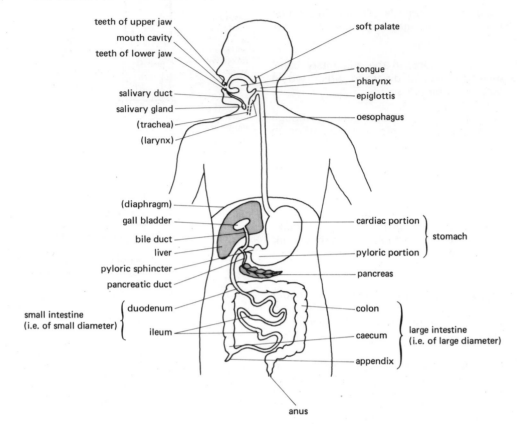

Mouth cavity

Food may be bitten off by the incisors and then chewed (masticated) by the cheek teeth (see p. 118).

Mastication is important because it breaks food into smaller particles:

(a) giving a much larger surface area for enzyme action;

(b) preventing discomfort during swallowing.

The sight, taste and smell of food stimulates the production of **saliva** down the ducts from three pairs of **salivary glands**—in the cheeks in front of the ears, under the tongue, and under the lower jaw. Saliva contains:

(i) the enzyme **ptyalin**.

(ii) **mucin**, to soften and lubricate food, and make food particles adhere to each other.

(iii) **salts**, e.g. sodium chloride and bicarbonate, to provide a near-neutral medium for ptyalin.

A hard palate separates the mouth from the nasal cavity so that mammals, unlike other animals, may retain food here without interfering with breathing. Tongue movements mix the food particles with saliva, then the particles are collected into a ball (**bolus**) prior to swallowing.

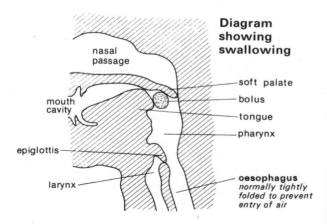

Diagram showing swallowing

The **tongue** forces the bolus upwards and backwards across the pharynx. The soft palate is pushed across the nasal opening and the **epiglottis**, like a trap-door, shuts the opening into the larynx. Aided by the contraction of pharynx muscles, the bolus is forced into the oesophagus.

Oesophagus

A narrow folded muscular tube; the bolus forces the folds open and **peristalsis** begins (see p. 104). Liquids travel quickly by gravity. Food passes along the oesophagus through the neck and thorax into the stomach.

Stomach

A storage sac of capacity one litre approximately, enabling mammals to eat only intermittently; it lies across the upper abdomen. The bolus first enters the cardiac portion and meets **gastric juice** secreted by glands in the stomach walls. The **hydrochloric acid** present in gastric juice:

(a) provides the correct acid medium for the gastric enzymes (rennin and pepsin).

(b) kills putrefying bacteria which enter with food.

(c) stops the activity of ptyalin.

Gastric juice also contains **mucus**, which further lubricates the food and protects the stomach walls from the action of enzymes and acid. (Mucus is produced for similar purposes in the small and large intestines.)

Food may remain in the stomach for as long as three hours, depending on the time needed for the completion of gastric digestion. Simple molecules such as glucose and alcohol may be absorbed through the stomach wall into the bloodstream. Fats such as butter are **melted** by the heat and the acid food mass (**chyme**) is continually churned by the complex arrangement of muscle layers in the wall.

The **pyloric sphincter** is a ring-shaped muscle band at the far end of the stomach. When it relaxes and contracts rhythmically, chyme is squeezed through a narrow opening in the ring by a series of small jets into the duodenum.

Duodenum

Acid food passing into the duodenum causes its cells to produce the hormone **secretin** which passes into the bloodstream. On reaching the pancreas secretin causes the production of juice down the pancreatic duct. A similar mechanism applies in relation to the production of bile from the gall bladder. As man feeds only at intervals, these hormones play an important part in the body's economy; they allow bile and pancreatic juice to be poured into the duodenum only when food is ready to be digested there.

Bile is greenish watery and alkaline, lacking enzymes but containing:

(i) **salts** (e.g. organic salts and sodium bicarbonate) which:

    (a) change acid chyme to alkaline **chyle**, so providing a suitable medium for the remaining digestive enzymes.

    (b) **emulsify** fats, i.e. change the pools of fat into small suspended droplets with a much larger surface area for subsequent enzyme action.

(ii) **excretory materials** (see p. 157), largely derived from the breakdown of haemoglobin from worn-out red corpuscles; these colour the waste food.

Pancreatic juice is also alkaline, and contains enzymes to continue digestion; it enters about halfway along the duodenum, which soon becomes the ileum.

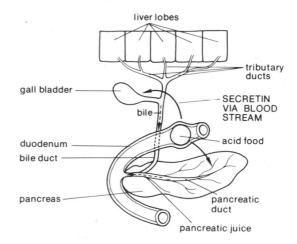

Diagrammatic representation of duodenum region, showing action of secretin

Ileum

A very long coiled tube about twenty feet long. The walls secrete enzymes which complete digestion.

The **end-products of digestion** are monosaccharides (carbohydrates), fatty acids and glycerol (fats), and amino-acids (proteins). The ileum is the main region for **absorption** of these end-products,

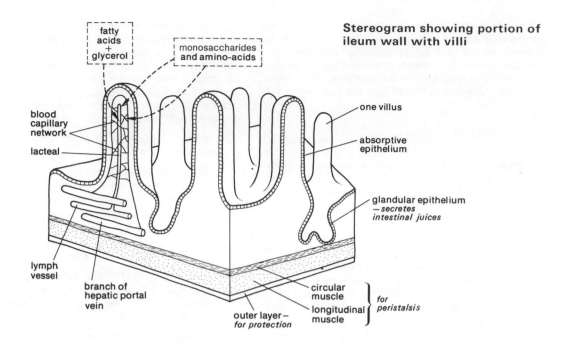

Stereogram showing portion of ileum wall with villi

having a very large surface area, because of

(a) its considerable length

(b) its inner surface, which contains inward finger-like projections called **villi**.

The end-products are absorbed through the thin lining into the capillary system, as shown in the diagram.

The food-remains after absorption are extremely fluid. About 0.5 litre (500 cm³) per day in man passes into the large intestine.

Large intestine

The **caecum** and **appendix** are very reduced in size compared with the rabbit, and have little importance in man. The food-remains which pass into the **colon** consist of (i) undigested food, (ii) roughage, (iii) mucus, (iv) alimentary secretions, (v) bacteria, (vi) dead cells from the canal lining, (vii) waste bile pigments, (viii) water. The colon extracts as much water as possible and passes this into the blood; **water conservation** of this kind is extremely necessary in land-living animals.

This leaves semi-solid cylinders of waste food (**faeces**) which pass into the **rectum** for storage. Faeces may remain here for more than twenty-four hours and are finally **egested** by muscular action through the anus.

THE DESTINY OF ABSORBED FOODS

1 Monosaccharides—carried by hepatic portal vein to liver. Depending on the immediate needs of the body some pass through the liver and are carried to the cells in the bloodstream. Excess (i.e. more than 80 mg per 100 cm³ of blood) is converted to glycogen and stored under the stimulus of the hormone insulin (see p. 174); this hormone is brought to the liver in the bloodstream from the pancreas. (Some glycogen may be stored in muscle tissue.) Later, glycogen may be reconverted to monosaccharide as a result of hormone stimulus (e.g. by adrenalin) and distributed around the body. Some monosaccharides are converted into fat for storage.

2 Fatty acids and glycerol—are reformed into fat before entering the lymph system eventually to be emptied into the jugular vein at base of neck; the fat is utilized at once or stored (e.g. under skin, around kidneys).

3 Amino-acids—carried by hepatic portal vein to liver. Depending on the immediate needs of the body, some pass through the liver and are carried to the cells by the blood. Excess cannot be stored; they are **deaminated**, i.e. nitrogen is removed from the molecule, leaving compounds composed basically of carbon, hydrogen and oxygen only. These compounds may be consumed to provide energy or stored as glycogen or fat. The released nitrogen is at first in the form of ammonia; the liver makes the ammonia non-poisonous by

changing it into urea which is carried by the blood to the excretory organs:

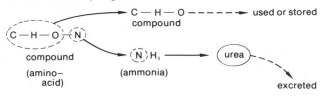

The liver

A five-lobed organ, the largest gland in the body, situated under the dome of the diaphragm (partly over the stomach and duodenum). It has a double blood supply—the hepatic artery (bringing oxygenated blood directly from the heart) and the hepatic portal vein (bringing blood rich in absorbed foods from the stomach and intestine). Blood leaves and returns towards the heart along the hepatic vein (see p. 153).

Principal functions:

1 Storage of glycogen (and its reconversion to monosaccharide when needed).

2 Alteration of the chemical structure of stored fats to make these fats available for metabolism.

3 Deamination of amino-acids and the production of urea.

4 Production of bile and its storage in the gall bladder (see p. 107).

5 Production of heat—the metabolic activities of the liver produce heat, so the temperature of blood leaving the liver is higher than that entering.

6 Production of blood proteins, e.g. prothrombin, fibrinogen, which are passed into the blood as it flows through (see p. 151).

7 Storage of iron—from the breakdown of haemoglobin, released when the red corpuscles are destroyed.

8 Storage of vitamins (e.g. vitamin A).

9 Detoxication—many poisonous substances in the blood either foreign to the body or produced naturally in it, are rendered harmless as the blood flows through.

The alimentary canal in other mammals

In common with other free-living animals, mammals may be classified according to their diets:

(a) **herbivorous**—those that ingest material only from plants.

(b) **carnivorous**—those that ingest material only from animals.

(c) **omnivorous**—those that ingest material from both plant and animal sources.

Differences in the structure and functioning of the alimentary canal are associated with these different sorts of diets. In particular there are differences in the arrangement of teeth and jaws (see p. 116). Herbivores tend to have proportionately longer alimentary canals than carnivores or omnivores; vegetation is more difficult to digest than is flesh.

In the rabbit the alimentary canal shows a conspicuous difference from that of man. The caecum is very large and ends in a prominent

appendix. Because the rabbit is herbivorous much of its diet consists of plant cell walls, i.e. of cellulose. In man there are no enzymes to digest cellulose and the plant cell walls constitute roughage; this loss of potential food is not serious, as man being omnivorous receives adequate carbohydrates from other food materials.

The rabbit also produces no cellulose-digesting enzymes itself. However the presence of certain bacteria in the caecum and appendix prevent what would otherwise be starvation. These bacteria secrete enzymes which digest the cellulose; a proportion of the digested products are absorbed by the bacteria themselves but the digestive processes are so extensive that sufficient is left for absorption by the rabbit.

A similar arrangement is found in **ruminants** (e.g. cows, sheep); these have a four-chambered stomach where cellulose-digestion occurs. Periodically, partly-digested food is brought back into the mouth for further mastication ('chewing the cud').

Other forms of nutrition

Many plants (e.g. fungi) lack chlorophyll and cannot feed themselves by photosynthesis. Similarly many animals are not free-living, so that such plants and animals use alternative methods of feeding. Three of the most common methods are outlined below:

1 **Saprophytism.** A saprophyte (Gk. *sapros*, rotten; *phyton*, plant) is a plant which feeds exclusively on dead organic matter, i.e. material from a living source but no longer alive. The plant exudes enzymes on to the organic matter, digests it externally, and then absorbs the simpler digested foods (see p. 15).

2 **Parasitism.** A parasite (Gk. *para*, beside; *sitos*, food) is a plant or animal which obtains food from a living source (called a host). Ideally the host remains alive so that the parasite may later take more food from the same source. Usually parasites are considerably smaller than their hosts and are of different species (see p. 28).

3 **Symbiosis** (Gk. *symbioun*, to live together). This is an association between two living organisms of different species (plant-plant, animal-animal, plant-animal) where both partners derive benefit, frequently nutritional benefit. In its caecum (see preceding section) the rabbit benefits from the digestion of cellulose by the symbiotic bacteria; the same bacteria benefit from a continual supply of cellulose brought to them by the rabbit.

SUMMARY OF DIGESTIVE ENZYMES

REGION OF ALIMENTARY CANAL	NAME OF ENZYME	DIGESTIVE CHANGE	OTHER FEATURES
MOUTH CAVITY	PTYALIN (≡SALIVARY AMYLASE)	Starch→maltose (polysaccharide→ disaccharide)	Enzyme continues activity down oesophagus until acidified in stomach
STOMACH	RENNIN	Milk protein→curds	Enzyme most abundant in young children; the curds are easier to digest, being semi-solid
	PEPSIN	Proteins→peptones	To prevent self-digestion of the stomach pepsin is inactive until it meets hydrochloric acid
DUODENUM	AMYLOPSIN (≡ PANCREATIC AMYLASE)	Starch→maltose	
	STEAPSIN	Emulsified fats→fatty acids + glycerol	(These are end-products of fat digestion)
	TRYPSINOGEN	Has no action until activated by *enterokinase* secreted by small intestine	
ILEUM	ENTEROKINASE + TRYPSINOGEN (from pancreas) gives TRYPSIN	Peptones→peptides	Enterokinase probably prevents self-digestion of pancreas and small intestine
	MALTASE	Maltose→glucose (disaccharide→monosaccharide)	(These monosaccharides are the end-products of carbohydrate digestion)
	SUCRASE	Sucrose→glucose and fructose	
	LACTASE	Lactose→glucose and galactose	
	EREPSIN	Peptides→amino-acids	(Amino-acids are the end-products of protein digestion)

☐ = carbohydrase   ┈ = lipase   ┄┄ = protease

CARBOHYDRATES

Compounds containing three elements only—carbon, hydrogen, oxygen.
Ratio of hydrogen : oxygen atoms per molecule 2 : 1

SUGARS (Sweet; soluble in water)		POLYSACCHARIDES (Not sweet; insoluble in cold water)
MONOSACCHARIDES Simple sugars	**DISACCHARIDES** Compound or complex sugars	
Formula—$C_6H_{12}O_6$	General formula—$C_{12}H_{22}O_{11}$	General formula—$(C_6H_{10}O_5)n$ where n = 300 approximately (Very large molecules, formed by linking monosaccharides together.)

Example: Glucose (dextrose or grape sugar) Fructose (fruit sugar) Galactose	*Common Food Source:* Fruits Fruits, honey Milk	*Example:* Sucrose (cane-beet-sugar) Lactose (milk sugar) Maltose (malt sugar	*Common Food Source:* Household sugar Milk Germinating barley	*Example:* Starch Cellulose Glycogen ('animal starch')	*Common Food Source:* Potatoes, bread Green vegetables (Living liver tissue only)

TESTS :	STARCH TEST :
All Monosaccharides and some Disaccharides will reduce the copper sulphate in Fehling's solution A or in Benedict's reagent to cuprous oxide— they are sometimes called 'reducing sugars' 1 Add Fehling's solution A, then Fehling's solution B to a glucose solution or material containing glucose, in a test-tube. Mix thoroughly. Gently boil, to obtain an orange-red precipitate (cuprous oxide) 2 Add Benedict's reagent instead of Fehling's solutions A + B. Repeat test 1, to obtain yellow or orange precipitate Non-reducing sugars, such as sucrose, give no precipitate with tests 1 and 2 unless they are first converted into reducing sugars in dilute hydrochloric acid and then neutralised with sodium hydroxide solution	Add iodine solution to starch powder or to a cold solution of starch. The starch becomes stained an intense blue-black Similar test with— Cellulose→blue stain Glycogen→red-brown stain

IMPORTANCE :	IMPORTANCE :
1 Broken-down during respiration to provide energy (17 kilojoules per 1 g of carbohydrate respired) 2 Basic food materials synthesised by green plants (during photo-synthesis) 3 Major way in which food substances are distributed inside living organisms	1 Food stores, e.g. starch in plants (cotyledons, tubers etc.), glycogen in animals (liver, skeletal muscles) 2 Structural materials, e.g. cellulose for plant cell walls

→

direction of build-up of simpler to more complex molecules inside living cells

←

direction of break-down of more complex into simpler molecules during digestive processes

FATS

Compounds containing three elements only—carbon, hydrogen, oxygen.
These three elements are present in varying proportions, and the number of oxygen atoms per molecule is relatively small

Example of formula—$C_{58}H_{98}O_6$ (tripalmitin)

OILS (Liquid at 20°C)	FATS (Solid at 20°C)
Examples: Cod liver oil—animal oil Maize oil Olive oil }—vegetable oil Linseed oil	Examples: Butter } Lard }—animal fats Margarine—vegetable fat
TEST: Add oil (e.g. olive, castor oil) to a small quantity of water in a test-tube. Shake the test-tube contents with red dye Sudan III. Allow to stand. The oil, stained red, floats on the water; the water remains colourless	TEST: 'Grease-spot test'. Rub fatty material into the centre of a clean sheet of paper. Hold paper to the light. A translucent stain appears where the fat was rubbed into the paper

IMPORTANCE

1   Sources of energy, producing more than twice as many calories as carbohydrate when respired (over 35 kilojoules per 1 g of fat), at the same time producing useful water for metabolism

2   Food stores—in seeds (e.g. endosperm of castor bean) and in mammals (e.g. under skin, around kidneys) where it acts as heat insulator and protective cushion

3   Structural materials, for protoplasm and cell membranes

4   Source of fat-soluble vitamins (see p. 114)

> Excess carbohydrates can be converted into fats; proteins may be converted into fats if the diet contains excessive carbohydrate.

PROTEINS

Compounds always containing at least four elements—carbon, hydrogen, oxygen, nitrogen.
Many proteins contain sulphur and phosphorus

Example of formula— $C_{254}H_{317}O_{75}N_{65}S_6$ (insulin)

Proteins are built from smaller units called amino-acids, joined together like a string of beads:

Many proteins are large molecules, composed of more than 100 constituent amino-acids

FIRST-CLASS (contain constituent amino-acids essential for growth of animals)	SECOND-CLASS (lack constituent amino-acids essential for growth of animals)
Examples: Most proteins from animal sources, e.g. proteins contained in lean meat, fish, cheese. Also contained in soya bean, certain nuts	Examples: Most proteins from vegetable sources, e.g. proteins contained in wheat flour

TESTS:

1   Add small quantity of sodium hydroxide to a protein solution in a test-tube. Then add very dilute copper sulphate solution, drop by drop, to obtain violet colour

2   Add nitric acid to solid protein (e.g. finger-nail) or solution of protein in a test-tube. Warm gently to obtain a yellow colour. Cool and add a few drops of strong ammonium hydroxide to obtain a bright orange colour

IMPORTANCE:

1 For growth: proteins are essential components of protoplasm and its structures

2 For the manufacture of essential chemicals, e.g. enzymes and hormones, for use in metabolism

3 Sources of energy: when respired, proteins produce over 20 kilojoules per 1 g (this is especially important in carnivores, whose carbohydrate intake is very low)

MINERAL SALTS

Required by plants and animals for a variety of reasons. Generally taken in as separate ions, i.e. calcium phosphate, for example, is not taken in as such, but as calcium (Ca^{++}) and phosphate (PO_4^{---}) ions. The following is a selection of the more important chemical elements; some may not be absorbed by living cells in 'mineral' form but rather as organic compounds, e.g. animals receive much of their nitrogen when they ingest protein.

ELEMENT	IMPORTANCE IN:		COMMON FOOD SOURCE FOR MAN
	FLOWERING PLANT	MAMMAL	
N	For synthesis of proteins and other complex chemicals		Protein foods, e.g. lean meat, fish
S	For synthesis of proteins and other complex chemicals	especially for hair, nails, horny layers of skin	
P	For synthesis of proteins and other complex chemicals and for structures concerned with heredity	for bone and tooth structure, muscle contraction	Protein foods, especially milk
K	For photosynthesis: for tissue fluids	For metabolism inside cells; for transmission of nerve impulses	
Na		For tissue fluids, including blood; for kidney functioning; for transmission of nerve impulses	Household salt
Mg	Constituent of chlorophyll	For bone and tooth structure	
Ca	For cell wall structure	For bone and tooth structure; for muscle contraction; for clotting of blood	Milk
Fe	For internal respiration / for chlorophyll manufacture	constituent of haemoglobin (lack causes anaemia, reduced oxygen-carrying capacity of blood)	Liver; certain vegetables (e.g. watercress)
I		For proper functioning of thyroid gland (lack causes simple goitre)	Sea-foods; household salt
Cl		For functioning of gastric juice; for tissue fluids, including blood	Household salt
Fl		For tooth structure (enamel) and hardness of bone	Milk; drinking water
Mn	For internal respiration / as a growth factor and for development of bone		

TABLE OF VITAMIN SUBSTANCES

VITAMIN	SOL-UBILITY	NAME OF DEFICIENCY DISEASE	NATURE OF DISEASE, CAUSED BY LACK OF VITAMIN	COMMON FOOD	OTHER FEATURES
A	Fat-soluble	Night-blindness	Reduced powers of vision in conditions of dim light (vitamin A is necessary for proper functioning of retina)	(a) directly from animal fats and oils e.g. milk, butter, egg yolk, fish liver oils, liver (b) indirectly from carotene in fresh green vegetables	1 Carotene, an orange pigment in green plants (masked by the chlorophyll) is converted into vitamin A by the cells in the intestine wall of mammals 2 Too much vitamin A produces unpleasant side-effects 3 Stable—not easily destroyed by heat
		Xerophthalmia	Damage to epithelial tissue, particularly cornea which becomes opaque and ulcerated leading to blindness. Linings of respiratory system become less resistant to invasion by bacteria and viruses		
B a complex of sub-stances including B_1 (thiamine)	Water-soluble				Vitamin B complex is present in all living cells
		Beri-beri	Inflammation of nerves leading to muscular weakness, paralysis, loss of sensation, mental deficiency and death	Whole cereal grains, whole-meal bread, yeast, liver, egg yolk, peas, beans	1 Beri-beri particularly prevalent in S.E. Asia where standard of living is low and staple diet is polished rice (i.e. the rice husk which actually contains vitamin B_1 is removed leaving only the 'kernel' for consumption, which is deficient in vitamin B_1) 2 Vitamin may be destroyed by heat
B_2 (riboflavin)			Cracking and flaking of skin leading to progressive skin infections	Fresh green vegetables, yeast, milk, liver, meat, eggs	Stable—not easily destroyed by heat
Nicotinic acid		Pellagra	1 Disorders of digestive system (loss of appetite, diarrhoea etc.) 2 Disorders of central nervous system (loss of memory, depression etc.) 3 Dermatitis (cracking and flaking of skin)	Whole cereal grains, yeast, liver, eggs, milk, cheese	Disease common where standard of living is low
C (ascorbic acid)	Water-soluble	Scurvy	1 Acute disorders of connective tissues—pains in muscles and joints, loss of weight, progressive haemorrhages (i.e. loss of blood both from external sites such as gums, nose, and from internal capillaries)	Citrus fruits (e.g. lemons, grape-fruits, oranges), blackcurrants, tomatoes, fresh green vegetables, potatoes, rose-hips	1 Scurvy used to be a common fatal disease of sailors on long oceanic voyages where fresh fruit and vegetables soon became unavailable 2 Unstable—quickly decomposed by oxidation at high temperatures

VITAMIN	SOL-UBILITY	NAME OF DEFICIENCY DISEASE	NATURE OF DISEASE, CAUSED BY LACK OF VITAMIN	COMMON FOOD	OTHER FEATURES
			2 Possibly reduced resistance to infection—vitamin C may be concerned with antibody production 3 Disorders of teeth 4 Inability of wounds to heal		3 Most mammals (but not man) synthesise vitamin C 4 Milk contains very little, so that supplementary vitamin C may be needed in infancy
D	Fat-soluble	Rickets	Bones lose their mineral contents (e.g. calcium phosphate), become soft and liable to fracture or give insufficient support—stresses cause distortion of bones leading to deformities	Animal fats and oils, e.g. milk, eggs, butter, fish liver oils	1 Vitamin D is manufactured by the action of ultra-violet light on skin (see p. 90) 2 Rickets particularly prevalent in (a) young mammals where bones are growing very rapidly (b) mammals, particularly man, living in industrial areas where fog and soot particles filter ultra-violet light out of atmosphere
E	Fat-soluble		Importance to man not conclusively proved. In other mammals—lack causes sterility. Male rats fail to produce fertile sperm and lose interest in females: in pregnant females, abortion may occur or offspring may be born dead	Fresh green vegetables, whole grains, egg, milk, butter	
K	Fat-soluble		Reduced ability of blood to clot; as a result, increased haemorrhages	Fresh green vegetables, tomatoes, liver	Vitamin K is synthesised by bacteria normally present in intestine

1 Plan one day's adequate and balanced diet for men and women of different ages and different occupations, with the aid of reference books on domestic science and nutrition.

2 Find out about the commercial production of milk in Great Britain from farm to consumer. In particular, find out what diseases are likely to be transmitted by milk and the ways in which milk is kept free from contamination.

3 Enzymes are found in plants as well as in animals. Find out about the enzymes of plants and their function in the plant's metabolism.

4 Examine prepared sections of stomach (to see gastric glands), small intestine (to see villi) and liver.

5 Read further about unusual methods of nutrition —symbiosis, commensalism and insectivorous plants.

6 Dissect a small mammal in order to display the alimentary canal from oesophagus to anus.

24 TEETH IN MAMMALS

Teeth are developments of the skin covering the jaws. The teeth of mammals are different from similar structures in many other animals; they are set into sockets along the jaw-bones.

Characteristically mammals possess *three* different types of tooth, each type adapted to a particular function:

1 **Incisors**

2 **Canines**

3 Cheek teeth – **Premolars**
                – **Molars**

Dental formulae:

The number and arrangement of teeth in the skull of an animal is called the **dentition**. As this varies considerably in different mammals it is convenient to have a shorthand method of showing dentition in the form of a **dental formula**. A common way of writing a dental formula can be shown, using an adult man as example:

$$\text{i.}\ \frac{2}{2}\ \ \text{c.}\ \frac{1}{1}\ \ \text{pm.}\ \frac{2}{2}\ \ \text{m.}\ \frac{3}{3} = 32$$

The initials i., c., pm., and m. refer to incisors, canines, premolars and molars respectively. The total at the end refers to the number of teeth in the

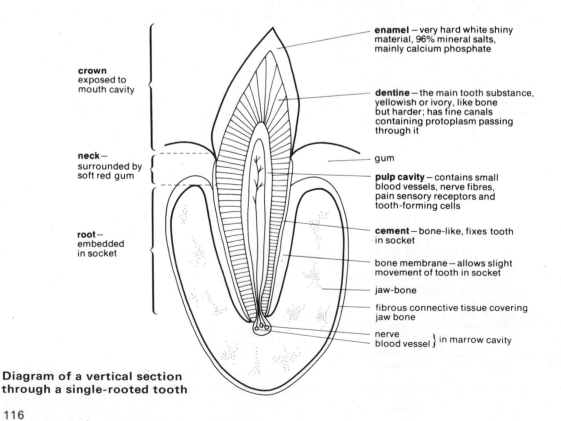

crown
exposed to
mouth cavity

neck –
surrounded by
soft red gum

root –
embedded
in socket

enamel – very hard white shiny material, 96% mineral salts, mainly calcium phosphate

dentine – the main tooth substance, yellowish or ivory, like bone but harder; has fine canals containing protoplasm passing through it

gum

pulp cavity – contains small blood vessels, nerve fibres, pain sensory receptors and tooth-forming cells

cement – bone-like, fixes tooth in socket

bone membrane – allows slight movement of tooth in socket

jaw-bone

fibrous connective tissue covering jaw bone

nerve
blood vessel } in marrow cavity

**Diagram of a vertical section
through a single-rooted tooth**

1 Incisor

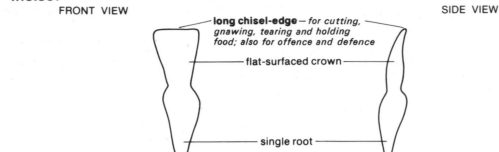

FRONT VIEW   SIDE VIEW

long chisel-edge — *for cutting, gnawing, tearing and holding food; also for offence and defence*

flat-surfaced crown

single root

2 Canine

conical pointed crown — *especially used for stabbing and tearing flesh*

single root

3 Cheek tooth

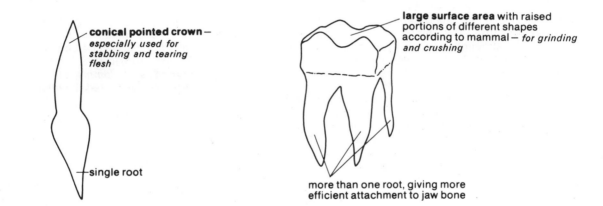

large surface area with raised portions of different shapes according to mammal — *for grinding and crushing*

more than one root, giving more efficient attachment to jaw bone

whole skull. The figures after each initial are the numbers of teeth of each particular type—the figures above the horizontal line deal with the teeth in the upper jaw, figures below the line with teeth in the lower jaw. However these individual figures are for *one side* of the skull only (right or left) so that they must finally be doubled to reach the overall total.

Milk and permanent dentitions

Each mammal normally has two sets of teeth in its lifetime. The first set, or **milk dentition**, serves during the time when the mammal is young. Some mammals, like the rabbit, are born with teeth; others, like man, begin to cut them shortly after birth.

The first incisors appear at about six months the milk dentition being complete by the end of the third year. The first sign of the permanent dentition is marked by the appearance of the 'six year' molars. From then on the milk dentition is gradually replaced by the rest of the **permanent dentition**. The eruption of the remaining molars, which are not represented in the milk dentition, is completed by the appearance of the **wisdom teeth.** This may not occur until the twentieth year or later. Hence in man:

Milk dentition:

Incisors + Canines + Premolars
             shed

Permanent dentition:

Incisors + Canines + Premolars + Molars

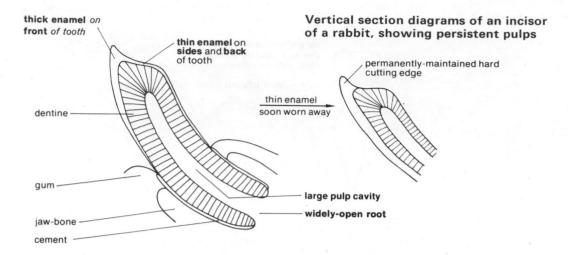

thick enamel *on front* of tooth

thin enamel on sides and back of tooth

Vertical section diagrams of an incisor of a rabbit, showing persistent pulps

dentine

permanently-maintained hard cutting edge

thin enamel soon worn away

gum

large pulp cavity

jaw-bone

widely-open root

cement

Open and 'closed' roots

In man, when the tooth is fully developed the entrance to the pulp cavity becomes very narrow. This means that the blood supply to the tooth is reduced because the blood vessels become constricted. Although enough food and oxygen enter the tooth to keep it alive there is now not enough for growth.

In many mammals, particularly those which eat harsh vegetation, the root remains widely **open** and the tooth continues to grow throughout life. Such teeth are said to have **persistent pulps**. The crown height is kept constant by the abrasive action of the food.

TEETH AND DIET

The dentition of mammals is closely related to the food they eat. There is also a relationship between diet and the way in which the two jaws move against each other; this in turn is affected by the type of movable joint between upper and lower jaws (i.e. by their **articulation**).

Man is **omnivorous** (L. *omnis*, all; *vorare*, to devour) and has an unspecialized dentition.

Gnawing mammal (e.g. rabbit)

Permanent dentition:

$$\text{i.}\ \frac{2}{1}\quad \text{c.}\ \frac{0}{0}\quad \text{pm.}\ \frac{3}{2}\quad \text{m.}\ \frac{3}{3} = 28$$

The roots are open.

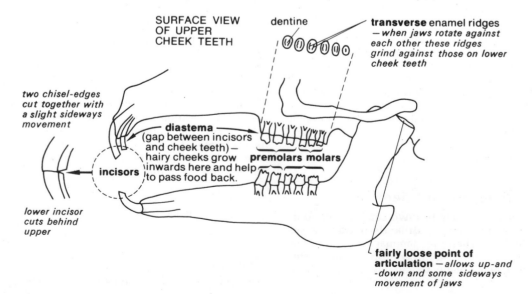

SURFACE VIEW OF UPPER CHEEK TEETH

dentine

transverse enamel ridges — *when jaws rotate against each other these ridges grind against those on lower cheek teeth*

two chisel-edges cut together with a slight sideways movement

diastema (gap between incisors and cheek teeth) — hairy cheeks grow inwards here and help to pass food back.

premolars molars

incisors

lower incisor cuts behind upper

fairly loose point of articulation — *allows up-and -down and some sideways movement of jaws*

Carnivorous (i.e. flesh-eating) mammal (e.g. dog)

Permanent dentition :

$$\text{i.} \ \frac{3}{3} \quad \text{c.} \ \frac{1}{1} \quad \text{pm.} \ \frac{4}{4} \quad \text{m.} \ \frac{2}{3} = 42$$

The roots are not open.

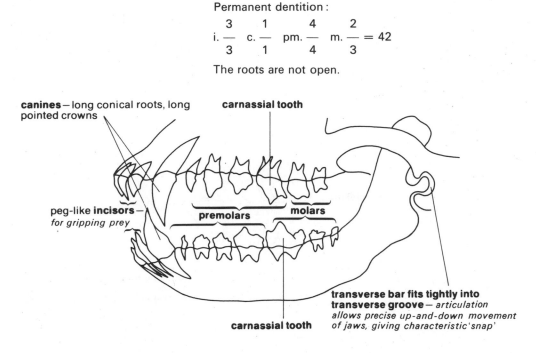

canines — long conical roots, long pointed crowns

carnassial tooth

peg-like **incisors** — *for gripping prey*

premolars

molars

carnassial tooth

transverse bar fits tightly into transverse groove — *articulation allows precise up-and-down movement of jaws, giving characteristic 'snap'*

Carnassial teeth give a characteristic shearing action, used for slicing through flesh and for crushing bones :

SIDE VIEW

upper carnassial bites outside lower carnassial - edges slide past each other like scissor blades

SURFACE VIEW OF CROWN

shearing ridge

point

The other cheek teeth of carnivores are generally less well developed.

Teeth in mammals

Herbivorous (i.e. vegetation-eating) mammal (e.g. sheep)

Permanent dentition:

$$i. \frac{0}{3} \quad c. \frac{0}{1} \quad pm. \frac{3}{3} \quad m. \frac{3}{3} = 32$$

The roots are open.

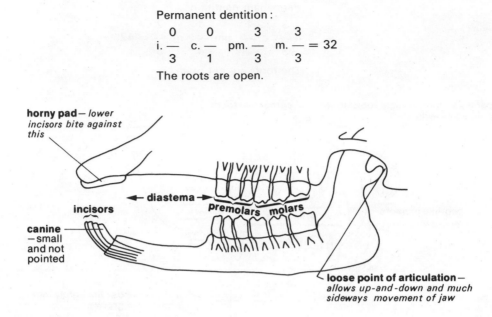

horny pad — *lower incisors bite against this*

diastema

incisors

premolars **molars**

canine — *small and not pointed*

loose point of articulation — *allows up-and-down and much sideways movement of jaw*

Molar crowns

VERTICAL SECTIONS

SURFACE VIEW

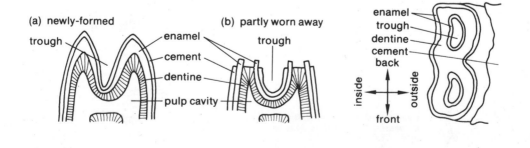

(a) newly-formed

trough

enamel

cement

dentine

pulp cavity

(b) partly worn away

trough

enamel
trough
dentine
cement
back

inside

outside

front

---

1 Examine as closely as possible the feeding of different types of mammals, e.g. cat, rabbit, sheep. Try to relate the movements of the mouth and jaw to the nature of the teeth which these animals possess.

2 Read further about the dentition of a range of mammals, e.g. whales, elephants, bats, etc.

Respiration is a property of all living protoplasm. It is the process whereby living organisms produce energy for their metabolic processes. This energy may be required for movement, synthesis of essential chemicals for use in metabolism (i.e. for secretion), and in birds and mammals for heat production. The energy is released from the breakdown of food materials, and waste products are released at the same time. Most plants and animals have developed special methods for the disposal of this waste, e.g. the respiratory and excretory systems of mammals.

DEFINITIONS OF RESPIRATION

At present there is confusion about some of the terms used in connexion with respiration. It is advisable to know these terms and also which of them have the same meaning.

Organisms may respire **aerobically** or **anaerobically**; many organisms can respire in either way according to circumstances.

Aerobic respiration—the usual method of most organisms: molecules of gaseous oxygen are used.

Anaerobic respiration—this method is used, for example, by many bacteria: chemical compounds containing oxygen atoms (such as sugars) are used.

Where organisms respire aerobically distinctions are also made between external and internal respiration:

1 External respiration—*the passing of oxygen molecules into, and carbon dioxide molecules out of an organism.* It is a purely *physical* or *mechanical process.*

It may be an obvious and active process as in the **breathing** movements of insects and mammals; or it may be more passive as in simple organisms and flowering plants.

In *Amoeba* for example, external respiration occurs by simple diffusion (see p. 7). In flowering plants oxygen diffuses through the stomata (in green leaves and stems) and through lenticels (in bark). In insects (see p. 45) movements of the skeleton draw air in and out of the openings called spiracles on the body surface, and in order to reach the tissues the air passes along tubes, tracheae, inside the body. In mammals movements of the ribs and diaphragm draw air into the lungs; the blood acts as the carrier of gases between lungs and tissues.

Where an organism lives in water the oxygen and carbon dioxide are in solution, but even with land-living organisms the gases must be dissolved before the body can use or dispose of them. Oxygen dissolves in the moisture on the skin of the frog, in the mucus which lines the walls of the air-sacs (alveoli) in the mammalian lung and in the water which forms a film on the outer surfaces of the cells inside a leaf.

2 Internal respiration—this is often referred to as *'tissue'* respiration. The phrase is misleading because it implies that the process occurs in tissues only, whereas it occurs in all living cells. It would be better called 'cellular' respiration. It is *the oxidation of simple chemical substances* (*usually carbohydrates*) *within the cells to liberate energy.* Because it is a *chemical* process, it can be represented by an equation:

$$C_6H_{12}O_6 + 6O_2 \longrightarrow 6CO_2 + 6H_2O + ENERGY$$

simple   oxygen   carbon   water
carbohydrate   dioxide   (vapour)
(glucose)

The chemical substances shown in this equation are the beginning and end products only. The oxidation of carbohydrate in fact takes place in a series of complex biochemical reactions; many of the stages are brought about by different enzymes. The energy is not all released at once but rather in small amounts at various stages; larger amounts of energy would have virtually an explosive effect in cells and could not be tolerated.

Various aspects of the above equation can be demonstrated experimentally. Although there is no satisfactory simple **experiment which shows that aerobic organisms need oxygen for**

Respiration

respiration, the following should be performed:

1

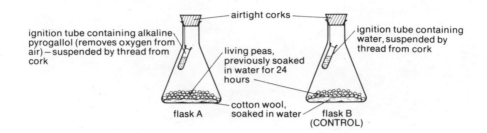

ignition tube containing alkaline pyrogallol (removes oxygen from air) — suspended by thread from cork

airtight corks

living peas, previously soaked in water for 24 hours

ignition tube containing water, suspended by thread from cork

cotton wool, soaked in water

flask A

flask B (CONTROL)

2 Leave for 1 week to 10 days.

Result → Peas in B germinate; peas in A do not.

Strictly this experiment shows that oxygen is necessary for germination, but as peas must respire in order to germinate, it suggests that oxygen is necessary for respiration.

Experiment to show that carbon dioxide is produced in respiration

ANIMAL EXPERIMENT

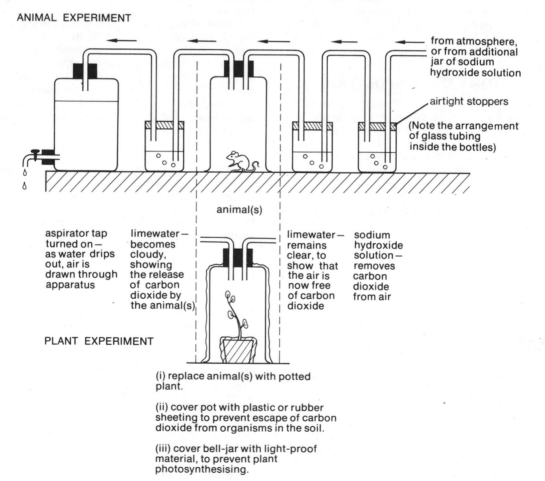

from atmosphere, or from additional jar of sodium hydroxide solution

airtight stoppers

(Note the arrangement of glass tubing inside the bottles)

animal(s)

aspirator tap turned on — as water drips out, air is drawn through apparatus

limewater — becomes cloudy, showing the release of carbon dioxide by the animal(s)

limewater — remains clear, to show that the air is now free of carbon dioxide

sodium hydroxide solution — removes carbon dioxide from air

PLANT EXPERIMENT

(i) replace animal(s) with potted plant.

(ii) cover pot with plastic or rubber sheeting to prevent escape of carbon dioxide from organisms in the soil.

(iii) cover bell-jar with light-proof material, to prevent plant photosynthesising.

The release of energy from respiration can be shown experimentally since some of the energy is released as heat:

Experiment to show that heat energy is produced by respiration

1 Assemble apparatus pictured below and immediately take readings from thermometers A and B.

2 Take further thermometer readings at regular intervals—ideally one reading per day for 4 or 5 days.

Result

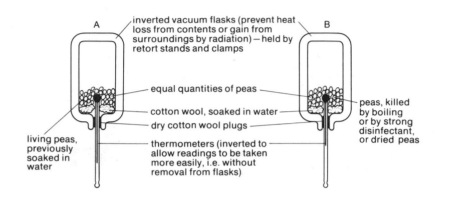

A · inverted vacuum flasks (prevent heat loss from contents or gain from surroundings by radiation) — held by retort stands and clamps · B

equal quantities of peas

cotton wool, soaked in water

dry cotton wool plugs

living peas, previously soaked in water

peas, killed by boiling or by strong disinfectant, or dried peas

thermometers (inverted to allow readings to be taken more easily, i.e. without removal from flasks)

The final reading should be several degrees higher in thermometer A than in B. Normally there should be *no* temperature change in B, but fluctuations in room temperature may cause some rise or fall. This experiment illustrates very clearly the importance of a **control** (apparatus B) in biological investigations.

The Production of Water Vapour

It is possible to infer that water vapour is produced in respiration by breathing out on to a flat glass surface such as a mirror. Droplets of a colourless liquid will condense on the surface. When tested with anhydrous copper sulphate (a white powder) or dry cobalt chloride paper (blue), blue crystals of copper sulphate are formed or the paper changes to pink, respectively. These tests prove that the droplets contain water but the experiment does not conclusively prove that the water has been formed as a waste product of respiration; a large proportion of it will be water evaporated from the linings of the respiratory system.

Anaerobic respiration

This form of respiration is relatively inefficient because it releases less than $\frac{1}{25}$ of the amount of energy released by the aerobic method. Plants such as yeast (see p. 18) produce ethyl alcohol and carbon dioxide as waste products; animals produce lactic acid.

Experiment to demonstrate anaerobic respiration

1

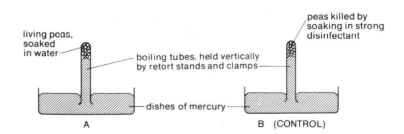

living peas, soaked in water

boiling tubes, held vertically by retort stands and clamps

peas killed by soaking in strong disinfectant

dishes of mercury

A   B (CONTROL)

Respiration

2 Inspect apparatus at daily intervals over following 4–5 days

Result

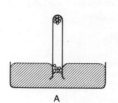

A
mercury level drops — tube
filled with colourless gas

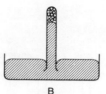

B
no change

Float sodium hydroxide pellets into tube A. Mercury level returns to original position, showing that gas (carbon dioxide) has been absorbed.

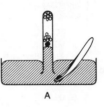

A

This experiment shows that pea seeds when surrounded by glass and mercury only (i.e. no air present) will respire anaerobically and produce carbon dioxide.

Respiration experiments with yeast

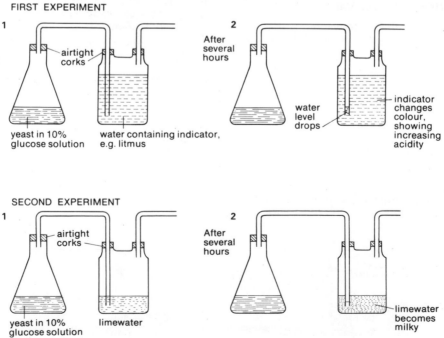

FIRST EXPERIMENT

1   airtight
     corks

yeast in 10%
glucose solution

water containing indicator,
e.g. litmus

2   After
     several
     hours

water
level
drops

indicator
changes
colour,
showing
increasing
acidity

SECOND EXPERIMENT

1   airtight
     corks

yeast in 10%
glucose solution

limewater

2   After
     several
     hours

limewater
becomes
milky

26 EXCHANGE OF GASES

IN A MAMMAL

Mammals respire aerobically; they will die within minutes if not supplied with oxygen gas. Certain parts of the body, e.g. the brain, are more sensitive to oxygen lack than are others. Muscles will temporarily continue contracting and relaxing without a supply of oxygen; they gradually build-up an 'oxygen debt'. A limit to the debt which can be tolerated occurs when too much waste material (lactic acid) has accumulated from anaerobic respiration. Much of the energy required during sprinting comes from anaerobic respiration; at the end of the race the athlete's breathing rate is very fast because extra oxygen gas is required to pay off the oxygen debt, in fact to break down the lactic acid to carbon dioxide and water vapour.

Mammals, being active complex creatures, need a large and rapid supply of oxygen. They require:

(a)   a sufficiently large surface area for absorption of oxygen

(b)   an efficient excretory method for large quantities of carbon dioxide.

Also, oxygen gas must be dissolved in liquid before it can pass into living membranes.

Many animals absorb oxygen (and excrete carbon dioxide) through the entire body surface. This is not possible in mammals because:

(i)   being large animals the ratio of surface area to body volume (see p. 205) is not high enough.

(ii)   as an adaptation to life on dry land mammals have dry impervious skins, so reducing water loss (water passes out from the sweat glands only).

Special organs, **lungs**, have been developed. These provide a very large moist surface area (over 100 m² in man) inside the body.

As the lungs are situated in one region of the body only—in the thorax—they are far removed from many of the tissues. The blood system (see p. 145) is used to transport oxygen to the various tissues and to bring carbon dioxide back to the lungs. This, in turn, means that dissolved respiratory gases have to pass between lung cavity and blood. This diffusion of gases is efficient only because the lung membranes are extremely thin.

The respiratory system

The internal position of the lungs some distance away from the body surface requires a respiratory system of tubes to connect lungs with atmosphere.

There are two lungs, one arranged on each side of the heart. Lungs are spongy elastic lobed organs occupying a considerable volume of the chest cavity. They are almost completely encircled by the rib cage (laterally) and the diaphragm (below). Two **pulmonary blood vessels** enter and leave each lung near to the position of entry of the breathing tube, the **bronchus**; the pulmonary **artery** brings deoxygenated blood from the right ventricle of the heart and the pulmonary **vein** carries oxygenated blood back to the left auricle. The artery divides into a vast network of capillaries inside the lung and these capillaries unite to form the vein.

Inside each lung the bronchus subdivides many times into smaller tubes called **bronchioles** which themselves subdivide, finally to end in **air-sacs** (called **alveoli**; singular, alveolus) that resemble clusters of grapes. Each human lung contains several million alveoli. Exchange of gases takes place only in the alveoli, through the very thin moist walls that separate the air in the alveoli from the blood.

V.S. Mammalian lung tissue, showing air-spaces

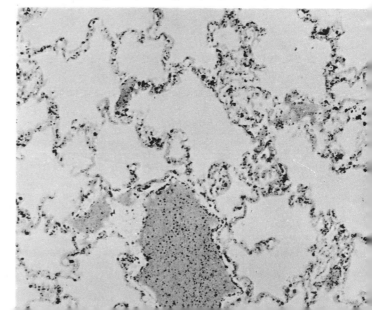

Exchange of gases

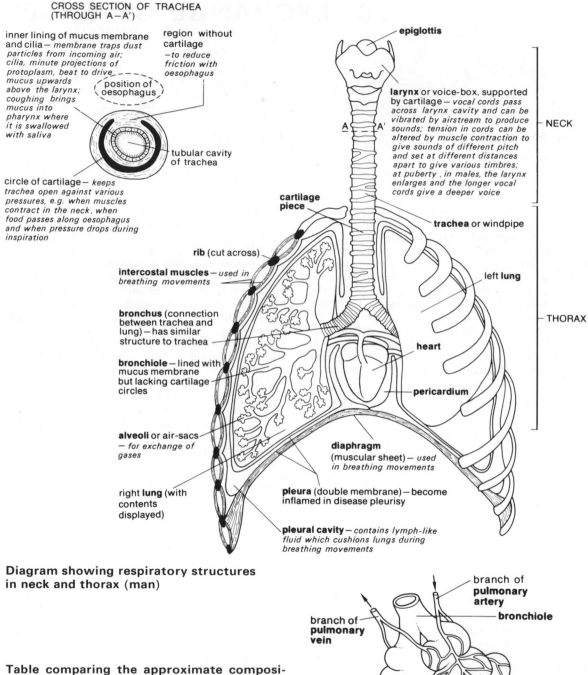

CROSS SECTION OF TRACHEA
(THROUGH A–A')

inner lining of mucus membrane and cilia — *membrane traps dust particles from incoming air; cilia, minute projections of protoplasm, beat to drive mucus upwards above the larynx; coughing brings mucus into pharynx where it is swallowed with saliva*

region without cartilage — *to reduce friction with oesophagus*

position of oesophagus

tubular cavity of trachea

circle of cartilage — *keeps trachea open against various pressures, e.g. when muscles contract in the neck, when food passes along oesophagus and when pressure drops during inspiration*

epiglottis

larynx or voice-box, supported by cartilage — *vocal cords pass across larynx cavity and can be vibrated by airstream to produce sounds; tension in cords can be altered by muscle contraction to give sounds of different pitch and set at different distances apart to give various timbres; at puberty, in males, the larynx enlarges and the longer vocal cords give a deeper voice*

NECK

A — A'

cartilage piece

trachea or windpipe

left **lung**

THORAX

rib (cut across)

intercostal muscles — *used in breathing movements*

bronchus (connection between trachea and lung) — has similar structure to trachea

bronchiole — lined with mucus membrane but lacking cartilage circles

heart

alveoli or air-sacs — *for exchange of gases*

pericardium

diaphragm (muscular sheet) — *used in breathing movements*

right lung (with contents displayed)

pleura (double membrane) — become inflamed in disease pleurisy

pleural cavity — *contains lymph-like fluid which cushions lungs during breathing movements*

Diagram showing respiratory structures in neck and thorax (man)

Table comparing the approximate composition of air entering and leaving the lungs of man

	Nitrogen	Oxygen	Carbon Dioxide	Water Vapour
Inspired air	78%	20%	0.03%	variable- near 1%
Expired air	75%	15%	4.0%	near 6%

branch of **pulmonary artery**

branch of **pulmonary vein**

bronchiole

capillary network

alveoli

Diagram showing cluster of alveoli with their blood supply

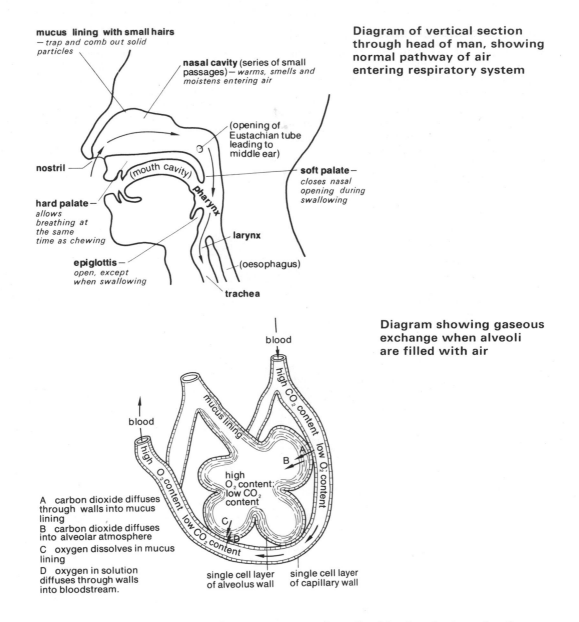

mucus lining with small hairs
— *trap and comb out solid particles*

nasal cavity (series of small passages) — *warms, smells and moistens entering air*

Diagram of vertical section through head of man, showing normal pathway of air entering respiratory system

(opening of Eustachian tube leading to middle ear)

(mouth cavity)

pharynx

soft palate — *closes nasal opening during swallowing*

nostril

hard palate — *allows breathing at the same time as chewing*

larynx

(oesophagus)

epiglottis — *open, except when swallowing*

trachea

Diagram showing gaseous exchange when alveoli are filled with air

blood

high CO_2 content

low O_2 content

mucus lining

blood

high O_2 content

high O_2 content; low CO_2 content

A

B

low O_2 content

A   carbon dioxide diffuses through walls into mucus lining

B   carbon dioxide diffuses into alveolar atmosphere

C   oxygen dissolves in mucus lining

D   oxygen in solution diffuses through walls into bloodstream.

low CO_2 content

C

D

low CO_2 content

single cell layer of alveolus wall

single cell layer of capillary wall

Experiment to demonstrate that man breathes out more carbon dioxide than he breathes in

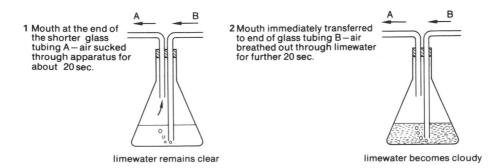

A   B

1 Mouth at the end of the shorter glass tubing A — air sucked through apparatus for about 20 sec.

A   B

2 Mouth immediately transferred to end of glass tubing B — air breathed out through limewater for further 20 sec.

limewater remains clear

limewater becomes cloudy

Breathing movements

Oxygenation of the blood in the lungs and the expulsion of carbon dioxide can be achieved only by active ventilation of the lungs. Breathing movements are due to:

(i) contraction and relaxation of the muscles of the diaphragm (Gk. *dia*, across; *phragma*, partition) which are attached to the sternum, posterior ribs and vertebral column.

(ii) movements of the rib cage caused by the obliquely-connecting intercostal (L. *inter*, between; *costa*, rib) muscles pulling on the ribs.

In most mammals the diaphragm movements are more important than movements of the rib cage because the rib cage is required to support the thorax more when all four legs are on the ground. Also, it is said that men and infants use their diaphragms more than their rib cages; women use their diaphragms less.

When the diaphragm contracts, it flattens and moves downwards; this produces the increase in the volume of the thoracic cavity necessary to draw air into the lungs.

Vertical section diagram through diaphragm showing movements

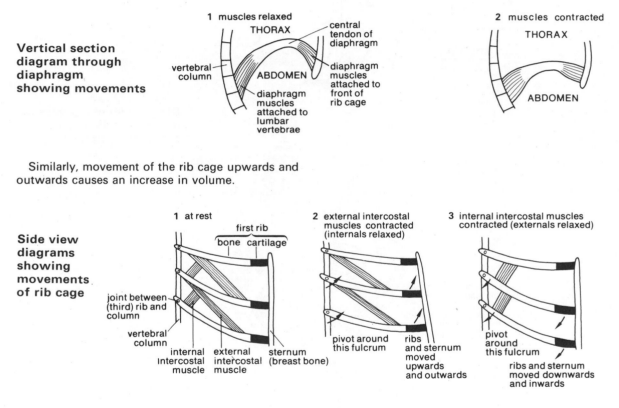

Similarly, movement of the rib cage upwards and outwards causes an increase in volume.

Side view diagrams showing movements of rib cage

This movement of the rib cage can be compared with the raising of a bucket handle from the resting position: the outward curvature of the handle compares with the outward curvature of each pair of ribs.

Overall, when air is drawn into the body, the thorax increases in length and diameter; the first is achieved largely by the diaphragm, the second by the rib cage.

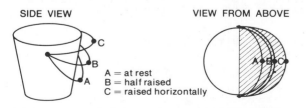

SIDE VIEW   VIEW FROM ABOVE

A = at rest
B = half raised
C = raised horizontally

A   Inspiration (Breathing in)

When air is drawn into the body, the thorax increases in diameter and length; the first is achieved largely by the rib cage, the second by the diaphragm.

1   Two simultaneous actions:

   (a)   external intercostal muscles contract—ribs pivot against thoracic vertebrae—rib cage and sternum move upwards and outwards

   (b)   diaphragm muscles contract—diaphragm (dome-shaped when at rest) flattens downwards, about 1·5 cm in man, pressing on organs of upper abdomen.

2   Pleural membranes (attached to ribs and diaphragm) follow movements of ribs and diaphragm, allowing elastic lungs to expand.

3   Volume occupied by lungs increases—air pressure inside lungs falls.

4   Atmospheric pressure forces air along respiratory tubes into lungs until pressure inside lungs equals that in atmosphere.

B   Expiration (Breathing out)

1   Two simultaneous actions:

   (a)   external intercostal muscles relax, internal intercostals contract—rib cage and sternum pivot down and in to original position.

   (b)   diaphragm muscles relax—diaphragm domes up to original position.

2   Ribs and diaphragm press on pleural fluid—pleural fluid compresses lungs.

3   Air pressure rises inside lungs.

4   Air forced out of lungs until lungs (assisted by their own elastic recoil) have deflated to their original volume.

Expired air contains less oxygen and more carbon dioxide and water vapour than does inspired air (see table on p. 126 and experiment on p. 127). The lungs extract only about $\frac{1}{5}$ of the volume of oxygen available in the inspired air.

Lung Capacity

When fully inflated, the total capacity of each lung in an adult man is approximately 2.5 litres. The air which passes in and out at each breath is called **tidal air**. During normal breathing this is seldom more than one tenth of the total lung capacity, and some of this tidal air does not even reach the alveoli; it remains in the air passageways of the lung.

The **vital capacity** of the lungs is the maximum volume of air which can pass in and out during forced breathing. In man this is normally just over $1\frac{1}{2}$ litres per lung, though athletes may achieve larger volumes as a result of training.

These figures show that the lungs can never be completely emptied of air, but always retain some **residual air** which stagnates inside the lungs between breaths (it is not, of course, always the *same* air which stagnates).

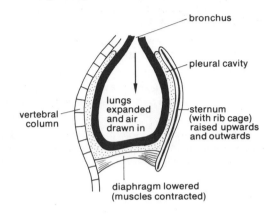

Diagram of section through thorax after inspiration

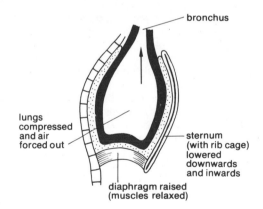

Diagram of section through thorax after expiration

Control of Breathing

Although breathing can be consciously altered (e.g. during speech), both the rate and depth of breathing are normally controlled unconsciously by the medulla oblongata of the hind-brain (see p. 170). The medulla is extremely sensitive to the carbon dioxide concentration of the blood that it receives (it is not sensitive to the oxygen concentration). A rise in blood carbon dioxide concentration, e.g. after strenuous exercise, stimulates the medulla; messages are sent from the brain along nerves to the intercostal muscles and diaphragm. Breathing movements become faster, so that carbon dioxide is removed from the blood more quickly and the blood receives a better oxygen supply. (Also the heart beats faster to speed the passage of these gases between respiring tissues and lungs.) This is another example of **homeostasis** (see p. 90).

There are in addition special movements associated with the respiratory system:

1   **yawns**—prolonged inspiration; may be caused by excessive accumulation of carbon dioxide in the blood.

2   **coughs**—short, forced expirations; to clear mucus and foreign particles from the upper end of trachea, etc.

3   **hiccoughs**—short, violent inspirations; caused by spasmodic involuntary contractions of diaphragm.

IN A FLOWERING PLANT

Exchange of gases is a much more passive process here than in a mammal. It takes place by simple diffusion, and breathing movements are not apparent.

Cells inside the plant continually produce carbon dioxide from their respiration and they continually use up oxygen. This means that there is more carbon dioxide inside the plant than outside, and more oxygen outside the plant than inside. In this way a diffusion gradient of these two gases is maintained. The process is not entirely straightforward because in sunlight plant respiration *appears* to cease; oxygen is given off from, and carbon dioxide is taken in for photosynthesis. Actually cell respiration continues at the same time as photosynthesis.

The roots of plants obtain their oxygen by diffusion from the soil air (see p. 254) and carbon dioxide is liberated in return. The gases are dissolved in the soil water. The older parts of the root may obtain oxygen from the above-ground part of the plant by way of the air passageways which extend down between the cortex cells of the stem.

The above-ground parts of plants have minute surface pores which connect internal parts with external atmosphere:

(a)   **lenticels**—found in woody stems, i.e. branches of shrubs and trees; they are small gaps in the corky bark, appearing as craters at the surface.

(b)   **stomata**—found in soft green (i.e. herbaceous) stems and in leaves; they are microscopic pores surrounded by sausage-shaped guard-cells which open and close the pores by altering their shapes (see also p. 60).

Diagram of cross-section through lenticel

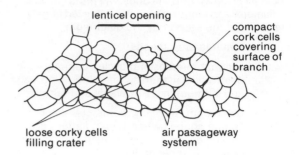

lenticel opening

compact cork cells covering surface of branch

loose corky cells filling crater

air passageway system

The pathway taken by oxygen diffusing into a leaf is illustrated below (carbon dioxide from respiration travels similarly but in the reverse direction):

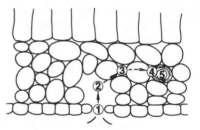

1   entry through open pore of stoma.

2 and 3   movement along intercellular air-space passageways.

4   dissolving on moist outer film of cell.

5   entry into cell (in solution) through cellulose wall.

1   Examine sections of lung tissue (to see alveoli, bronchioles and blood supply) and of trachea (to see cartilage) microscopically.

2   Compare and contrast the breathing methods of a range of animals—protozoa, worms, insects, fishes, amphibians and birds.

3   Devise a method for determining the vital capacity of your own lungs.

4   Suggest why it is important that residual air should remain in the lungs.

5   Dip a recently-picked leaf into very hot water and watch for the emergence of air bubbles which demonstrate the presence of stomata.

27 ABSORPTION

Absorption of water

OSMOSIS

Plants absorb water by their roots from the soil by osmosis. *Osmosis is the passage of water molecules from a weaker solution to a stronger solution through a semipermeable membrane.* (A **semipermeable membrane** is one which allows certain substances—usually small molecules such as water—to pass through it, whereas other substances—usually large molecules such as starch—cannot.)

So far no single theory has been put forward which completely explains the causes of osmosis.

Experiments on osmosis and absorption should either be performed by the student individually or else be seen as demonstrations.

Experiment A: Simple demonstration of osmosis as a physical phenomenon

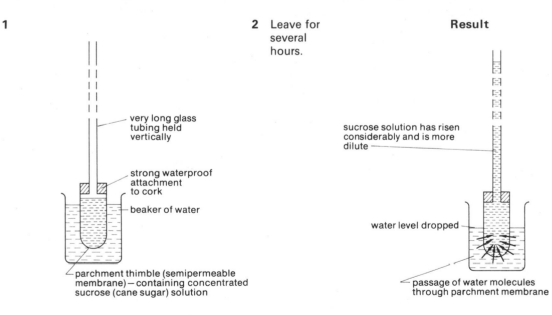

1

very long glass
tubing held
vertically

strong waterproof
attachment
to cork

beaker of water

parchment thimble (semipermeable
membrane) — containing concentrated
sucrose (cane sugar) solution

2 Leave for
several
hours.

Result

sucrose solution has risen
considerably and is more
dilute

water level dropped

passage of water molecules
through parchment membrane

The sucrose solution has risen in the tube against the force of gravity—the pressure which osmosis has generated in the experiment to make this possible is the **osmotic pressure**. (A control experiment set up with the fluids reversed, i.e. with water in the parchment thimble and strong sucrose solution in the beaker, will produce the opposite result—fluid leaves the thimble and the level rises in the beaker.)

Absorption

Further experiments demonstrating osmosis

Experiment B:

1 Two fresh eggs with shells removed by standing in acetic acid or very dilute hydrochloric acid for several hours.

2

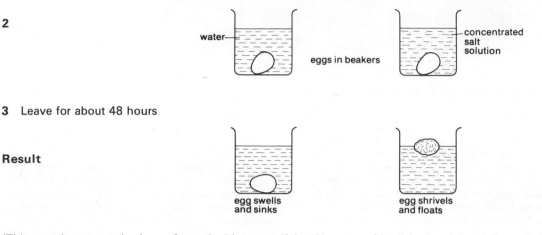

water — / eggs in beakers / concentrated salt solution

3 Leave for about 48 hours

Result

egg swells and sinks   egg shrivels and floats

(This experiment can also be performed with grapes if the skins are unblemished and the stalks are left intact.)

Experiment C: To demonstrate osmosis in living tissue

(a) Plant tissue (The Potato Osmometer)

1 Cut a large potato into halves.

2 Remove about 1 cm of peel from circumference of each half adjacent to the cut; bring one half to the boil.

3

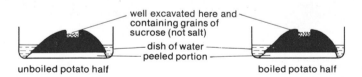

well excavated here and containing grains of sucrose (not salt) / dish of water / peeled portion

unboiled potato half   boiled potato half

4 Cover dishes and leave for 24 hours

Result

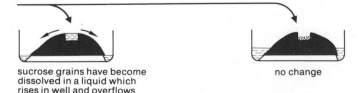

sucrose grains have become dissolved in a liquid which rises in well and overflows   no change

Both potato halves must be partly peeled because the corky potato covering is impermeable to water. The living cells of the unboiled potato provide a series of semipermeable membranes separating the sucrose grains (which soon form a strong solution by dissolving in the sap of the damaged potato cells lining the well) from water in the dish. Water passes from the dish upwards against gravity into the well by osmosis. In the boiled potato the cells are killed and the semipermeable mernbranes are destroyed. No osmosis takes place.

(b) Animal tissue

1
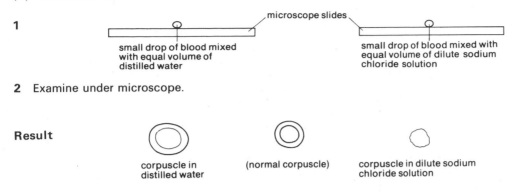

microscope slides

small drop of blood mixed with equal volume of distilled water

small drop of blood mixed with equal volume of dilute sodium chloride solution

2 Examine under microscope.

Result

corpuscle in distilled water

(normal corpuscle)

corpuscle in dilute sodium chloride solution

Root-hairs and their functions

The part of the root largely responsible for absorbing water from the soil is the zone of root-hairs a few millimetres behind the root tip. The zone usually extends along the root for about two centimetres. The root surface is densely clothed by these minute projections, perhaps several hundred per square centimetre. Damp soil tends to adhere to this zone when a plant is uprooted. The surface (epidermis) cells of the root which produce these hairs—one hair from each cell—are called the **piliferous layer** (L. *pilus*, hair; *ferre*, to carry).

In many plants, the root hairs live for a few days only, after which they collapse and die, and are rubbed off. The epidermis cells that bore them may become thickened with cork and impermeable to water; alternatively the cells may be cast off by friction with the soil. The root continues to grow through the soil by producing new cells just behind the root tip. These cells then rapidly expand in length. As fast as the old root-hairs die they are replaced by hairs developed between the present root-hair zone and the zone of elongation.

The appearance of the root-hair zone varies under different conditions. When the soil is dry the zone is densely clothed and the individual hairs are very long.

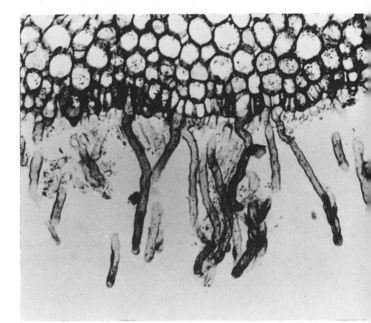

T.S. through root of bean, showing epidermis with root-hairs (piliferous layer)

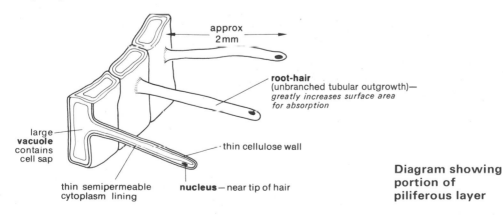

approx 2 mm

root-hair (unbranched tubular outgrowth)— *greatly increases surface area for absorption*

large **vacuole** contains cell sap

thin cellulose wall

thin semipermeable cytoplasm lining

nucleus—near tip of hair

Diagram showing portion of piliferous layer

**Diagram of
vertical section
through root
at apex and
root-hair region**

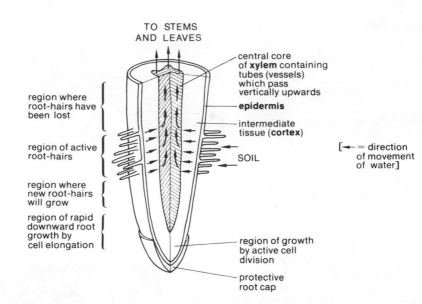

TO STEMS
AND LEAVES

region where
root-hairs have
been lost

region of active
root-hairs

region where
new root-hairs
will grow

region of rapid
downward root
growth by
cell elongation

central core
of **xylem** containing
tubes (vessels)
which pass
vertically upwards

epidermis

intermediate
tissue (**cortex**)

SOIL

[← = direction
of movement
of water]

region of growth
by active cell
division

protective
root cap

The water absorbed by the root-hairs travels across intermediate tissues (**cortex**) in the root until it reaches the xylem in the **vascular strand** at the root centre. Here, very long tubes in the **xylem** carry the water upwards to the stem and leaves.

The cellulose walls of all cortex cells are freely permeable to water and dissolved substances, but the cytoplasm lining provides a semipermeable membrane. Water is able to travel into and across the root by osmosis.

Although soil water may contain dissolved salts, this is only a relatively weak solution; the sap in the root-hair cells, which may contain certain accumulated salts and food such as sugar, is always more concentrated.

As a result of transpiration (see p. 139) there is a continual upward current of fluid in the xylem tubes. This current is used as a starting-point to explain water movement across the root, reading from left to right on the following diagram:

1 water drawn out from sap into transpiration stream
2 sap in cell A becomes more concentrated
3 water drawn through by osmosis
4 sap in cell B at first weaker than that in cell A becomes more concentrated as it loses water
5 water drawn through by osmosis
6 sap in cell C at first weaker than that in cell B becomes more concentrated as it loses water
7 water drawn through by osmosis

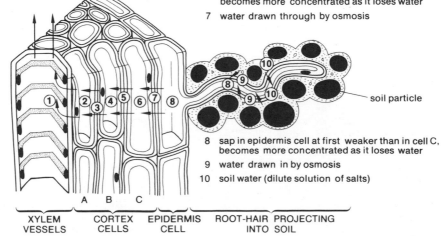

water passing upwards in
transpiration stream

soil particle

8 sap in epidermis cell at first weaker than in cell C, becomes more concentrated as it loses water
9 water drawn in by osmosis
10 soil water (dilute solution of salts)

A B C

XYLEM
VESSELS

CORTEX
CELLS

EPIDERMIS
CELL

ROOT-HAIR PROJECTING
INTO SOIL

Experiment to demonstrate absorption of water by roots

1 Remove plant (e.g. groundsel) carefully from soil, causing minimum damage to roots.

2

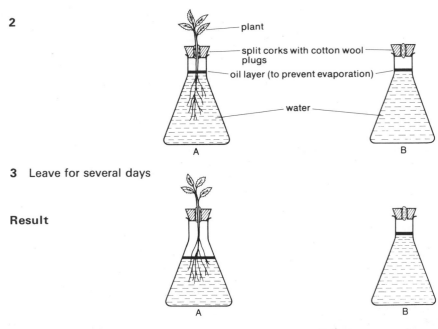

3 Leave for several days

Result

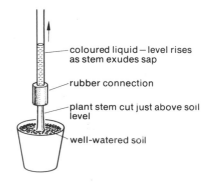

Root pressure is believed to assist the upward movement of water in the xylem. Sap is exuded under quite considerable pressure when a stem is cut across just above soil level, particularly in spring. The pressure is often sufficient to carry the sap several feet upwards but it is not sufficient by itself to move sap to the top of a tall tree. The cause of root pressure is not easily explained.

Experiment to demonstrate root pressure.

Absorption of mineral salts

Those regions of the root responsible for absorbing water from the soil by osmosis are also those which absorb salts dissolved in the soil water. Salts are not absorbed as molecules but as **ions**, i.e. potassium nitrate does not enter as such, but as K^+ and NO_3^-, and they may enter at different rates.

Salts may sometimes pass in by **diffusion**; this implies movement of ions from a region of higher to one of lower concentration. Although the soil water is only a weak solution of salts, diffusion is possible because immediately after being absorbed the ions may be:

(a) carried away to other parts of the plant;

(b) used-up as a result of being converted into other chemicals inside the plant.

In these ways a relatively even weaker solution of the absorbed ions would be maintained inside the plant than outside.

Frequently however the concentration of particular salts is greater in the plant than in the soil water. In this case the plant uses **energy from respiration** to draw these ions into its cells against the diffusion gradient. As a result, if the oxygen supply to the roots is insufficient, absorption of the salts ceases.

Absorption

Experiment to demonstrate absorption of dissolved substances by roots

1   Remove suitable plant (e.g. balsam) carefully from soil, causing minimum damage to roots.

2

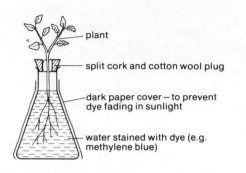

plant

split cork and cotton wool plug

dark paper cover — to prevent dye fading in sunlight

water stained with dye (e.g. methylene blue)

3   Leave for several days.

4   Cut cross-sections and longitudinal sections of stem.

Result

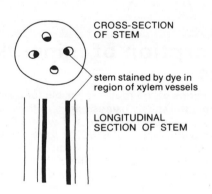

CROSS-SECTION OF STEM

stem stained by dye in region of xylem vessels

LONGITUDINAL SECTION OF STEM

28 TURGIDITY, PLASMOLYSIS AND WILTING

When a piece of plant tissue is placed in a very strong sucrose solution, water leaves the sap vacuole by osmosis through the semi-permeable cytoplasm lining. The volume of the vacuole decreases and the cytoplasm lining appears to shrink away from the cellulose wall. This phenomenon is called **plasmolysis** (Gk. *plasma*, shape; *lysis*, loosing, breaking) and has a characteristic appearance. It can, perhaps, best be seen if a piece of epidermis is carefully stripped from a red rose petal; the petal's colour is caused by pigment dissolved in the vacuole sap of the epidermal cells. The presence of dissolved pigment gives a much clearer indication of the size of the vacuole.

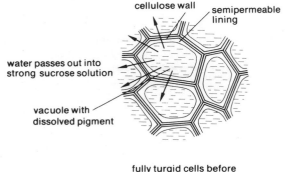

cellulose wall
semipermeable lining

water passes out into strong sucrose solution

vacuole with dissolved pigment

fully turgid cells before plasmolysis

cytoplasm lining pulled away from cellulose wall

spaced filled with strong sucrose solution

plasmolysed cells

A similar process can be observed when a drop of blood is placed in a drop of strong sugar or salt solution on a slide under a microscope (see p. 133); the red corpuscles lose water by osmosis, shrink and become wrinkled.

The cytoplasm and vacuole can often be restored to their original condition by immersing the strip of plasmolysed epidermis in water. The vacuole contents, having lost water previously, will now be much more concentrated and water will enter the vacuoles from the surroundings.

As the water enters, the vacuole volume increases and sap presses outwards against the cytoplasm lining and cell wall:

PHASE 1

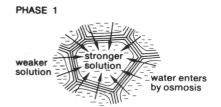

weaker solution
stronger solution
water enters by osmosis

PHASE 2

vacuole enlarges— water presses outwards against lining

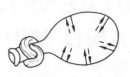

—→ outward pressure of air against
rubber causing inflated condition
of balloon

—→ inward pressure of rubber wall,
opposing air pressure

—→ outward pressure of water against
cellulose wall, causing inflated
condition of cell

—→ inward pressure of cellulose wall,
opposing water pressure

The cell is behaving rather like a rubber balloon. The rubber wall of the balloon is elastic and so, to some extent, is the cellulose cell wall. Air is blown into the balloon, water enters the cell by osmosis.

In both balloon and cell the outward pressure is balanced by an inward pressure caused by the elastic nature of the wall. It is possible to force so much air into a balloon that the wall pressure is

overcome and the elastic wall bursts. In a plant cell, however, the pressure of water acting outwards against the cell wall (called the **turgor pressure**) is never sufficient to burst the cellulose; instead a point is reached when no more water can enter the cell, or rather, the cellulose wall presses water out of the cell as fast as it enters. When this happens, the cell is fully inflated and said to be **turgid**.

FULLY TURGID CELL

—→ outward pressure (**turgor pressure**)
of water

—→ inward pressure (**wall pressure**)
strong enough to oppose any
further entry of water into cell

The importance of turgidity

Flowering plants are supported in two main ways:

1 By the xylem vessels, which are strengthened by deposits of hard woody material (**lignin**) on their walls. The deposits are often ring-shaped or spiral (see diagram on p. 55). In this way the vascular tissue provides a skeletal framework for the plant. This can be clearly seen in a leaf, where the vein arrangement helps to hold the leaf flat.

2 By turgidity, i.e. by pressure of water inside otherwise unstrengthened cells.

In roots and stems, the turgidity of the cortex cells (see pp. 55 and 57) assists in keeping the plant firm and erect. Similarly the turgidity of mesophyll cells plays an important part in keeping a leaf flat. The cells are arranged side-by-side and when fully turgid press against each other, keeping the tissue as a whole tense. (See adjacent diagram.)

At times when more water is being lost from a plant than is being absorbed by the roots, the vacuole volumes of its cells decrease and the cellulose walls become limp (**flaccid**). If the process continues, soon the entire plant sags. This is called **wilting**. It is most likely to occur on

sunny days when the rate of transpiration (see next section) is high and the soil is dry or frozen.

Seedlings are particularly prone to wilt especially when they are being transplanted ('**pricked-out**'). The root system needs time to grow and to make contact with the soil water in its new situation yet the seedlings continue to lose water by transpiration. Since the xylem in young plants is not fully strengthened, they rapidly wilt.

Watering plants with salt water produces a similar effect; water is forced out of the tissues by osmosis into the new, stronger external solution in the soil.

— — —→ = turgor pressure against cell walls

29 TRANSPIRATION

Transpiration is the loss of water vapour by land plants.

Water evaporates from all the above-ground parts of a plant (leaves, green and woody stems, flowers) but the greatest loss takes place through the stomata of leaves. These are usually more plentiful on the underside, in the lower epidermis.

Elsewhere the leaf epidermis is covered by a waxy cuticle; this prevents water loss except when the air outside the leaf is extremely dry. The following experiments associated with transpiration are to be carried out by the student. Simpler versions of the potometer can be assembled if the apparatus of the type described on p. 141 is unavailable.

Experiment to demonstrate transpiration

1

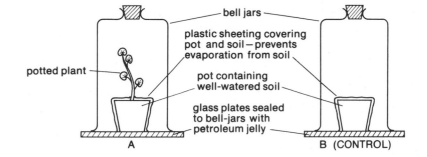

bell jars
plastic sheeting covering pot and soil – prevents evaporation from soil
potted plant
pot containing well-watered soil
glass plates sealed to bell-jars with petroleum jelly

A   B (CONTROL)

2 After about 30 minutes:

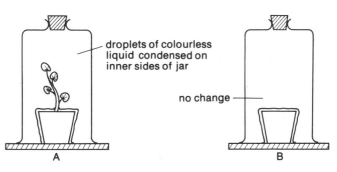

droplets of colourless liquid condensed on inner sides of jar

no change

A   B

3 Test liquid with *either*

anhydrous copper sulphate: result → white powder changes to blue crystals

or

anhydrous cobalt chloride: result → blue powder changes to pink crystals

These tests show that the liquid is water.

Transpiration

Experiment to show that more water is lost from the under surface than from the upper surface of most leaves

1   As many leaves as possible (preferably still attached to plants), each treated as follows:

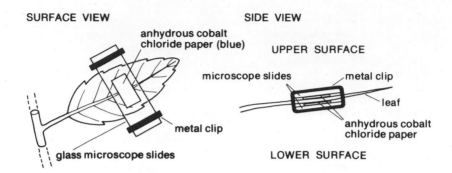

2   Leave for several minutes

Result

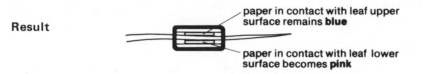

paper in contact with leaf upper surface remains **blue**

paper in contact with leaf lower surface becomes **pink**

The effect of blocking the stomatal pores with waterproof material can be shown as follows:

1   A series of similar leaves have their surfaces variously smeared (see diagram) with petroleum jelly.

2   The leaves are then suspended by threads attached to their stalks from a line fixed at each end to a retort stand.

3   The suspended leaves are left for a few days and observed at regular intervals.

Result

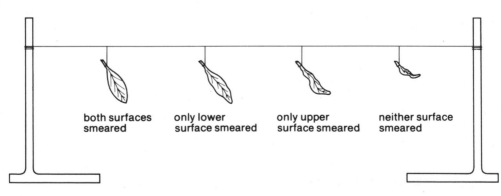

both surfaces smeared

only lower surface smeared

only upper surface smeared

neither surface smeared

The mechanism of transpiration

The cells in a healthy leaf are fully turgid (see p. 137). This means that the leaf behaves as a reservoir of water and the spaces which surround the cells in the leaf are saturated with water vapour. Even when the external atmosphere is extremely humid the air outside the leaf is always less saturated with water vapour than that inside. It follows then that whenever the stomata are open, water will escape from the leaf. This phenomenon is transpiration; it sets up a continuous sequence of events in the leaf that result in movement of water (the **transpiration stream**) throughout the plant :

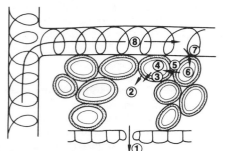

1   diffusion of water vapour through stomatal pore from intercellular air space into less saturated air around leaf

2   evaporation of water from thin water film on mesophyll cell surface into intercellular air space

3   water seeps through cellulose wall from vacuole into outer film

4   vacuole contents, having lost water, become more concentrated

5   water drawn into vacuole by osmosis from adjacent cells

6   vacuole contents of adjacent cells, having lost water, become more concentrated

7   water drawn into adjacent cells by osmosis from xylem vessel in leaf vein

8   water drawn along xylem vessel in vein and from stem vein into leaf

The rate of transpiration

This depends on external factors, i.e. climatic conditions outside the leaf; these factors are concerned basically with the degree of saturation of the air surrounding the leaf :

(a)   **Temperature**—transpiration increases at higher temperatures. Any given volume of air requires more water vapour to saturate it at higher than at lower temperatures.

(b)   **Humidity**—transpiration decreases at higher humidities. If the air surrounding the leaf is already considerably saturated with water vapour (for example, as a result of evaporation from soil and objects after heavy rain) the tendency for water vapour to escape from the leaf will be less.

(c)   **Air movement**—transpiration increases with wind speed.

The transpiration rate is also affected by the stomata, which generally open more fully in the day and close at night : the rate increases as the stomata open. It is possible that the amount of water in the leaf affects the rate; transpiration may decrease as the leaf loses water because the osmotic concentration of the vacuoles in the leaf cells rises and the cells hold on to their water with greater force.

Measurement of the transpiration rate

No simple experiment can be devised to *measure* the rate of transpiration. Those experiments which often are supposed to do this in many cases measure only the rate of water absorption; although this may be *proportional* to the rate of water loss from the shoot, the rate of absorption is not necessarily the *same* as the rate of transpiration.

However the **potometer** can be used to *compare* rates of transpiration.

Potometer

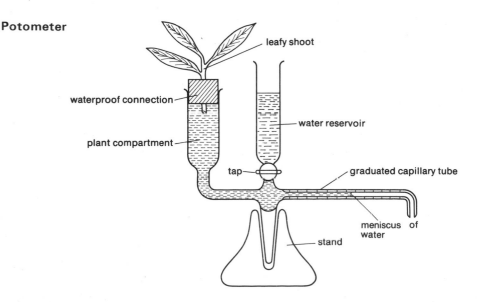

Setting up a Potometer

1 Cut from shrub (e.g. cherry laurel) a leafy shoot larger than that required for potometer and place cut end in sink of water.

2 With tap of potometer closed, fill reservoir with water.

3 With open end of capillary tube under water, open tap and completely flood apparatus with water.

4 Cut a portion of suitable size off shoot under water; this prevents the entry of air bubbles into the water column in the xylem vessels.

5 Place cut end of shoot through watertight rubber (or cork) connection, tightly fitting into plant compartment.

6 Turn tap off and remove capillary tube end from water.

As the leafy shoot removes water from the apparatus a meniscus will move along the capillary tube. The distance moved in unit time gives an indication of the rate of transpiration.

At any time the potometer can be refilled with water by opening the tap below the reservoir. Rates of transpiration can be compared by noting the rate of meniscus movement when the potometer is moved to different situations, e.g. near a warm radiator, close to the breeze generated by an electric fan. With more than one potometer the rates of transpiration of different leafy shoots can be compared under the *same* conditions, e.g. a laurel (evergreen) shoot can be compared with a rose (deciduous) shoot (with what results?).

The importance of transpiration

1 Water movement

The transpiration stream transports water absorbed by the roots upwards in the plant. A very small proportion of this water is taken to the green parts and used in photosynthesis (see p. 95). It also passes into the cell vacuoles and makes cells turgid; this water pressure inside the cells acts as a skeleton and keeps them in shape (see p. 137).

2 Mineral salt transport

The water sucked upwards in the transpiration stream along the xylem vessels carries with it dissolved mineral salts. These salts which have been absorbed from the soil are needed to make proteins, etc., by combination with foods made during photosynthesis (see p. 99). Much of this synthesis of complex chemicals takes place in the leaves. Unlike a mammal, a plant has no central pump such as a heart; in order to reach the leaves, mineral salts are sucked upwards in the transpiration stream. The force to do this, against gravity, is generated by those factors (sunlight, etc.) which cause transpiration.

3 Cooling

Heat energy (latent heat of evaporation) is required to evaporate water from the plant's surface. Transpiration therefore cools the leaf. This protection, which resembles sweating in mammals (see p. 90) is particularly important around mid-day.

Some plants, e.g. cacti, are adapted to retain water by reducing transpiration. Their tissues are able to tolerate quite high temperatures (above 60°C) in direct sunlight.

4 The problem of water loss

Transpiration is a considerable liability to many plants. At midsummer the soil may be dry and the sunlight intense. In these conditions plants lose large quantities of water by transpiration and are unable to replace this water from the soil. The cells wilt; the plant collapses and eventually dies.

Plants require a large porous surface (provided by the stomata) so that they can obtain enough carbon dioxide gas for photosynthesis. Unfortunately water vapour is able to leave through the same pores used for entry of carbon dioxide: the plant has little control over this loss.

Xerophytes

These are plants which can survive conditions of water shortage, and possess adaptations that tend to reduce transpiration:

(a) In pine, the stomata are sunk below the leaf surface: the humidity of the air trapped outside the stomata is increased.

(b) In sea buckthorn, the leaf bears branched hairs which reduce movement of air across the leaf surface.

(c) In many cacti, the leaves are modified to form spines: this reduces the surface area from which transpiration can take place.

Transpiration and leaf-fall

Deciduous leaves are those that are shed regularly. In temperate regions this occurs at the end of the growing season, in autumn. **Evergreen** leaves are shed at irregular intervals at any time in the year.

Generally evergreen leaves possess fewer stomata per unit area than are present in deciduous leaves; also the waxy cuticle of the epidermis is thicker. As a result, an evergreen leaf will lose less water in a given time than will a deciduous leaf of the same area; this can be demonstrated by a potometer (see p. 141).

Because of the problems of water loss caused by transpiration, leaves of the deciduous type must be shed before winter. On sunny but frosty days in mid-winter, the soil water is often frozen and cannot be absorbed. However, the temperature in and around the leaves may be quite high, so that the transpiration rate is also high. Leaves would soon wilt and the plant die.

The withering of leaves on dead branches is not an active process and therefore differs from leaf-fall. Details of the process of leaf-fall vary in different plants; the following is fairly typical:

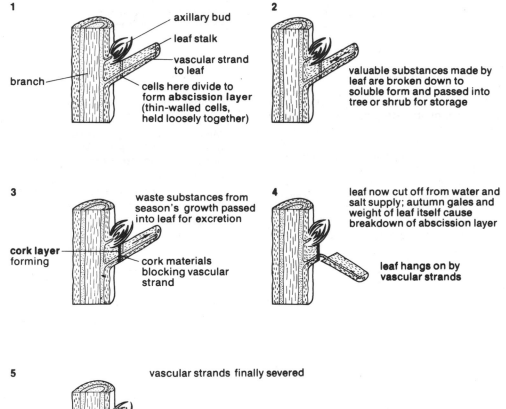

1

axillary bud

leaf stalk

vascular strand to leaf

branch

cells here divide to form **abscission layer** (thin-walled cells, held loosely together)

2

valuable substances made by leaf are broken down to soluble form and passed into tree or shrub for storage

3

waste substances from season's growth passed into leaf for excretion

cork layer forming

cork materials blocking vascular strand

4

leaf now cut off from water and salt supply; autumn gales and weight of leaf itself cause breakdown of abscission layer

leaf hangs on by vascular strands

5

vascular strands finally severed

cork scar — seals wound from water loss, entry of disease-producing organisms etc.

Transpiration

Diagram summarising water movement in a plant

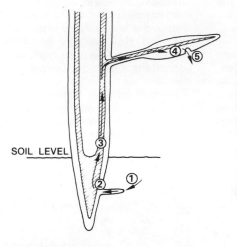

SOIL LEVEL

1 by osmosis from soil water film into root-hair cell vacuole

2 osmotic gradient across root to vascular strand

3 movement upwards along xylem tubes in transpiration stream

4 osmotic gradient across mesophyll cells

5 evaporation from mesophyll cells into intercellular spaces and transpiration through stomata

---

1 Carefully uproot a young seedling and examine the root-hair zone in a drop of water on a slide under a microscope.

2 Find out about the importance of osmosis to animals.

3 Place strips of epidermis from the concave (inner) surface of the scale leaves of an onion in strong sugar solution and in distilled water. Examine the strips under a microscope and compare the appearance of the cells after about 10 minutes.

4 Measure the length of a stick (petiole) of rhubarb and then re-measure when the epidermis has been removed. Account for any differences in measurements.

5 Why is it more efficient to hoe a garden in the morning than in the evening?

6 To what extent are transpiration and sweating similar processes?

30 TRANSPORT IN MAMMALS

Blood is the major transporter in a mammal; it acts as a communicating link between all parts of the body. Blood systems were probably a necessary first stage in the evolution of large compact animals; because of their small surface area in relation to volume (see p. 205), cells near the centre of the body are too far away from contact with the atmosphere. Oxygen cannot reach these cells by diffusion alone: neither can waste products be removed by diffusion alone. A special respiratory tissue, blood, evolved to carry respiratory gases between external atmosphere and respiring tissues.

The human body contains about 5.5 litres of blood, approximately 10 per cent of the body weight.

THE COMPOSITION OF BLOOD

Blood can be regarded as a tissue, but here the basic part of the tissue is a fluid; cells (the various types of which differ from each other in detail) are suspended in this fluid. Thus blood has two main components:

1 **Plasma**—a pale straw-coloured fluid occupying about one half of the total blood volume. Approximately 90 per cent of plasma is water, the remainder dissolved materials.

   (a) **blood proteins**, e.g. prothrombin and fibrinogen.

   (b) **foods** (end-products of digestion), e.g. glucose, amino-acids and fatty substances in process of transportation.

   (c) **mineral salts**, in the form of ions, e.g. Na^+, Cl^-, Mg^{++}, Ca^{++}, HCO_3^- (bicarbonate).

   (d) **excretory materials**, in a mammal, especially urea.

   (e) **hormones**, in minute quantities.

   (f) **gases**, small traces of oxygen, carbon dioxide and nitrogen dissolved in the plasma.

   (g) **antibodies** and **antitoxins** (see p. 150).

The proportions of these dissolved materials vary in:

   (i) different circumstances—e.g. plasma may contain more foods after a heavy meal and more of the hormone adrenalin when the animal is frightened; certain antibodies will be present only when the body is subjected to specific diseases.

   (ii) different blood vessels—more urea is present in the renal artery than in the renal vein (see p. 158) and more dissolved foods are to be found in the hepatic portal vein after a meal (see p. 109 and p. 153) than in other blood vessels.

Serum is the name given to blood plasma which has had its protein fibrinogen removed; in this form the plasma cannot clot, so it can be stored in hospital blood banks for transfusions.

2 **Cells** (Corpuscles)—see table on next page.

BLOOD VESSELS

Blood flows around the body in tubes called blood vessels. Except for the smallest of these tubes (capillaries), blood vessels are organs, i.e. they are composed of different tissues. The movement of blood in the vessels depends on the activity of a central pump, the heart.

Arteries and **veins** are the two principal types of blood vessels. Blood is pumped from the heart into and along arteries and is emptied back into the heart by way of veins. The blood flows *in one direction only* along any particular blood vessel; this is because of valves in the heart and veins.

Both arteries and veins have three-layered walls:

   (a) inner lining

   (b) middle lining—contains muscle and elastic tissue. The contraction and relaxation of the muscle tissue, particularly in arteries, changes the diameter of the blood vessel: this alters the blood pressure and rate of flow. The elastic tissue allows the vessel to expand with the

BLOOD CELLS

	RED CORPUSCLES (ERYTHROCYTES)	WHITE CORPUSCLES (LEUCOCYTES)	PLATELETS (THROMBOCYTES)
APPEARANCE	Biconcave circular discs without nuclei; contain a solution of the pigment haemoglobin (a complex chemical containing iron) inside an elastic envelope. In slow-moving blood, red corpuscles often adhere together like piles of dishes and are said to be 'in rouleaux' 0·008 mm SURFACE VIEW SIDE VIEW	Several different types, all with nuclei—three common types are illustrated below: TYPE A — large spherical nucleus, 0·008 mm, thin rim of clear cytoplasm TYPE B — 0·02 mm, kidney-shaped nucleus, relatively clear cytoplasm TYPE C — 0·01 mm, very granular cytoplasm, amoeboid outline, irregularly lobed nucleus	Non-nucleate fragments of protoplasm 0·003 mm — irregular outline, clusters of granules
FREQUENCY (in cu mm of human blood)	Five million (Ratio of reds : whites = 600 :1 approx.) The number varies at different times during the day; also, more red corpuscles are present where there is less oxygen, at higher altitudes ('acclimatisation')	Very variable—approximately 8000→10 000 (Ratio of types A :B :C = 25 :3 :72)	Very variable—approximately 250 000
SITE OF FORMATION	In the red bone marrow (especially at upper end of femur and in ribs and vertebrae) at the rate of about 1 million per second	A and B—in the lymph system (see p. 149) C—in the red bone marrow	In the red bone marrow
LENGTH OF ACTIVE LIFE	Approximately three months; they are then destroyed in the liver and spleen	Very variable—usually from two to three weeks (some much shorter during infections) (see p. 150)	Unknown
FUNCTIONS	Transport of respiratory gases. Oxygen diffuses from the air in the lungs into the haemoglobin—unstable oxyhaemoglobin is carried around in the corpuscles—the oxygen is released at and diffuses into the respiring tissues, where the concentration of oxygen is very low	A—perhaps concerned with formation and/or storage of antibodies. B and C—attack and engulf foreign particles in blood stream (see p. 150)	Clotting of blood (see p. 151)

Blood can be examined microscopically by making a blood smear. A mounted needle is sterilised by heating to red heat, and used to collect a small amount of blood from finger or ear-lobe. The blood is then allowed to drop, without being squeezed, onto a clean dry microscope slide. The drop is spread by drawing the edge of a cover-slip across it. In order to see white corpuscles more clearly the smear is allowed to dry and then covered for a few minutes by a suitable stain. The stain is washed off and a cover-slip placed over the smear.

flow of blood and then to recoil and squeeze the blood onwards.

(c) outer lining—fibrous connective tissue, to resist excessive expansion of the wall.

Arteries progressively divide and subdivide as they carry blood from the heart into the various body organs. In each organ, the artery divides into smaller, thinner-walled vessels called **arterioles**. Arterioles themselves divide into a network of the smallest of all blood vessels called **capillaries**. These eventually unite into larger blood vessels, **venules**, which carry blood out of the organ. Venules in turn unite to form veins, and more of these unite as blood is brought back to the heart. In this way blood circulates around the body between heart and organs.

BLOOD FLOW

1 Along arteries

Here blood flows between heart and organs under considerable pressure; this pressure is maintained, even for example against the force of gravity, in two ways:

(a) by the force of contraction of the ventricles at each heart-beat (see p. 155), squeezing blood from the heart into the arteries.

(b) by blood which has entered the arteries pressing outwards against the artery wall— the elastic tissue stretches and then recoils

inwards to squeeze blood further onwards. (The blood cannot return to the heart because of more blood being forced along behind it and because of the semilunar valves where blood leaves the heart).

These two processes result in the characteristic pulsing of blood, which therefore gives an exact measure of the heart-beat; it can be felt where arteries pass near the skin surface, e.g. at the wrist.

2 Along veins

Here blood flows between organs and heart at low pressure; much of the pressure that was present in the arterial flow has been lost as the blood flowed along the capillaries. Blood cannot therefore return to the heart 'under its own steam' but instead relies on three main methods:

(a) the largest veins (e.g. in the limbs) pass closely adjacent to the muscles of the skeleton. When these muscles contract they become shorter and thicker and press on blood in the veins.

Simple diagram showing the pathway of circulating blood (dimensions of blood vessels are those commonly found in man)

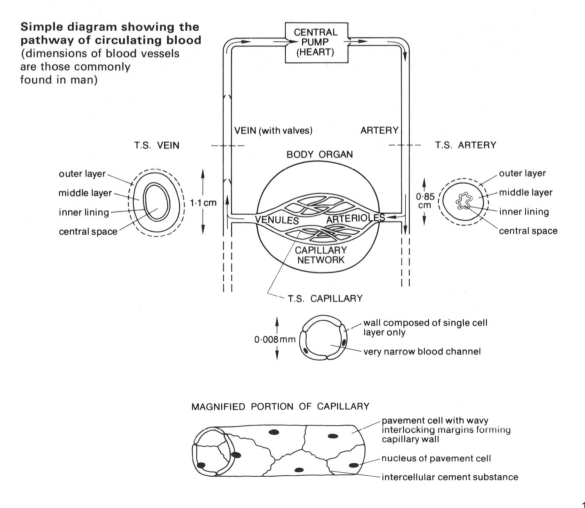

147

TABLE SHOWING GENERAL DIFFERENCES BETWEEN ARTERIES AND VEINS

	ARTERIES	VEINS
1	Carry blood away from heart	Carry blood towards heart
2	Carry oxygenated blood	Carry deoxygenated blood
	(except in pulmonary circulation—see p. 152)	
3	Blood flows rapidly under high pressure (90 mm Hg for example, in man)	Blood flows slowly under low pressure (5 mm Hg for example, in man)
4	Blood flows in pulses	Blood flows more smoothly
5	Walls have thick middle layer of muscle and elastic tissue	Walls have thin middle layer of muscle and elastic tissue
6	Valves not present	Valves present
7	Blood flows along a proportionately smaller space	Blood flows along a proportionately larger space
8	Cross-sectional outline rounded	Cross-sectional outline more irregular or oval

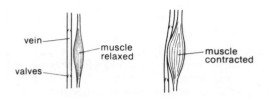

The blood is squeezed either forwards towards or backwards away from the heart. When the blood flows backwards it fills pocket-like valves in the vein walls: these valves close the vein when they are full and blood is trapped.

(The importance of limb muscles in maintaining blood flow can easily be demonstrated by sitting completely still for several minutes; the 'pins-and-needles' which result are caused by insufficient blood flow to the nerves in the limbs, which become deprived of food and oxygen.)

(b) the pull of gravity—from all those parts of the body above the heart.

(c) inspiration movements (see p. 128)—as well as drawing air into the lungs, these movements suck blood along veins into the chest cavity and thence into the heart.

3 Along capillaries.

Every organ in the body has its own capillaries; these minute tubes form a complex network so that each living cell is near the flow of blood. Capillaries are so narrow that the flow of red corpuscles along them often is impeded; when the capillaries finally unite to form venules leaving an organ the blood flows somewhat faster again.

Capillaries allow material to leave the bloodstream and pass into the surrounding spaces containing fluid; these spaces actually bathe the tissues. Being narrow, capillaries have a large surface to volume ratio; this, combined with thin (one-cell-thick) walls and the slow steadier flow of blood along them, provides a very efficient arrangement for permeation. Substances in solution—foods, oxygen (from the red corpuscles), hormones, etc.—diffuse from the bloodstream into the tissue fluid spaces and thence into the cells; dissolved waste materials—urea,

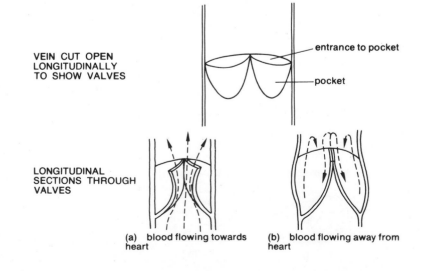

VEIN CUT OPEN LONGITUDINALLY TO SHOW VALVES — entrance to pocket — pocket

LONGITUDINAL SECTIONS THROUGH VALVES

(a) blood flowing towards heart

(b) blood flowing away from heart

carbon dioxide, etc.—diffuse from the cells back into the bloodstream. Certain white (but not red) corpuscles can also move between blood and tissue fluid spaces; apparently they are able to form very thin pseudopodia and pass down narrow channels of protoplasm in the cells of the capillary walls.

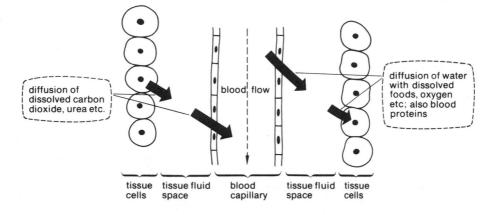

THE LYMPH SYSTEM

The fluid which has diffused from the blood into the spaces outside the capillaries is called **lymph**. It is a pale straw-coloured liquid very similar to blood plasma; it lacks red corpuscles but because it contains blood proteins it can clot. Lymph is in a sense a 'go-between', just as a retailer acts between wholesaler and customer.

The lymph system deals with this lymph and is an accessory part of the blood system. It begins as fine lymph capillaries situated actually in the spaces bathing the tissue cells. The capillaries are closed at one end. Lymph diffuses into them from the tissue spaces and the capillaries empty at their opposite ends into larger tubes called **lymphatics**. Two very large lymphatic ducts eventually return the collected lymph into the bloodstream at the base of the neck; one empties into the right, and the other into the left jugular vein.

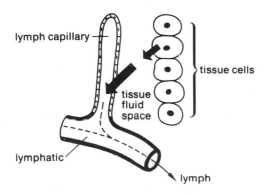

Lymphatics resemble most veins in that:

(**a**)   they carry fluid towards the heart.

(**b**)   they have thin relatively non-muscular walls.

(**c**)   they contain valves.

(**d**)   the flow of fluid is steady and at low pressure.

(**e**)   the fluid is deoxygenated

Unlike veins, lymphatics possess swellings at intervals along their length. These are **lymph nodes** and are glandular in function. Each node is composed of connective tissue and cells that produce white corpuscles. In many infections lymph nodes become swollen and painful; this is because the disease is being limited and fought locally in the gland by the activities of its white corpuscles (see p. 150. Lymph nodes are particularly plentiful in the arm pits, groin and neck. The tonsils and even the spleen (see p. 146) are essentially very large lymph nodes.

The lymph system has the following main functions:

1   to collect blood plasma which has diffused out from the blood capillaries, and return it to the bloodstream.

2   to remove some of the excretory materials produced by metabolising cells.

3   to remove and distribute secretions of gland cells —e.g. the hormone thyroxine is carried away from the thyroid gland in the lymph, perhaps in preference to the blood capillary network.

4   to collect absorbed fatty materials from lacteals in the villi of the small intestine (see p. 108).

5   to produce certain white corpuscles and fight infections.

Functions of the circulatory system

Since blood continually circulates in the body it provides a rapid fluid method of communication between all the various body regions. The cells of mammals are specialized (see p. 85) and carry out precise complex functions; these functions can be maintained only if their immediate surroundings are kept within certain narrow limits, i.e. at constant temperature, constant acidity, etc. Blood and lymph bathing the cells play an important part in maintaining this constant internal environment.

1 Transport

(a) **oxygen**—carried as oxyhaemoglobin in red corpuscles from lungs to respiring tissues (see table on p. 146).

(b) **carbon dioxide**—carried from respiring tissues to lungs in these ways:

    (i) dissolved in plasma.

    (ii) as sodium bicarbonate, after associating with sodium ions in the plasma.

    (iii) as carbamino-haemoglobin, as a result of the carbon dioxide entering the red corpuscles and reacting with haemoglobin.

        Carbon dioxide diffuses from the respiring tissues into the tissue fluid spaces and then into the blood and lymph. In the lungs, the gas diffuses from the capillaries to the alveoli, where the concentration of carbon dioxide is low.

(c) **simple foods**—the end-products of digestion are absorbed by the villi in the small intestine (see p. 108). Glucose and amino-acids are carried dissolved in the plasma to the liver and then to the respiring and growing tissues, according to the requirements of the body. Fatty materials are transported in the plasma to places of fat storage, etc., when lymph vessels empty their contents into the blood-stream at the neck (see p. 149).

(d) **excretory substances**—apart from carbon dioxide, this is mainly nitrogen-containing waste. Excretions from all metabolizing cells diffuse into the bloodstream and are carried to the liver, where they are converted into urea; the plasma transports dissolved urea from liver to kidneys for excretion (see p. 157).

(e) **hormones**—carried in plasma from endocrine organs which secrete them. Some, e.g. insulin (see p. 108 and p. 174) affect specific 'target organs'; others, e.g. adrenalin (see p. 173) produce more general and widespread effects.

(f) **heat**—carried from main heat-generating organs (liver, muscles) to all regions of body as means of heat distribution and maintenance; also to the skin as a means of temperature regulation (see p. 90).

(g) **antibodies** and related substances (see 'Prevention of disease').

2 Prevention of disease

Foreign organisms such as bacteria may enter the body through the lining of the respiratory system, through the alimentary canal and through breaks in the skin. Many of these organisms are **pathogens**, i.e. capable of producing disease. Blood has several components which protect the body from infection:

(a) **white corpuscles**—those white corpuscles that are amoeboid are sometimes called phagocytes, i.e. they are capable of ingesting living bacteria, dead cells and other foreign particles in the blood and tissue fluid spaces; they do this rather as *Amoeba* ingests its food (see p. 6).

(b) **antibodies**—these are complex blood proteins found in the plasma. They are almost certainly secreted or stored by those white corpuscles that are formed in the lymph system. Antibodies are said to act against pathogens in the circulation in three main ways:

    (i) they cover the surface of the pathogen so that the phagocyte can ingest it more easily.

    (ii) they make the surface of the pathogens 'sticky' so that they clump together; the pathogens are then unable to invade the tissues but remain localized at one place in the circulation.

    (iii) they actually dissolve the surface of the pathogen and disintegrate it.

(c) **antitoxins**—these are a special type of antibody. They are secreted into the plasma but act against the excretions of pathogens rather than against the pathogens themselves. These excretions are called **toxins**; they are largely responsible for the characteristic symptoms (fever, rash, etc.) of a particular disease. Each type of bacterium produces a different toxin, which can be neutralized by a specific antitoxin. Where an infection is extensive, e.g. at a septic cut, bacteria at first outnumber the white corpuscles; the bacteria secrete toxins which kill large numbers of white corpuscles. The dead corpuscles accumulate and form pus.

(d) clotting—this process prevents excessive loss of blood when blood vessels are damaged. It is extremely complex, but the essential stages are as follows:

(i) when the skin is cut and blood flows out, the **platelets** become exposed to the air; they disintegrate and liberate an enzyme **thrombokinase** into the plasma. Apparently some thrombokinase is also set free from the damaged blood vessel walls.

(ii) thrombokinase in the presence of calcium ions, found in blood, can now change the plasma protein **prothrombin** to an active enzyme **thrombin**.

(iii) thrombin reacts with the soluble plasma protein **fibrinogen** and changes it to insoluble **fibrin** which is precipitated as strands in a meshwork.

(iv) corpuscles become trapped in the mesh, dry, die and harden to form a scab under which the wound can heal and which prevents entry of foreign particles.

(The blood of people who suffer from the hereditary disease **haemophilia** lacks substances necessary to complete the clotting process.)

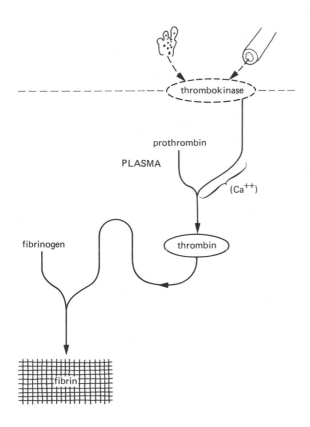

(e) inflammation—where part of the body becomes infected, e.g. when a wood splinter is left in the skin, the blood vessels may dilate. More warm blood reaches this part: the local rise in temperature kills many bacteria. At the same time, the blood capillaries become increasingly permeable, allowing antibodies and white corpuscles to reach the infected part more easily.

(f) immunity—pathogens liberate protein substances called **antigens** into the bodies of the organisms which they invade. Each type of pathogen produces its own special type of antigen, and this provokes the production of a specific antibody against it.

The first time that an animal is infected by a particular pathogen (even if the infection is not sufficient to produce disease symptoms), specific antibodies are manufactured in response. Normally, its various protective activities (inflammation, the activities of white corpuscles, and the formation of antibodies and antitoxins) enable the animal to survive. Afterwards, some of the specific antibodies remain in the animal, which somehow 'remembers' how to make them—the ability to form the particular antibody is even passed on to daughter cells after cell division. In any subsequent infection of the same type, specific antibodies will be readily available to deal with it. Because of this, it is rare for anyone to suffer from measles, mumps or chickenpox more than once.

This process is called **natural immunity**, because it has not been brought about by medical treatment. (Many viruses, such as those that cause colds and influenza, exist in different forms called 'strains'; although an animal may develop natural immunity from several of these strains, other 'new' strains can cause disease symptoms.)

Animals can also develop artificial immunity. This was first demonstrated by Edward Jenner (1749–1823) and other scientists towards the end of the eighteenth century. Pathogens can be treated so that they are killed or weakened. When they are then introduced into the body by vaccination or inoculation, they produce no disease symptoms at all or only mild symptoms. Even so, their antigens stimulate the formation of sufficient antibodies to give immunity from future attacks of the same disease. (Sometimes, closely-related pathogens can be used, with identical results.) This sort of immunity is called active artificial immunity, because the animal makes antibodies for its own protection.

There is always a time lapse between inoculation and the production of antibodies. As a result, active immunization is unsatisfactory where treatment is needed quickly. If an animal already has antigens in its blood (e.g. after a snake-bite), or is suspected of having already contracted a disease (e.g. tetanus bacteria may have entered a wound which has been in contact with infected soil), passive artificial immunity may be a more effective treatment. Here, blood serum is taken from another animal—often, a horse—which has contracted the particular disease, or else has been artificially stimulated by inoculation of the appropriate antigens. The serum contains the correct antibodies and is inoculated into the patient. It gives only temporary immunity, because the patient's body has not produced its own antibodies.

Another example of antigen-antibody reaction occurs when different blood groups are mixed. Certain of these groups are not compatible: if mixed in error during a transfusion, they provoke a reaction—the red cells stick together and block the blood capillaries. Allergic reactions are caused by antigens: the proteins of inhaled pollen-grains act as antigens which provoke a reaction ('hay-fever') from antibodies.

In a similar way, the body recognizes organs from other animals, even those of the same species, as 'foreign' protein. This leads to great difficulties in organ transplants when, for example, a man receives a kidney from another man. Unless the natural immunity reactions of the body are suppressed by drugs, the transplanted organ is likely to be rejected.

THE CIRCULATION OF BLOOD

The idea that blood circulates between heart and organs was first put forward by **William Harvey** near the start of the seventeenth century; previously blood was thought to have an ebb and flow movement in the body.

Mammals have a **double circulation**, i.e. there is a major circulation from heart to all regions of the body except the lungs, and a less extensive circulation to the lungs alone. This means that blood passes through the heart twice before it returns to the same part of the body. Thus the heart is partitioned down the middle: deoxygenated blood destined for the lungs on the right side is completely separated from oxygenated blood destined for other parts of the body on the left. Much of the blood pressure has been lost by the time the blood returns to the heart from the body, so contraction of the right side forces blood to the lungs and through the intricate lung capillaries. On returning to the heart, contraction of the left side boosts the pressure so that the blood has sufficient force to pass to the rest of the body.

Simple diagram showing double circulation

152

Diagram showing blood circulation to major parts of a mammal

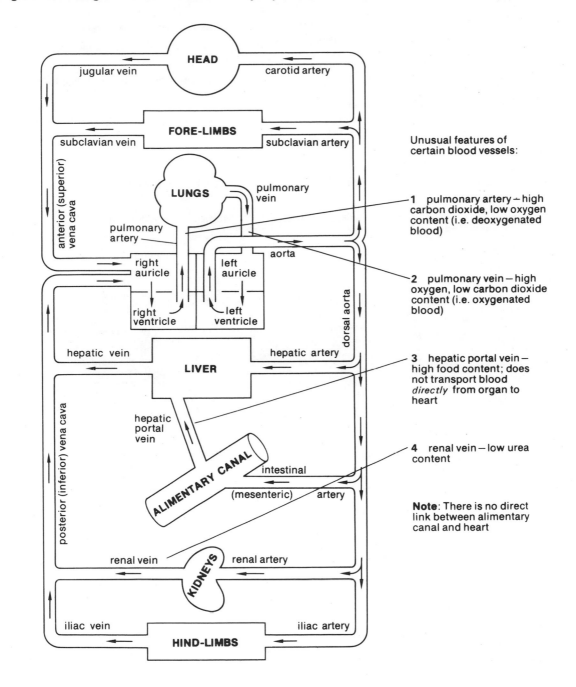

Unusual features of certain blood vessels:

1 pulmonary artery – high carbon dioxide, low oxygen content (i.e. deoxygenated blood)

2 pulmonary vein – high oxygen, low carbon dioxide content (i.e. oxygenated blood)

3 hepatic portal vein – high food content; does not transport blood *directly* from organ to heart

4 renal vein – low urea content

Note: There is no direct link between alimentary canal and heart

The heart

The heart can be regarded as a hollow muscle. The central spaces contain the blood; contractions of the muscle pump the blood and maintain the circulation.

The heart is a roughly pear-shaped organ situated in the thoracic cavity between the lungs, slightly displaced to the left. It has its own blood supply—the **coronary artery** and **vein** pass over the heart surface and divide many times. The right side is completely partitioned from the left by a **septum**.

Each half has two interconnecting chambers:

(a) **auricles (atria)**—thin-walled; *receive* blood from veins.

(b) **ventricles**—thick-walled; *despatch* blood into arteries.

Each auricle or atrium is above or in front of its respective ventricle.

Diagram showing a longitudinal section through the heart of a mammal

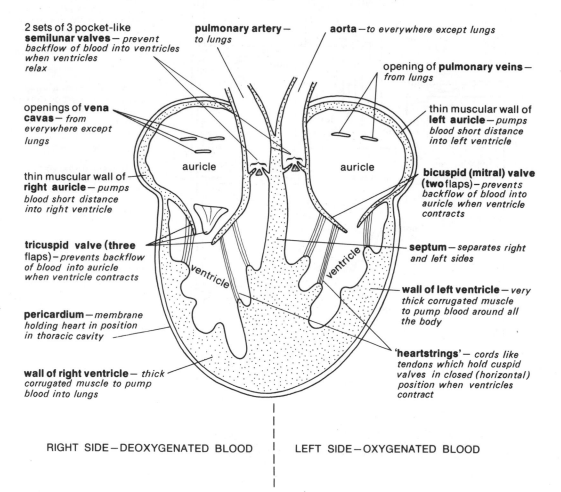

2 sets of 3 pocket-like **semilunar valves**— *prevent backflow of blood into ventricles when ventricles relax*

pulmonary artery— *to lungs*

aorta—*to everywhere except lungs*

opening of **pulmonary veins**— *from lungs*

openings of **vena cavas**— *from everywhere except lungs*

thin muscular wall of **left auricle**—*pumps blood short distance into left ventricle*

thin muscular wall of **right auricle**— *pumps blood short distance into right ventricle*

bicuspid (mitral) valve (two flaps)— *prevents backflow of blood into auricle when ventricle contracts*

tricuspid valve (three flaps)—*prevents backflow of blood into auricle when ventricle contracts*

septum—*separates right and left sides*

wall of left ventricle— *very thick corrugated muscle to pump blood around all the body*

pericardium—*membrane holding heart in position in thoracic cavity*

'heartstrings'— *cords like tendons which hold cuspid valves in closed (horizontal) position when ventricles contract*

wall of right ventricle— *thick corrugated muscle to pump blood into lungs*

auricle

auricle

ventricle

ventricle

RIGHT SIDE—DEOXYGENATED BLOOD

LEFT SIDE—OXYGENATED BLOOD

Heart of mammal—external features (ventral view)

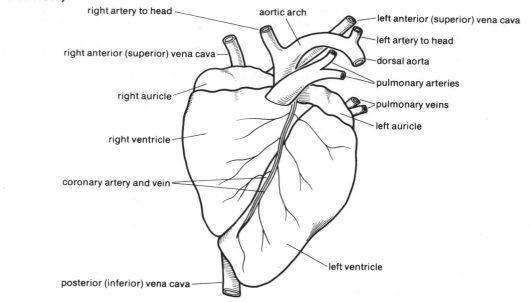

right artery to head

aortic arch

left anterior (superior) vena cava

left artery to head

right anterior (superior) vena cava

dorsal aorta

pulmonary arteries

right auricle

pulmonary veins

left auricle

right ventricle

coronary artery and vein

left ventricle

posterior (inferior) vena cava

The action of the heart

Blood passes through the heart in a basically two-phase sequence; the two phases alternate continuously.

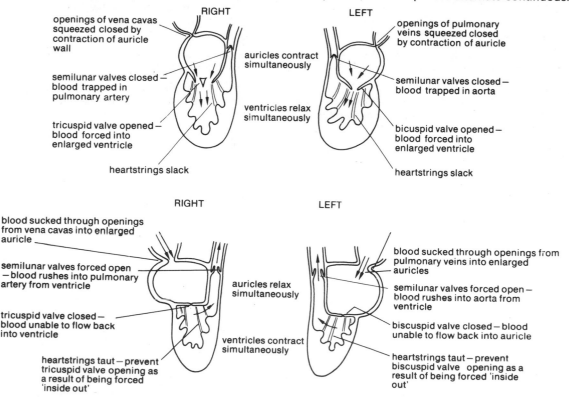

RIGHT

LEFT

openings of vena cavas squeezed closed by contraction of auricle wall

openings of pulmonary veins squeezed closed by contraction of auricle

auricles contract simultaneously

semilunar valves closed—blood trapped in pulmonary artery

semilunar valves closed—blood trapped in aorta

ventricles relax simultaneously

tricuspid valve opened—blood forced into enlarged ventricle

bicuspid valve opened—blood forced into enlarged ventricle

heartstrings slack

heartstrings slack

RIGHT

LEFT

blood sucked through openings from vena cavas into enlarged auricle

blood sucked through openings from pulmonary veins into enlarged auricles

semilunar valves forced open—blood rushes into pulmonary artery from ventricle

auricles relax simultaneously

semilunar valves forced open—blood rushes into aorta from ventricle

tricuspid valve closed—blood unable to flow back into ventricle

biscuspid valve closed—blood unable to flow back into auricle

ventricles contract simultaneously

heartstrings taut—prevent tricuspid valve opening as a result of being forced 'inside out'

heartstrings taut—prevent biscuspid valve opening as a result of being forced 'inside out'

Each heart-beat produces two characteristic sounds—'*lubb*, *dup*'—a prolonged dull thud followed by a more abrupt sound of higher pitch. The former is caused by the closure of the cuspid valves when the ventricles contract; the latter occurs when the semi-lunar valves close.

The rate of heart-beat in man

The average heart-beat rate is 75 beats per minute, i.e. each cycle is completed in 0.8 second. The rate increases at higher temperatures and during exercise (up to 200 beats per minute) and emotional disturbances such as fright. The heart also beats faster during fevers, but other disease conditions may cause the rate to decrease. The rate is slower during sleep. In children the heart beats somewhat faster and there may be an increase again towards old age.

The heart beat of an average man pumps out 5 litres of blood per minute; this increases to 8 litres per minute during strenuous exercise. Consequently the blood flow to the tissues, and hence the supply of food and oxygen, varies in relation to the immediate requirements of these tissues.

The heart will continue beating even if its nerve supply is removed. The heart muscles begin their rhythmic contractions before birth and continue automatically in this way until death. However, stimulation from the nervous system is necessary to control the *rate* of heart-beat. The hormone adrenalin (see p. 173) also has an effect on the beat rate.

The pressure of blood in the aorta following contraction of the left ventricle in a 20-year-old man is approximately 120 mm Hg. Quite substantial variations from this figure do not necessarily indicate poor health. Blood pressure increases with age.

The muscle in the walls of the heart (**cardiac muscle**) is found nowhere else in the body. It does not easily fatigue. (See p. 195.)

1   Make a smear of blood as directed on p. 146 and examine microscopically. Devise some method of finding an approximate ratio of red to white corpuscles.

2   Examine prepared sections of arteries and veins microscopically and note the differences.

3   Devise a method for comparing the rates of your heart-beats when sitting quietly, after moderate exercise, and after strenuous exercise.

4   Place a living tadpole in a small drop of water on a slide under a microscope. Examine the thin skin of the tail to see movement of blood through the skin capillaries.

5   Dissect a sheep's heart (obtainable from butchers' shops) and display its main structures including the arteries and veins connected to it.

6   Read about human blood groups. Special cards can be obtained so that you can determine your own blood group.

7   Read about the work of Edward Jenner and about the history and importance of artificial immunity.

31 EXCRETION

Excretion is the removal of the waste products formed by metabolism.

Metabolism is the name given to all the complex chemical processes which perpetually take place in living cells. These processes produce waste materials (excretions) that are poisonous in varying degrees and have to be removed from the body as soon as possible. The excretions of living organisms are essentially of two types:

(a) **carbonaceous**, i.e. containing carbon—in particular the gas carbon dioxide, formed directly as a waste product of respiration (see p. 121). If allowed to accumulate, carbon dioxide poisons cells on account of its acidity; it dissolves in the tissue fluids, forming carbonic acid.

(b) **nitrogenous**, i.e. containing nitrogen—in particular ammonia (in many aquatic organisms), uric acid (in insects, reptiles and birds) and urea (in mammals). These substances are especially poisonous, and accumulations would soon result in death. The urea which is excreted by mammals originates either from excess amino-acids, which are deaminated in the liver (see p. 109), or from the breakdown of protoplasm in damaged or worn-out cells.

It is a characteristic feature of most plants that, unlike animals, they produce carbonaceous but not nitrogenous waste under normal conditions of metabolism.

Excretion should not be confused with two similar-sounding processes, egestion and secretion. *Egestion is the removal of undigested materials from the processes of digestion.* Except in single-celled creatures, such as *Amoeba*, these materials have never entered the cells or taken part in metabolism. In mammals they have entered the body at the mouth and left it at the anus; they have hardly been altered during their passage along the alimentary canal.

Secretion is the production of useful chemical substances by living cells. Secretions have definite functions; they can be passed from the cells that manufacture them into ducts (e.g. salivary, pancreatic secretions) or into the lymph or blood-stream (e.g. endocrine secretions—see p. 173).

It is sometimes difficult to distinguish between excretions and secretions. Sweat, for example, may be regarded as either; its water component is a secretion because it serves the positive function of body temperature regulation, whereas substances dissolved in the water (such as small traces of urea) are clearly waste products of metabolism.

Excretion in mammals

Mammals possess four main excretory organs:

1 **Lungs**—excretion of carbon dioxide and water vapour from respiration (see p. 125).

2 **Liver**—excretion of bile pigments derived from the breakdown of haemoglobin in worn-out red corpuscles (see p. 109).

3 **Skin**—excretion of a watery solution of salts and urea by means of the sweat glands (see p. 90).

4 **Kidneys**—excretion of relatively large quantities of urea and other materials.

The lungs and kidneys are the most important excretory organs. All except the kidneys have been described elsewhere in this book.

THE KIDNEYS

The kidneys are deep red, dense, compact organs situated in the upper part of the abdominal cavity, one on each side of the vertebral column. They may be asymmetric, the right further forwards than the left, so that the right kidney lies quite close to the diaphragm.

Each kidney is covered by a tough translucent membrane which fixes it to the dorsal abdominal wall. In most mammals the kidney has a characteristic shape, concave on its inner and convex on its outer surface. The tissues around the kidneys are places where fat may be stored, so that the kidneys are often partly embedded in fat. The kidneys are otherwise protected only by the muscle layers of the lumbar region, and the fat assists in cushioning the kidneys from impacts.

Diagram showing the position in the abdomen of the excretory system

The positions of kidneys are variable. In man the left kidney is higher, in rabbit the right, while in cat the two kidneys are at the same level

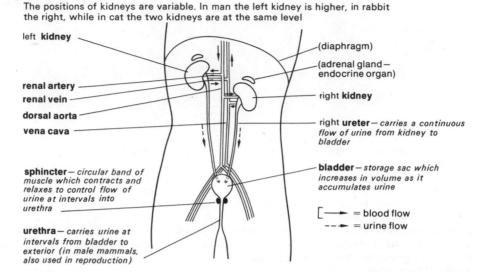

left **kidney**

(diaphragm)

(adrenal gland— endocrine organ)

renal artery

renal vein

right **kidney**

dorsal aorta

vena cava

right **ureter**— *carries a continuous flow of urine from kidney to bladder*

sphincter— *circular band of muscle which contracts and relaxes to control flow of urine at intervals into urethra*

bladder— *storage sac which increases in volume as it accumulates urine*

⟶ = blood flow

---► = urine flow

urethra— *carries urine at intervals from bladder to exterior (in male mammals, also used in reproduction)*

Three tubes are connected to each kidney at the centre of the concavity:

(i) **renal artery**—bringing oxygenated blood from the dorsal aorta into the kidney for purification.

(ii) **renal vein**—removing purified but de-oxygenated blood from the kidney into the vena cava.

(iii) **ureter**—a narrow white muscular tube which drains urine away to the bladder.

Kidney structure

A vertical section through a kidney shows that there are two main zones, **cortex** and **medulla**. The cortex follows the outer convex surface and the medulla lies towards the centre of the kidney; each zone has a quite different appearance under the microscope (see photograph opposite). The ureter is swollen to form the **pelvis** at the point where it emerges from the concave surface of the kidney. The inner border of the medulla region is extended into several conical **pyramids** which project towards the pelvis.

Three main types of tissue are found in a kidney:

(i) **blood capillaries**, formed by the division of the renal artery many times into a complex arrangement inside the kidney; these capillaries eventually unite to form the renal vein.

(ii) **kidney tubules**, (more than one million per human kidney), closely associated with the blood capillaries; these microscopic tubules are closed at their inner ends in the cortex and finally open into the pelvis at the tips of the pyramids.

(iii) **connective tissue**, binding the capillaries and tubules together.

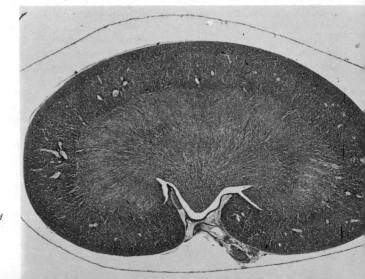

V.S. Mammalian kidney, showing outer cortex (containing rounded Bowman's capsules), central medulla and a pyramid

Diagram of a vertical section through a kidney

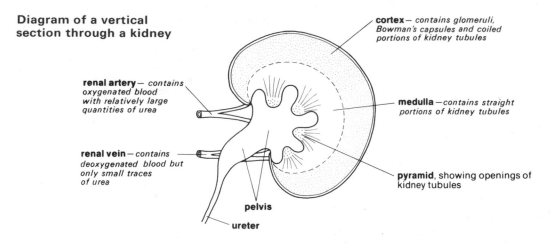

cortex — *contains glomeruli, Bowman's capsules and coiled portions of kidney tubules*

renal artery — *contains oxygenated blood with relatively large quantities of urea*

medulla — *contains straight portions of kidney tubules*

renal vein — *contains deoxygenated blood but only small traces of urea*

pyramid, showing openings of kidney tubules

pelvis

ureter

The kidney tubules

Each tubule has four portions, all joined together. It begins as a bowl-shaped **Bowman's capsule**; the walls of the bowl are hollow, the hollow being lined by a single layer of cells.

The other three portions are much more tubule-like, and again have one-cell-thick walls. In two of these the tubule is twisted around itself forming the **first** and **second coiled tubules**. Between these coiled portions the tubule forms the **loop of Henlé**, so called because two straight pieces of tubule are linked together in a U-shaped loop. At the end of the second coiled tubule several kidney tubules from one region of the kidney join together to form a **collecting duct** which passes towards the pelvis. Bowman's capsule and the coiled tubules are in the cortex; the loop and the collecting duct are in the medulla.

The bowl of Bowman's capsule contains a **glomerulus**, i.e. tight knot of blood capillaries formed from the branching of the renal artery. After leaving the glomerulus the blood vessel forms a complex network of capillary loops around the other parts of the kidney tubule. It is these capillaries that unite to become the renal vein.

Kidney functioning

Approximately one quarter of the blood distributed by each heart-beat passes through two kidneys.

The force of the heart-beat drives blood along the branches of the renal artery into the glomerulus. The blood vessel leaving each glomerulus is narrower than that entering; this creates a bottle-neck in the glomerulus capillaries. Blood pressure then drives water and dissolved substances in the glomerulus through the thin barrier which separates

Diagram showing Bowman's capsule

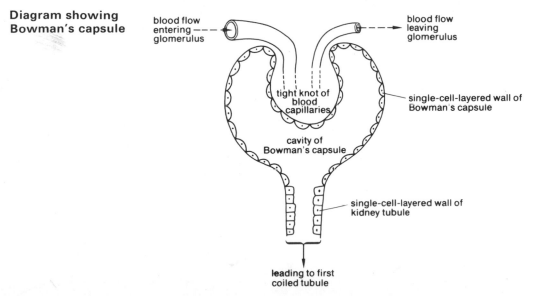

blood flow entering glomerulus

blood flow leaving glomerulus

tight knot of blood capillaries

single-cell-layered wall of Bowman's capsule

cavity of Bowman's capsule

single-cell-layered wall of kidney tubule

leading to first coiled tubule

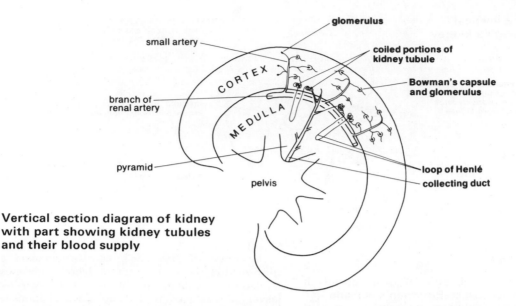

Vertical section diagram of kidney with part showing kidney tubules and their blood supply

the blood from the cavity of Bowman's capsule. In addition to losing urea and water the blood loses salts and foods, so that the blood that leaves the glomerulus contains mainly corpuscles and blood proteins together with some water.

The fluid in Bowman's capsule which has thus been ultra-filtered (i.e. filtered under pressure) contains valuable materials that the animal cannot afford to lose. These materials include much water, foods such as glucose, and many of the salts.

Diagram showing one kidney tubule isolated

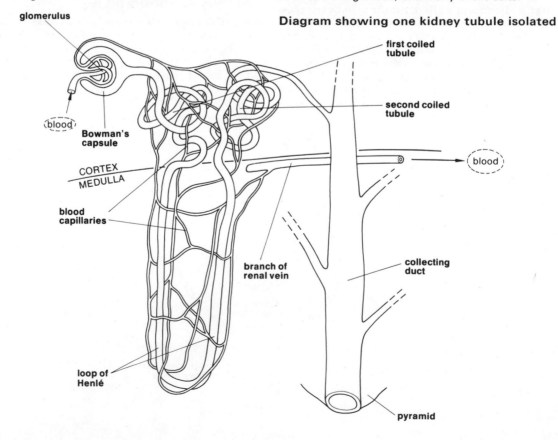

The cells lining the first coiled tubule use energy from respiration to absorb useful materials from the filtered fluid as it flows by them. As much as possible is extracted and passed into the blood which circulates in the closely-adjacent capillary network. In the disease **diabetes** (see p. 174) the blood reaching the kidney, and the filtered fluid also, contains abnormally large amounts of dissolved glucose; the cells of the coiled tubule are unable to extract such large quantities and glucose appears in the urine—urine is tested for glucose when diabetes is suspected.

The fluid passing from the first coiled tubule into the loop of Henlé is mainly a dilute solution of urea. Mammals are predominantly land-living animals and could not afford to lose the large amount of water that reaches the loop of Henlé. The function of this portion of the tubule is therefore to bring about the removal of as much water from the tubule as possible, according to the needs of the body. After excessive drinking the blood has a high water content and the kidney allows a very dilute urine to be excreted; after much sweating the blood water content is low and the urine is concentrated.

In the second coiled tubule it is possible that a 'tidying-up' is done. Further useful materials which may have escaped re-absorption in the first coiled tubule are passed back into the blood and excretory materials remaining in the blood are passed into the tubule.

Diagram showing kidney functioning

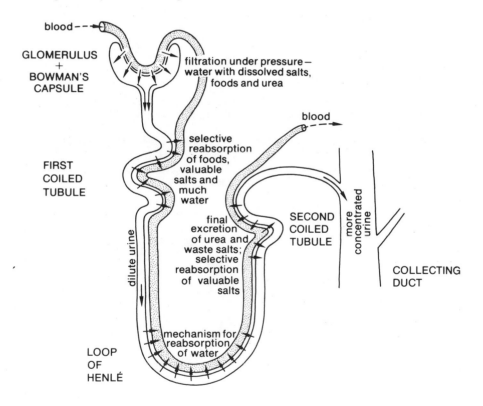

Urine, a concentrated solution of urea in water, together with certain dissolved salts and other excretory substances, flows down the collecting duct where it meets urine from other tubules in the kidney. It then flows out of a pyramid into the pelvis and trickles down the ureter in a continuous stream: the movement is caused by waves of muscle contraction in the ureter walls—peristalsis. (See p. 104.)

The **bladder** progressively fills with urine. Periodically the urine is emptied to the exterior along the urethra by conscious action; the muscles in the bladder wall contract.

Blood leaving the kidney tubules by the renal vein has lost almost all its urea but has regained its foods and valuable salt content. Because the tubules' re-absorption processes require an energy supply the blood also contains less oxygen but more carbon dioxide.

Summary of kidney functions

1 Excretion of nitrogenous waste, especially urea.

2 Osmoregulation, i.e. regulation of the water content of the body.

3 Excretion of salts and the regulation of blood acidity—many of the excreted salts are acid salts; this maintains the slight alkalinity of blood.

4 Excretion of other toxic substances (drugs etc.).

The functioning of the kidney is a good example of homeostasis (see p. 90).

	Concentrations of Principal Ingredients in:	
	1500 cm³ of Blood Plasma entering Kidney	Daily Output of Urine (1500 cm³) Per Man
Water	1370 g	1440 g
Urea	0.5 g	30 g
Glucose	1.5 g	0
Protein	110 g	0
Sodium ions	4.8 g	5.2 g
Chloride ions	5.4 g	9.0 g
Sulphate ions	0.05 g	2.7 g

Excretion in flowering plants

Plants do not regularly excrete nitrogenous waste and their methods of excretion are comparatively simple. Apart from waste gases from respiration and photosynthesis, plants generally make their waste insoluble and therefore harmless to living cells. Excretory materials are deposited inside cells or in the spaces between cells. They may be stored here permanently until death or removed when that region of the plant is discarded in some way. Normally the discarding is part of some other important plant process and the excretion is only a subsidiary function.

1 Waste carbon dioxide from respiration (at night) and waste oxygen from photosynthesis (during the day) diffuse out through the stomata (see p. 130). During the day the waste carbon dioxide is used for photosynthesis and the oxygen for respiration.

2 Waste materials may be passed into the leaves. In deciduous plants some of the waste accumulated during the season's growth passes into the leaves each autumn; leaf-fall then provides a convenient method of excretion (see p. 143). Similarly some excretion occurs when trees, e.g. plane, regularly shed their bark. Waste may also be excreted in the covering of fruits and seeds at dispersal.

3 The stems of trees and shrubs grow in thickness by means of an active growth zone (sap-wood) below the bark. As the stem grows outwards the central tissues progressively die, so producing the **heart-wood.** Waste materials are passed into the future heart-wood before the cells die.

4 Insoluble excretory crystals are deposited into the vacuoles of living cells in many herbaceous plants, e.g. silicates in the leaves of grasses. They remain here until death.

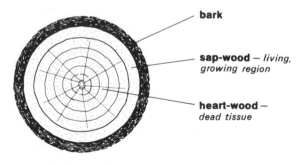

T.S. of tree trunk showing heart-wood

1 Obtain a large kidney from a butcher's shop and carefully cut it into halves vertically (as in the diagram on p. 159). Examine the cut surface and identify the main regions.

2 Dissect the abdomen of a small mammal to see the kidneys and other associated structures concerned with excretion. Note the close association of excretory and reproductive systems.

3 Test a sample of human urine for the presence of reducing sugars (see p. 111).

32 ANIMAL AND PLANT BEHAVIOUR

Both animals and plants react to changes in the world around them in ways which can easily be observed. Such reactions are *adaptive*, that is, they have as their end-result an improvement in the well-being of the organism concerned. The term **irritability** is sometimes used to describe this characteristic property of living things. The word was first used in this context by a doctor, Francis Glisson, towards the end of the seventeenth century. He applied it to the general property of living substance, or protoplasm, to react to change : for example the contraction of a muscle cell when suitably stimulated. The term is now more generally used to refer to the reactions of whole organisms rather than to their separate cells. Any situation, or change in a situation, which can clearly be shown to be responsible for a particular reaction in a living organism is called a **stimulus** (L. *stimulus*, a goad). The reaction it produces is called a **response**. It is not possible to define these two terms independently of each other. A stimulus can only be described by reference to a response. A response necessarily implies a stimulus. There are a number of physical factors in the environment which can act as stimuli to living things. These are described below.

Gravity This can provide a stimulus to nearly all living organisms. As a force, gravity is a constant but it provides a stimulus to animals, plants, or parts of plants when they are displaced from their normal attitudes. A cat, dropped the wrong way up, lands on all fours. A root laid on its side will turn and grow downwards.

Light In some cases light causes an avoiding reaction, as in earthworms. The aerial parts of flowering plants grow *towards* light. The total amount of light, over a period, may determine when flowering occurs, or, in animals, when they migrate.

Some animals possess *eyes*, which are devices for obtaining information about the outside world, using light as a medium.

Temperature Plants generally show no reaction to change in temperature although, like cold-blooded animals, their chemical processes are affected by temperature. Animals react in many ways to temperature change. It may be the stimulus to begin hibernation, migration, mating or nesting. Warm-blooded animals possess sense organs which act like thermostats. They act so as to oppose any change in internal temperature.

Touch All animals are sensitive to touch. Some plants, such as climbers, or insect-eating plants, are also sensitive to touch.

Chemical substances Some parts of plants, such as the roots and perhaps pollen tubes, are sensitive to certain chemical substances. In animals chemical stimulation may be either as **smell** or **taste**. To act as a stimulus to taste, substances must either be liquid, or soluble in water. For a land animal a perfectly dry material has no taste. To stimulate the sense of smell substances must be soluble, and for land animals they must also be volatile.

Water Land organisms often seem to be able to find water. Plant roots grow towards water. Amphibia, such as frogs, may often be found far from ponds at other times of the year but always seem to be able to find ponds at mating time.

Pressure Animals are sensitive to pressure changes both in water and on land. Higher animals possess *ears*, which are devices for sensing the variations in frequency of pressure changes, which are called sound. They thus provide another medium for receiving information about the world around.

33 NERVOUS SYSTEM

The study of behaviour and the mechanisms underlying it presents the biologist with some of his most complex problems. A great deal can be explained about the links between stimulus and response but this is by no means the whole story.

In order that a stimulus may produce a response three factors must be involved. These are:

(a) A **receptor** consisting of cells which are sensitive to the stimulus,

(b) An **effector** which may be a gland or a muscle,

(c) A **connection** between receptor and effector. This is usually the nervous system, although the link may also be a chemical one, where hormones are involved.

In a mammal the nervous system consists of the **brain** and **spinal cord** together with the **peripheral nerves** (see p. 168). These consist of nervous tissue which is made up of two kinds of cells:

(a) **Neurons** These are cells which can conduct signals. Such cells vary a great deal in size and form but they all share a number of common features:

  They possess **dendrites**. These are delicate twiglike outgrowths of the cell. They are in close contact either with sense cells or with parts of other nerve cells, and it is here that a nerve impulse begins.

  They possess one or more outgrowths which are called **axons**. These carry impulses away from the cell, and they end either in close contact with another nerve cell or with an effector organ.

  They become specialized as conducting cells very early during the development of the mammal and, because they are so specialized, cell division cannot occur in nerve cells.

  They may continue to grow as the individual grows but there is no increase in the total number present at birth. In fact the total number decreases throughout life as some cells die and are not replaced.

Nerve cells have a limited capacity for regeneration. If the axon is severed the part separated degenerates and disappears. The stump which remains will then sometimes grow out to replace the lost part.

(b) **Neuroglial cells** These play an important part in acting as a supporting and nutritive tissue for the neurons.

The nerve impulse

The nerve impulse, which is so characteristic of nerve cells, is an electro-chemical disturbance which spreads rapidly along the neuron from one end to the other. The following important points should be remembered:

  All nerve impulses are alike.

  They are accompanied by minute changes in the electrical conductivity of the cell.

  They use up energy, so that oxygen is consumed and carbon dioxide and heat are produced.

  A nerve impulse can be induced to travel either way along a nerve fibre, although it does not normally do so.

The synapse

As far as is known, nerve cells have no other function than that of transmitting impulses. The special properties of the nervous system as a whole can be explained only in terms of the way in which these cells are connected together. The links between nerve cells are called **synapses**. At a synapse the ends of one nerve fibre are in close contact with dendrites of another cell. They may touch but there is no physical connection between the two cells other than simple contact.

A synapse has three functions.

(a) It acts as a valve. Although a signal may travel either way along the fibre it can cross the synapse in one direction only.

(b) It behaves rather like a resistance in an electrical circuit. Signals will pass from the fibre of one neuron to the next only if their frequency is high enough.

(c) A synapse is literally a junction.

e.g.

When a nerve impulse reaches the end of a nerve fibre it causes the secretion of minute amounts of a chemical substance. The higher the frequency of the impulses, the more of the substance is formed. It is unstable and is attacked by an enzyme which decomposes it rapidly, but provided its concentration is high enough for a limited period it will cause the next neuron in the chain to fire.

Reflex action

The simplest form of nervous activity is a reflex action. A reflex is a simple, unlearned response to a stimulus. The following are examples of reflex action.

1 *Withdrawal reflexes*. These occur in response to any unpleasant or painful stimulus. The flexor muscles of the limbs are involved.

2 *Extensor reflexes*. These involve the extensor muscles and they play an important part in the maintenance of posture (see diagram on p. 167).

3 *Visceral reflexes*. 'Viscera' means all the internal organs, and all these reflexes refer to the smooth muscles of the body (see p. 196) and to glands.

Examples:

Erection of hairs on the skin as a result of cold or fear.

Constriction of the pupil of the eye in bright light.

Flushing or paling of the skin.

Sweating.

Salivating when food enters the mouth.

Posture reflexes play an important part in maintaining the body upright. Any tendency for the limb to collapse is the starting point for a series of events which end in stimulation of the extensor muscle which, by its contraction, prevents collapse.

The knee jerk, which can be elicited by a sharp tap just below the knee when legs are crossed, is used by doctors as a test for the general health of the nervous system. The tap stimulates a large number of pressure sense cells in the tendon, thus causing a vigorous contraction in the extensor and a jerk of the lower leg.

A **reflex arc** is the nerve pathway involved in a reflex action. It consists of at least two neurons linked by a synapse in the brain or spinal cord. Not all reflex actions are brought about by the nervous system. When food passes from the stomach into the duodenum it causes the secretion of pancreatic juice. The acid food stimulates the wall of the duodenum, which produces a hormone, secretin, which passes into the blood. When this reaches the pancreas it causes the pancreas to produce its secretion. (See also p. 107.)

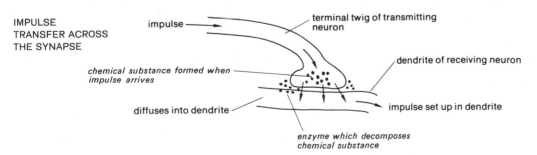

IMPULSE TRANSFER ACROSS THE SYNAPSE

impulse →

terminal twig of transmitting neuron

chemical substance formed when impulse arrives

dendrite of receiving neuron

diffuses into dendrite

impulse set up in dendrite

enzyme which decomposes chemical substance

A sensory neuron (conducts impulses from sensory cells in skin and muscle to the spinal cord)

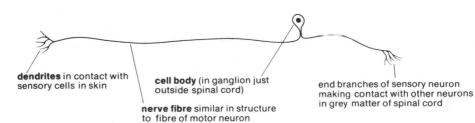

dendrites in contact with sensory cells in skin

cell body (in ganglion just outside spinal cord)

nerve fibre similar in structure to fibre of motor neuron

end branches of sensory neuron making contact with other neurons in grey matter of spinal cord

A motor neuron (transmits impulses from the spinal cord to a muscle or gland)

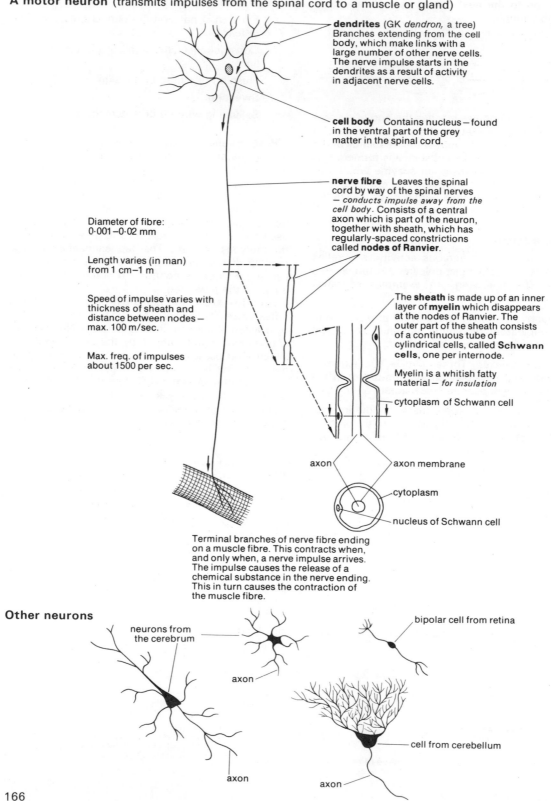

dendrites (GK *dendron,* a tree) Branches extending from the cell body, which make links with a large number of other nerve cells. The nerve impulse starts in the dendrites as a result of activity in adjacent nerve cells.

cell body Contains nucleus — found in the ventral part of the grey matter in the spinal cord.

nerve fibre Leaves the spinal cord by way of the spinal nerves — *conducts impulse away from the cell body.* Consists of a central axon which is part of the neuron, together with sheath, which has regularly-spaced constrictions called **nodes of Ranvier.**

Diameter of fibre: 0·001 – 0·02 mm

Length varies (in man) from 1 cm – 1 m

Speed of impulse varies with thickness of sheath and distance between nodes — max. 100 m/sec.

Max. freq. of impulses about 1500 per sec.

The **sheath** is made up of an inner layer of **myelin** which disappears at the nodes of Ranvier. The outer part of the sheath consists of a continuous tube of cylindrical cells, called **Schwann cells**, one per internode.

Myelin is a whitish fatty material — *for insulation*

cytoplasm of Schwann cell

axon

axon membrane

cytoplasm

nucleus of Schwann cell

Terminal branches of nerve fibre ending on a muscle fibre. This contracts when, and only when, a nerve impulse arrives. The impulse causes the release of a chemical substance in the nerve ending. This in turn causes the contraction of the muscle fibre.

Other neurons

neurons from the cerebrum

axon

bipolar cell from retina

axon

cell from cerebellum

axon

axon

Posture reflex
(involving two neurons)

SECTION THROUGH SPINAL CORD

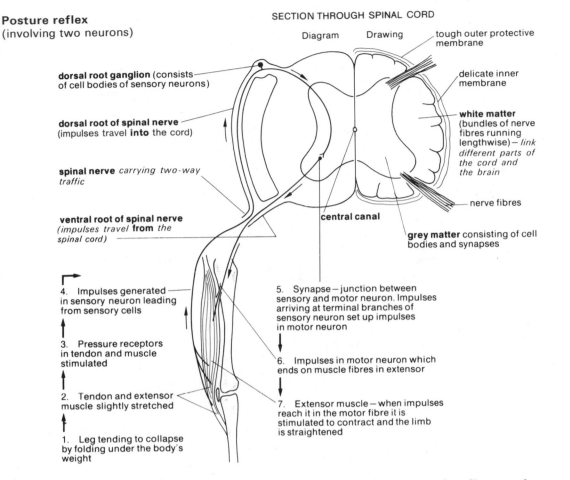

Diagram   Drawing

tough outer protective membrane

dorsal root ganglion (consists of cell bodies of sensory neurons)

delicate inner membrane

dorsal root of spinal nerve (impulses travel **into** the cord)

white matter (bundles of nerve fibres running lengthwise) – *link different parts of the cord and the brain*

spinal nerve *carrying two-way traffic*

nerve fibres

ventral root of spinal nerve *(impulses travel **from** the spinal cord)*

central canal

grey matter consisting of cell bodies and synapses

4. Impulses generated in sensory neuron leading from sensory cells

5. Synapse – junction between sensory and motor neuron. Impulses arriving at terminal branches of sensory neuron set up impulses in motor neuron

3. Pressure receptors in tendon and muscle stimulated

6. Impulses in motor neuron which ends on muscle fibres in extensor

2. Tendon and extensor muscle slightly stretched

7. Extensor muscle – when impulses reach it in the motor fibre it is stimulated to contract and the limb is straightened

1. Leg tending to collapse by folding under the body's weight

More complex reflex actions require more complex circuitry. For example, some reflex actions can be caused by more than one sort of stimulus. Breathing rate is altered by many different factors, for example, by skin temperature, the amount of carbon dioxide in the blood, the degree to which the lungs are stretched, the degree of body activity, fear and excitement. In this case there are several sensory pathways all sharing the same motor path.

The importance of reflex actions

Mammals, like many other animals, are capable of a variety of complex activities as they interact with the constantly changing world around them. Much of the activity however is purely routine, and can therefore be built to operate quite automatically and independently without the need to involve the whole organism. The reflex response to a particular stimulus is constant and predictable and therefore no 'decisions' have to be made. The complete auto-

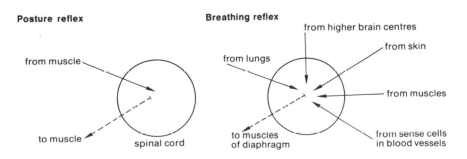

Posture reflex

from muscle

to muscle   spinal cord

Breathing reflex

from higher brain centres

from lungs

from skin

from muscles

to muscles of diaphragm

from sense cells in blood vessels

mation of all the protective and avoiding reactions, and of the mechanisms for internal regulation (homeostasis—see p. 90), leaves the higher centres of the nervous system free to deal with the more complex problems involved in coping successfully with the environment.

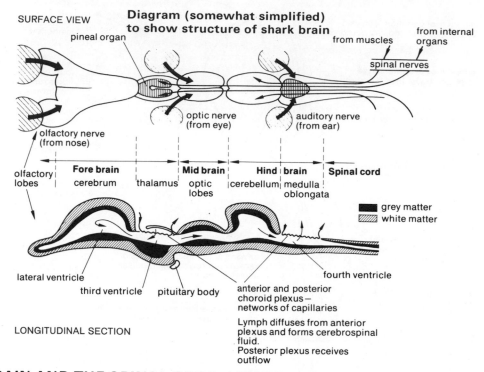

Diagram (somewhat simplified) to show structure of shark brain

LONGITUDINAL SECTION

Lymph diffuses from anterior plexus and forms cerebrospinal fluid.
Posterior plexus receives outflow

THE BRAIN AND THE SPINAL CORD (CENTRAL NERVOUS SYSTEM)

The structure of the central nervous system can be more easily understood in a simpler vertebrate than a mammal (see diagram). The brain and spinal cord together form a thick walled tube which consists of three parts :

(a) **An inner canal**. This is the central canal. In the brain it widens to form a series of spaces called ventricles. Central canal and ventricles contain cerebro-spinal fluid, which is similar to lymph (see p. 149). It undergoes a slow circulation in the ventricles, acting as a nutritive medium for the cells bordering the ventricles.

(b) **Grey matter**. This lies immediately external to the tube and consists of the cell bodies of neurons and their connections with each other. It can be thought of as an extremely complex switchboard. Its upper half contains cells which receive impulses from sensory cells transmitted by sensory neurons. The lower half contains cells whose fibres end among muscle and gland cells (i.e. motor neurons).

(c) **White matter**. This forms an outer sheath of nerve fibres running lengthwise along the brain and spinal cord, linking the different parts together. The white colour is due to the myelin which forms the insulation around the separate nerve fibres.

Paired spinal nerves leave the spinal cord at regular intervals. A typical spinal nerve contains both sensory and motor fibres surrounded by connective tissue. There are paired cranial nerves also, but their arrangement is rather specialized and need not concern us here. The brain is very much larger than the

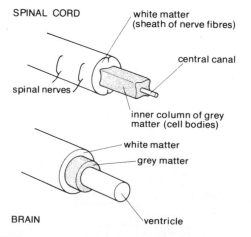

spinal cord. Its sensory input is also much greater, receiving fibres as it does from the special sense organs in the head.

The spinal cord has three main functions:

(i) It contains the circuitry, in the form of reflex arcs, for all those reflex actions which involve the muscles or glands of the body, as distinct from the head.

(ii) It provides connections between reflex arcs, both along and across the cord, so that a limited degree of co-ordination is possible.

(iii) It acts as a relay station for impulses passing between the brain and the sensory and motor apparatus of the body.

The brain also contains reflex arcs, those involved in reflex actions which occur in the head region. It also has an input from five main sources into its five main divisions (see diagram). In this respect it differs from the cord, for while both spinal cord and brain appear to be organized on a regional basis, the different regions of the cord receive impulses from similar sense cells in different parts of the body, whereas the different regions of the brain receive impulses from different sorts of sensory apparatus. Each of the five centres receives signals (information) ranging over a wide field. For example, the impulses passing along the thousands of fibres in the optic nerve to the mid-brain are 'information' about the whole of the visual field covered by the eye. The mid-brain therefore has the additional

The mammalian brain (rabbit)

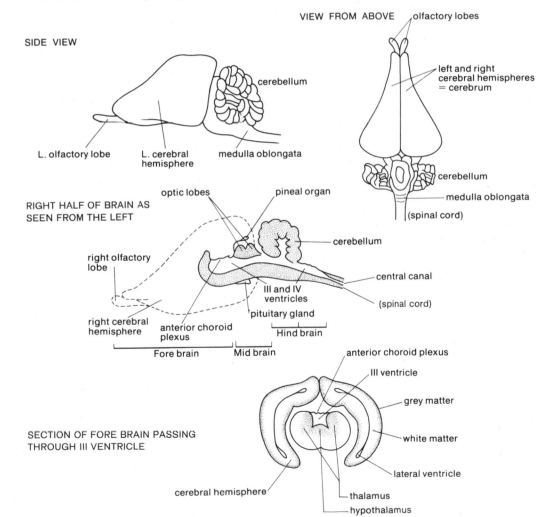

169

function of acting as a 'data processing' system. The co-ordination of this information requires an enormous amount of interconnection between the neurons which receive these impulses and this, in turn, means an increase in the number of connecting neurons in the brain. The third function of the brain provides for a limited amount of co-ordination between the different regions. The response of an animal to a particular stimulus, say, the smell of food, will depend partly on its internal state, whether it has eaten recently or not, and upon stimuli acting on its other sense organs. In the lower vertebrates this function of co-ordination is the task of the mid-brain, which receives nerve fibres from all the other brain centres.

The mammalian brain is built on the same plan as that of the shark, although it differs in a number of important ways:

1 It is much larger in relation to body size.

2 This increase in relative size is due to the proportionately greater size of the cerebrum and the cerebellum.

3 In these two regions the positions of grey and white matter are reversed. The grey matter is spread on the surface in the form of a thin sheet. In some mammals its area is further increased by infolding.

4 There are extra connections between the two halves of the brain.

5 The optic lobes, which are paired in other vertebrates, are further subdivided in mammals to form lobes called the corpora quadrigemina.

These differences are a reflection of the fact that there is much more centralization of control in the mammal as compared with other vertebrates. In lower animals only sensory impulses from organs of taste and smell end in the cerebrum. In the mammal, sensory fibres lead to the cerebrum from all the special sense organs as well as from the skin and the muscles. Only a few fibres lead from the eye to the optic lobes and they are concerned only with reflex actions involving the eye. Removal of or total damage to the cerebrum in a mammal results in part or total impairment of smell, vision and hearing. Furthermore, the mammal effectively becomes a mere automaton, showing none of the behaviour usually displayed by a healthy intelligent animal.

A great deal is known about brain function in mammals and the main facts are summarized below.

Olfactory lobes

Sensory nerve fibres from the organ of smell end here and link up with other neurons which terminate in the cerebral hemispheres.

Cerebrum (cerebral hemispheres)

Many specific areas on the surface are concerned with vision, hearing, touch, smell. Damage to some regions causes paralysis of the muscles. Large parts of the cerebrum however seem to have no function at all. As a whole the cerebrum is associated with all those elements of behaviour which distinguish one animal from another.

Thalamus

This is chiefly a relay station between the cerebrum and the rest of the brain.

Hypothalamus

A reflex control centre for many internal control mechanisms, for example, temperature regulation, water balance, sugar level of the blood. Many activities which may be described as instinctive are controlled here. Stimulation of various parts of the hypothalamus in experimental animals by means of implanted electrodes causes complicated activities associated with mating, feeding, nestbuilding, or preparation for sleep.

Mid-brain

Mostly forms a reflex centre. The upper part is involved when the pupil of the eye contracts in bright light. The lower part is a reflex centre for balancing reflexes.

Cerebellum

This is a large reflex centre for the co-ordination of muscle activity. It has both sensory and motor connections with the skeletal muscles of the body.

Medulla oblongata

This is the centre where many complex reflexes are controlled. Some examples are: reflex control of breathing, swallowing, salivation, chewing, coughing, sneezing and blinking.

Many fibres passing to and from the brain cross over in the spinal cord. Damage to parts on one side of the brain causes paralysis or loss of sensation on the opposite side of the body.

THE HUMAN BRAIN

A great deal is known about the functions of various parts of the human brain. This is because human beings are able to report back on their experiences. Experimental stimulation of parts of the surface of the cerebral hemispheres in a rabbit may give no information at all. If the same exercise is carried out on a man, as it may be if he is being operated on, under local anaesthetic, for the removal of a tumour, he is able to describe what happens to him. Much of the knowledge we have about the brain is summarized on the diagram.

The higher functions of the brain

1 Instinct

Many of the complex activities which animals display seem to originate from within, rather than as a response to a particular stimulus. They often appear to be carried out as if with a definite aim or purpose. Some of these activities have been studied in great detail and they appear in each case to be essentially the same for all members of the same species. They are displayed where no opportunity for learning the pattern has been possible. Such forms of activity are said to be instinctive. Some examples are courting, mating and nesting behaviour, the mutual grooming practised by monkeys, the stalking and hunting of cats, and the behaviour of animals which occupy a particular territory which they patrol and guard. The pattern of activity seems to be built in, although not entirely. In nest building, for example, not all the factors can be allowed for in advance. The animal

cannot know what sites are available, nor the availability of nesting materials. The internal 'machinery' may depend upon external stimuli acting as triggers to set it going, but the basic circuitry seems to be determined and inherited just as eye colour is. To borrow a phrase from the computer industry, the system is supplied with its circuitry and its programme complete and ready for use. It is thought that each pattern of instinctive behaviour is made up of a complex of reflex actions. The operation of one of these, triggered by some external stimulus, itself sets off one or more reflex actions forming the next step in the chain. The completion of the whole series of reflex actions depends on appropriate stimuli being presented at the right time, each of them acting like a switch closing a circuit. Each stimulus can be thought of as functioning like the signals in a railway system, the train waiting for a green before it can enter the next section of the track. It is certainly known that there are complex networks of neurons

View of human brain from the left side (showing left cerebral hemisphere, cerebellum, and medulla oblongata)

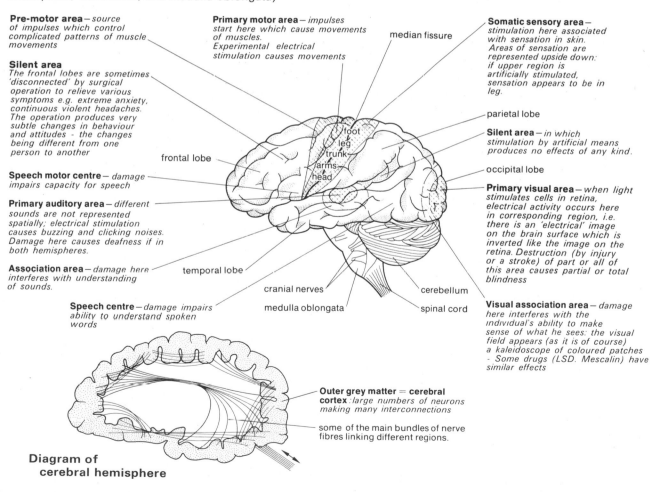

Pre-motor area — *source of impulses which control complicated patterns of muscle movements*

Silent area
The frontal lobes are sometimes 'disconnected' by surgical operation to relieve various symptoms e.g. extreme anxiety, continuous violent headaches. The operation produces very subtle changes in behaviour and attitudes - the changes being different from one person to another

Speech motor centre — *damage impairs capacity for speech*

Primary auditory area — *different sounds are not represented spatially; electrical stimulation causes buzzing and clicking noises. Damage here causes deafness if in both hemispheres.*

Association area — *damage here interferes with understanding of sounds.*

Speech centre — *damage impairs ability to understand spoken words*

Primary motor area — *impulses start here which cause movements of muscles. Experimental electrical stimulation causes movements*

median fissure

foot
leg
trunk
arms
head

frontal lobe

temporal lobe

cranial nerves

medulla oblongata

Somatic sensory area — *stimulation here associated with sensation in skin. Areas of sensation are represented upside down: if upper region is artificially stimulated, sensation appears to be in leg.*

parietal lobe

Silent area — *in which stimulation by artificial means produces no effects of any kind.*

occipital lobe

Primary visual area — *when light stimulates cells in retina, electrical activity occurs here in corresponding region, i.e. there is an 'electrical' image on the brain surface which is inverted like the image on the retina. Destruction (by injury or a stroke) of part or all of this area causes partial or total blindness*

cerebellum

spinal cord

Visual association area — *damage here interferes with the individual's ability to make sense of what he sees: the visual field appears (as it is of course) a kaleidoscope of coloured patches - Some drugs (LSD. Mescalin) have similar effects*

Outer grey matter = cerebral cortex :*large numbers of neurons making many interconnections*

some of the main bundles of nerve fibres linking different regions.

Diagram of cerebral hemisphere

in parts of the brain which can be stimulated by artificial means to cause complex activity in animals.

2 Learning

Mammals are not entirely like automatic machines built to run through fixed patterns of behaviour, however complex. They interact with the environment in a very positive way. One has only to observe the active curiosity of a young kitten to realise this. All animals are able to profit by experience to some extent, and their behaviour becomes modified as a result of learning. Mammals are particularly efficient at adapting their behaviour in this way. To use another metaphor, they behave like machines which are not completely programmed. Some of the programme is fed into the system as it goes along.

The conditioned reflex

At the beginning of the century Pavlov, the famous Russian biologist, carried out his classic work on conditioned reflexes. He discovered that the flow of saliva in dogs, normally caused by the taste of food in the mouth, could also be induced by a different stimulus, such as the ringing of a bell just before food was presented, if it was repeated enough times. Once the conditioned reflex had been established, i.e. reflex salivation to the stimulus of the bell, it could be abolished if the bell was sounded enough times without the presentation of food. Reflexes of this kind only occur in regions where complex nerve pathways already exist and also where they are closely interconnected. It is not possible to set up conditioned reflexes which pass only through the spinal cord; for example, the knee jerk cannot be caused by the sound of a bell. In man the great majority of reflexes are probably conditioned, for example, salivating at the sight of a juicy steak. A mother who tries to get her baby to take codliver oil by adding orange juice to it often creates in the baby a conditioned reflex which shows itself as a dislike of orange juice as well as of codliver oil. Many of our irrational fears, of spiders, or of the dark, arise as conditioned responses.

Learning by reinforcement

Any action on the part of an animal which produces a beneficial result is much more likely to occur again. Taking advantage of this fact it has been found possible to build up new patterns of behaviour in experimental animals by reinforcing (for example, by giving food), any random action which contributes to the desired pattern. The circus animal trainer has long known the value of reinforcement or reward in the training of circus animals. Much learning takes place in this way in human beings. When very young, the reward for appropriate behaviour may be the approval of the mother. Later in life it may be the approval of others, or the personal satisfaction of reaching a particular goal.

At present very little is known about the actual mechanisms involved in learning. Clearly, no structural rearrangement of the nerve cells can be involved. It is thought that the ease with which impulses can cross particular synapses may be increased, although exactly how this happens is uncertain. There is some evidence to suggest that chemical changes may actually occur within the neurons, as a result of impulses passing along them frequently.

The brain as a computer

More and more operations which were once thought to be possible only for man to do can now be carried out by complex machinery, the heart of which is a computer. The following is a typical example. A large airliner is normally brought in to land at the airport by an experienced pilot. He receives information of all kinds, from various instruments in the aircraft, from ground control, and from what he sees outside the plane. The whole process is highly complex and requires long training. It is also possible to land an aircraft completely automatically. In this case a computer takes the place of the pilot's brain. The information is fed to it by electrical signals and it controls the aircraft by means of electrical impulses directly operating the control system of the aircraft. Because they work so much faster, computers are more efficient for many purposes than brains, although the human brain is very much more versatile than any computer and is likely to remain so. Nevertheless, every time a piece of apparatus is designed which serves as a model of human brain function, however limited, another step is taken towards an understanding of the complex circuitry of the brain.

Measure your own reaction time. A simple way of doing this is as follows. Get one of your friends to hold a metre rule suspended by one end. With your hand resting on the edge of the bench, position finger and thumb on either side of the rule without actually touching it. Note the mark on the ruler. Your colleague releases the ruler without warning. You stop it as soon as it moves by pinching it between finger and thumb. Read off the mark. The difference in the two readings is a measure of the distance the ruler dropped before you reacted. If you don't study physics, get one of your friends who does to convert this distance into time. Compare your own reaction times at different periods of the day. Compare them with those of your colleagues. You can work out an approximate length for the nerve pathway involved and you know the speed of a nerve impulse. How long should an impulse take to travel from your eye to the muscles involved? Why do you think this time differs from your recorded reaction time? What could account for the difference?

34 CHEMICAL CO-ORDINATION

Much of the body's internal activity is subject to chemical control through the secretions of certain organs called ductless, or endocrine glands (endocrine, literally, inward secretion). Their secretions do not pass out by means of a duct but instead enter directly into the blood circulating through the organ. The substances secreted are called hormones (Gk. *hormoein*, to excite), and they produce effects in other organs and tissues. Chemical signals of this kind are necessarily slower than nerve impulses, since they can travel only at the speed of circulating blood. Just as the nervous system acts rather like a telephone system of 'private lines', so the endocrine system can be compared with a public address system. Endocrine organs play a very important role in the regulation and coordination of body activity. In some cases their secretions reinforce the action of the nervous system (e.g. adrenalin); in other cases reflex actions may have a chemical 'middle man' instead of a nervous reflex arc (e.g. secretin— see p. 107). Hormones regulate continuous or long term processes, and are usually stable compounds capable of exerting long-lasting effects. The system of endocrine organs functions as a coordinated whole, the secretions of one gland having effects on the activity of others. This provides a further example of homeostasis (see p. 90).

The **pituitary gland** is situated on the under side of the brain close to the hypothalamus and it is sometimes called the master gland because of its effect on the other glands. A number of pituitary hormones have been isolated and their chemical structure investigated. Most of them appear to be proteins. Among the widespread effects of pituitary hormones the following should be noted:

1   The activity of the thyroid gland is increased.

2   The proper functioning of the adrenal cortex depends upon pituitary secretion. In its absence degeneration of adrenal cortex occurs.

3   The ovary and testis are stimulated at puberty to produce sex cells and their own particular hormones.

4   Lactation depends on pituitary hormone.

5   Growth is retarded if the pituitary is underactive. If it is overactive growth is excessive, particularly in bones.

6   The blood sugar level is raised.

7   Blood pressure is raised.

8   The retention of water by the kidney is increased.

The **thyroid gland** is situated in the throat, lying in front of the larynx. It secretes an iodine-containing compound, thyroxine, which regulates the rate of body metabolism. Underactivity of the thyroid results in a general slowing down of the rate of metabolism. In the young this means that growth, both physical and mental, is retarded. Serious thyroid deficiency in children gives rise to a condition known as **cretinism**. Cretins do not live long and seldom develop the ability to look after themselves in even the most elementary way. Nowadays deficiency can be detected at a very early stage and its ill effects avoided by the administration of thyroid extract. Adults who suffer from thyroid deficiency are generally lethargic and dull. Sometimes fluid accumulates in the lymph spaces under the skin, causing swelling. This condition is known as **myxoedema**. Thyroid deficiency may be due to a diet deficient in iodine. The gland in this case enlarges to form a **goitre**. This condition, known as endopthalmic goitre, is much rarer than used to be the case, because in regions where iodine does not normally occur in sufficient amounts iodine is added to food. Table salt, for example, may be iodized. Overactivity of the thyroid causes a general rise in metabolic rate. Growth in the young is speeded up. Adults become nervous, tense and irritable. Treatment usually involves removal of part of the gland, which often enlarges to form a goitre. The goitre is often accompanied by the appearance of bulging eyes and is referred to as exophthalmic goitre.

The **parathyroid gland** consists of small patches of tissue embedded in the substance of the thyroid. Its secretion is essential in the control of calcium balance.

The **adrenal gland** lies immediately above the kidney in man. It is really two glands, the outer cortex and inner medulla.

The **cortex** produces a secretion which regulates the metabolism of sodium in the body. Addison's disease, caused by bacterial infection of the adrenal gland, leads to considerable loss of sodium from the body and sufferers exhibit a craving for salt, which they consume in large quantities.

The **medulla** produces a hormone called **adrenalin**. It is unstable and is rapidly destroyed so that its effects are not long-lasting. Its action on the body is similar to that produced by the nervous system supplying the viscera, i.e. it prepares the body for emergency action, for fight or flight. Its effects are widespread. It causes:

Increased heart and breathing rate.

Raised blood sugar level.

Redistribution of blood away from the skin and gut capillary networks to the heart and skeletal muscles.

Inhibition of digestive activity.

Erection of hairs on the skin.

Dilation of the pupil.

The **duodenum** is known to produce at least three hormones. Secretin was the first hormone to be identified as such. It is produced when acid food enters the duodenum from the stomach. It produces its effects on the pancreas, causing it to secrete its digestive juice.

The **pancreas** contains patches of tissue, the Islets of Langerhans, which produce a hormone called **insulin**. Insulin causes the lowering of the blood sugar level by its action on the tissues of the liver. If the pancreas fails to produce insulin the blood sugar rises and sugar is excreted by the kidney.

This condition is called diabetes mellitus, and can be countered by injection of insulin.

The **gonads** (ovary and testis) have been dealt with elsewhere (p. 230).

Oestrogen stimulates the development of egg cells and the development of secondary sexual characters. **Progesterone**, the pregnancy hormone, is produced by the empty follicle after the egg has been shed, and prevents further ovulation. **Testosterone** stimulates sperm development and the appearance of the secondary sexual characteristics in the male.

Two other organs in the body are thought to produce hormones:

The **thymus gland** is active in young mammals, gradually disappearing in the adult. It is the source of antibodies in the very young mammal. Later this function is taken over by the lymph glands under the stimulating action of a hormone produced by the thymus.

The **pineal organ** was believed by Descartes to be the seat of the soul. In primitive vertebrates it acts as an eye. In mammals it is believed to play some part in regulating the activity of the ovary and testis.

The endocrine organs

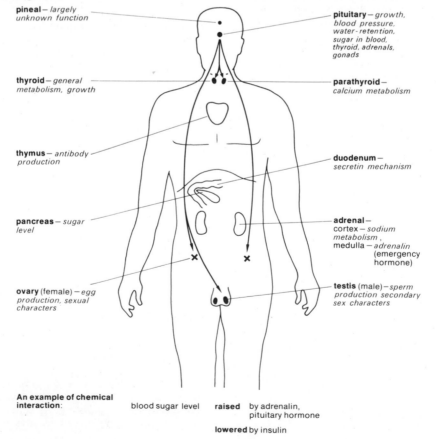

pineal — *largely unknown function*

pituitary — *growth, blood pressure, water-retention, sugar in blood, thyroid, adrenals, gonads*

thyroid — *general metabolism, growth*

parathyroid — *calcium metabolism*

thymus — *antibody production*

duodenum — *secretin mechanism*

pancreas — *sugar level*

adrenal — cortex — *sodium metabolism*, medulla — *adrenalin (emergency hormone)*

ovary (female) — *egg production, sexual characters*

testis (male) — *sperm production secondary sex characters*

An example of chemical interaction: blood sugar level   **raised** by adrenalin, pituitary hormone

lowered by insulin

35 MAMMALIAN SENSORY RECEPTORS

In mammals the ability to respond to a stimulus depends upon the existence of specialized cells called **receptors**. These are cells which, when suitably stimulated, generate and release small amounts of electrical energy sufficient to trigger off nerve impulses in the nerve endings of sensory neurons associated with them. There appear to be six types of sensory cell, according to the particular type of stimulus to which they respond. The stimuli are those caused by:

(a) **Mechanical stress or distortion** (Cells which react to this type of stimulus are found in the skin as pressure receptors, in the muscles and joints, and also in the ear.)

(b) **Temperature** (Temperature receptors are found in the skin.)

(c) **Pain** (Specialized receptors for pain are found in the skin.)

(d) **Smell** (Chemical receptors stimulated by smell are found in the lining of the nasal cavity.)

(e) **Taste** (Chemical receptors stimulated by taste are found in the mouth on the surface of the tongue.)

(f) **Light** (Light receptors are found in the lining layer of the eye.)

In some cases the receptor cells are distributed more or less regularly throughout the tissue in which they are found. This is the case with the receptors for touch, pain, temperature in the skin, and for the pressure receptors in the muscles and joints.

The receptors may be concentrated in particular regions: for example, the olfactory receptors are densely packed in the lining epithelium of the nasal cavity, and the taste sensitive cells are concentrated within taste buds.

Thirdly, the receptors may be concentrated and organized to form with other tissues, a specialized organ—for example, the eye or ear.

SKIN RECEPTORS

Pressure and touch

A number of different types of receptor have been identified in the skin which are sensitive to touch. Some of these are stimulated by light contact and appear to be associated with hair follicles. Others appear to be stimulated only by pressure. The general sensitivity of the skin varies from region to region, particularly in the ability to discriminate. For example, it is difficult to tell whether one is being prodded in the small of the back by two fingers, an inch apart, or by one. On the tip of the tongue it is possible to discriminate between two points of contact a millimetre apart. There is reason to believe that the touch receptors are fairly uniformly distributed over the surface of the body, so that the ability to discriminate must depend upon the number of nerve fibres per unit area. In regions such as the small of the back, which is relatively undiscriminating, the touch receptors share nerve fibres. In highly sensitive regions such as the tip of the tongue, possibly each touch receptor has its own nerve fibre.

Temperature

There are thought to be two types of temperature-sensitive receptors in the skin. One type is most active at temperatures about 30°C, the other is stimulated most effectively by temperatures of about 40°C. These correspond to cold and hot spots on the surface of the skin, which can be mapped by exploring the skin by means of cooled or heated rods (knitting needles) held close to the skin.

Pain

Special receptors are known to exist in the skin which respond to excessive stimulation, (mechanical, or heat or cold).

Pressure receptors in muscles, tendons and joints

Muscle spindle cells provide information about the degree of stretch of muscles. Tendon receptors give information about the tension set up in muscles and the sensory cells in the joints give information about the angles between bones. Signals from all three play an important part in providing information about the position of the limbs and the state of the muscles generally. In order to carry out any particular movement it is necessary to know the starting state. Such

sensory cells appear to signal information only when there is a change of state. If one lies perfectly still, for example in bed at night with the light out, after a few seconds it is impossible to tell where one's limbs are, without *moving* them. When nerve fibres leading *from* muscles are cut, so that sensation is lost, paralysis results, although the muscle may still contract by reflex stimulation.

THE SENSE OF SMELL

Less is known about the sense of smell than almost any other form of perception. In the eye, for example, all the light-sensitive cells are stimulated in basically the same way. It is the lens and the cornea which determine which cells shall be stimulated. The cells in the cochlea are all stimulated by mechanical distortion. It is the membrane upon which they stand which determines which cells shall be stimulated. In the nose there is no such 'screening' apparatus. The cells of the olfactory surface are directly stimulated by contact with the molecules of the substance which is being smelled.

Smell cannot be measured. Sound and light waves are forms of energy which can be measured. They can both be investigated objectively. Machines can be made which will emit sound of any particular frequency. It is not possible to make a machine which will produce smells. They are qualitatively different. Pictures and sounds can be converted into patterns of electrical energy and transmitted by radio, but not smells. It is not known what particular property of molecules it is that decides their smell. It has been suggested that shape may be important, that the molecules fit the sensitive surface in some way, rather as a key fits a lock. However, there is no necessary similarity between chemical properties and smell. Some substances are chemically similar but smell quite different. Others differ markedly in chemical structure but smell very much the same. It is possible to group smells to a limited extent. For example, lilac and lily of the valley are obviously more like each other in smell than either is like petrol or benzene.

There are no groupings upon which all experts are agreed. The following represents one provisional grouping :

Ethereal (fruity) ; resinous (camphor) ; fragrant (flower scent) ; burning smells ; spicy smells ; ripe cheese smells ; smells of rotting organic matter or excreta ; onion or garlic smells.

That it is possible to divide smells into groups suggests that perhaps olfactory cells can be divided into groups also. It is known that certain types of smell are detected only in certain parts of the sensory surface. In order that a substance shall have an odour it must be volatile i.e. it must vaporize. When it enters the airways of the nose it must then enter

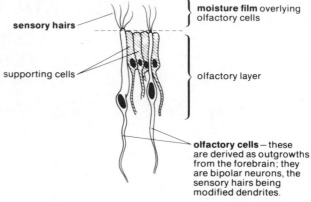

moisture film overlying olfactory cells

sensory hairs

supporting cells

olfactory layer

olfactory cells — these are derived as outgrowths from the forebrain ; they are bipolar neurons, the sensory hairs being modified dendrites.

into solution in the moisture film covering the sensory cells. The most volatile substances generally have the most powerful smells.

Among vertebrate animals the sense of smell is highly developed except among bony fishes, Amphibia and birds. These have practically no sense of smell. Smell plays an important part in the life of mammals. It is important, not only in the seeking of food and the avoidance of predators but also in mating behaviour. In man the sense of smell is easily fatigued. One soon becomes accustomed to a particular smell and then no longer notices it. The human male seems to be more sensitive to smell than the female. It is also probable that human beings have personal smells quite apart from smells which depend on particular diets or habits of hygiene.

TASTE

Taste, like smell, is a chemical sense, associated with the properties of substances in solution. Materials which are completely dry and completely insoluble have no taste. As far as is known, human beings are sensitive to four tastes only :

Sweetness is a property of sugars, although there are other substances which taste sweet, saccharin for example.

Sourness is a property of acids.

Saltiness is characteristic of the salts of strong acids.

Bitterness is a taste associated with a class of substances known as alkaloids. They are of plant origin and are often poisonous. Some, for example, quinine, are used medically.

Taste cells are found in groups called taste buds. As far as can be discovered, taste buds are of four different kinds corresponding to the four tastes, and they are mainly concentrated in the surface of the tongue in different localized regions. In animals

**Taste bud viewed
in section**

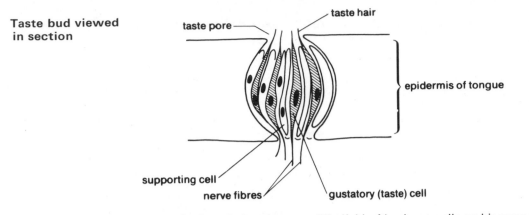

taste pore

taste hair

epidermis of tongue

supporting cell

nerve fibres

gustatory (taste) cell

which live in water they may be found elsewhere. In some fishes they may be found on the tail. In the ordinary way taste is assumed to be infinitely variable. The taste of food is in fact a compound of taste and smell. It is common knowledge that the sense of taste is blunted when one has a head cold.

Blindfold a friend, put a clip on his nose and then ask him to distinguish between a piece of apple and a piece of potato. Much of what is described as taste is often a reaction to the *feel* of food, for example the blandness of full cream or the stringy fibres in runner beans.

THE EYE

Most animals are sensitive to light, but it is only among the higher arthropods, some molluscs and the vertebrates that light serves as a medium for the transmission of information about the outside world. The structure of the eye is essentially the same in all vertebrates.

Eye reflexes

1 Blinking, caused by:
   (a) sudden movement near the eye,
   (b) foreign body entering eye,
   (c) an involuntary mechanism lubricating eye surface.

2 Accommodation. Contraction of ciliary muscle altering refractive power of lens for viewing objects at close range.

3 Constriction of pupil in bright light and dilation in dim light.

4 Secretion of tears in large amounts to wash out grit particles or insects.

The formation of images

The eye is in many respects like a camera. The light-proof case of the camera corresponds to the capsule of the eye. The film in the camera and the retina in the eye are both photosensitive. A real, inverted, diminished image is projected on to the film by means of a convex lens in the camera. In the eye there are two refracting bodies. The cornea together with the aqueous humour has the greater converging power, the lens rather less, since it is surrounded by materials which have very nearly the same refractive index as itself. Together the lens and cornea behave like a single converging (biconvex) lens. The camera can be focused for objects at different distances by actually moving the lens forward or backward, but the adjustment in the mammalian eye is made by altering the radius of curvature of the lens. A camera has a variable aperture: the iris, by altering the size of the pupil, acts in the same way.

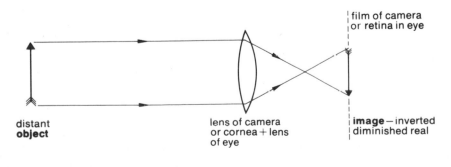

distant
object

lens of camera
or cornea + lens
of eye

film of camera
or retina in eye

image—inverted
diminished real

Vertical section through mammalian eye

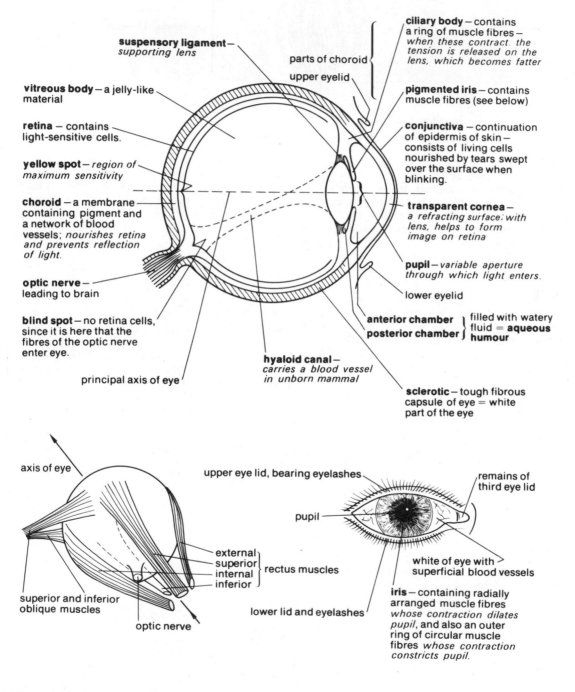

suspensory ligament — *supporting lens*

parts of choroid

upper eyelid

ciliary body — contains a ring of muscle fibres — *when these contract. the tension is released on the lens, which becomes fatter*

vitreous body — a jelly-like material

pigmented iris — contains muscle fibres (see below)

retina — contains light-sensitive cells.

conjunctiva — continuation of epidermis of skin — consists of living cells nourished by tears swept over the surface when blinking.

yellow spot — *region of maximum sensitivity*

choroid — a membrane containing pigment and a network of blood vessels; *nourishes retina and prevents reflection of light.*

transparent cornea — *a refracting surface; with lens, helps to form image on retina*

optic nerve — leading to brain

pupil — *variable aperture through which light enters.*

lower eyelid

blind spot — no retina cells, since it is here that the fibres of the optic nerve enter eye.

anterior chamber } filled with watery fluid = **aqueous humour**
posterior chamber }

principal axis of eye

hyaloid canal — *carries a blood vessel in unborn mammal*

sclerotic — tough fibrous capsule of eye = white part of the eye

axis of eye

upper eye lid, bearing eyelashes

remains of third eye lid

pupil

external
superior
internal } rectus muscles
inferior

white of eye with superficial blood vessels

superior and inferior oblique muscles

lower lid and eyelashes

optic nerve

iris — containing radially arranged muscle fibres *whose contraction dilates pupil*, and also an outer ring of circular muscle fibres *whose contraction constricts pupil.*

VIEW OF WHOLE EYE (RIGHT) FROM BEHIND AND ABOVE

FRONT VIEW OF RIGHT EYE (HUMAN)

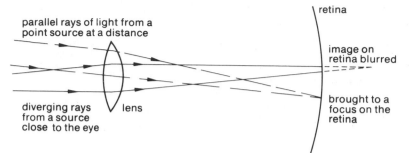

The unaccommodated eye

The eye, when it is directed towards a distant object, which means in practice any object more than about ten feet away, is said to be unaccommodated and the image formed is normally in focus on the retina. Any close object will appear blurred.

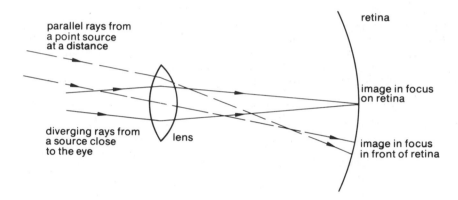

The accommodated eye

When the focus of the eye is adjusted so that near objects are clear it is said to be accommodated. Under these conditions distant objects are blurred.

In the unaccommodated eye the lens is flattened. This is because it is subject to tension exerted by the suspensory ligament. When accommodation occurs the iris and ring of ciliary muscle contract, the tension in the suspensory ligament is reduced and the lens bulges due to its natural elasticity. Its radius of curvature therefore decreases, and also its focal length.

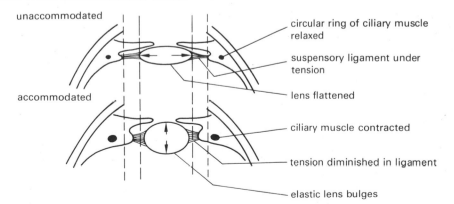

Fishes live in a medium (water) with the same optical properties as the cornea, which therefore plays no part in converging light rays. The lens is completely spherical and has much greater converging power. It cannot focus by change of shape. Focusing in a fish's eye involves physical movement back and forth of the eyeball, as it does in a camera.

179

Eye defects

1 Short sight (myopia)

A short-sighted person is unable to see objects clearly in the distance. Objects close to the eye may be seen clearly without the eye having to accommodate. This defect is due to the eyeball being too long from front to back. It can be corrected by the use of concave (diverging) lenses, except when examining objects at close range or reading small print.

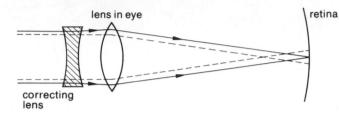

2 Long sight (hypermetropia)

A person suffering from long sight has to accommodate to see objects at a distance. For close range viewing the eye cannot accommodate sufficiently to bring objects into focus. This is caused by the eyeball being too short. It is corrected by the use of convex (converging) lenses which are worn continuously.

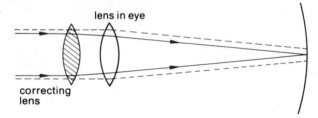

3 Astigmatism

The surface of the cornea, or more rarely the lens, is not spherical. Its radius of curvature is different in different planes. It causes dots to appear as lines. It can be corrected by using cylindrical lenses which make a correction in one plane only.

4 With increasing age the lens loses its elasticity so that accommodation becomes increasingly difficult. In this case convex lenses are employed for reading or other close work.

The retina

The retina is backed by a pigment layer which absorbs light. In nocturnal mammals the pigment layer is also a reflecting layer, which thus increases the sensitivity of the retina. The retina consists of closely-packed **light-sensitive cells**. They are activated by light energy. Although the reaction is much more rapid and sensitive, it is a photochemical effect similar to that causing the bleaching of curtains or wallpaper exposed to bright sunshine. There are two kinds of cell:

Cones are activated by bright light and possess very extensive nerve connections with the brain. They are particularly concentrated at the yellow spot which corresponds to the region of maximum visual acuity. In mammals which can discriminate colour there are thought to be three types of cone cell, each sensitive to one of the three primary colours.

Rods are sensitive to low-intensity illumination. They produce a picture of poor definition, since they are linked in groups to single nerve cells. None is present at the yellow spot. Both sorts of cells are modified neurons.

The mammalian retina is inverted. This means that the nervous connections from the rods and cones lie in front of these cells. Light has to pass through a network of nerve cells and fibres before it reaches the rods and cones. Where the nerve fibres converge to form the optic nerve there are no light-sensitive cells.

In some mammals, including man, the two eyes point in the same direction. They are then able to be used as range-finders. For distant objects their visual axes are parallel. The extent to which they converge to view near objects is a measure of the distance of such objects.

THE EAR

The ear consists of two parts which have distinct and different functions.

1 The Membranous Labyrinth

This is found in all vertebrates. It consists of the semi-circular canals, the utricle and the saccule. Its function is concerned with the position and movement of the head in space.

2 The Lateral Line System (found in fishes)

It consists of a number of fluid-filled canals running below the surface of the skin, opening to the surface at intervals. It contains sensory cells which are stimulated by pressure changes in the surrounding water.

The Cochlea (found in land vertebrates)
Associated with the cochlea is a middle ear chamber containing one or more bones, an eardrum, and, in mammals, an external ear. The cochlea is a device for analysing pressure waves (sound) and their variations in frequency (pitch).

Both the lateral line system and the cochlea are continuous with the labyrinth and are, in fact, outgrowths from it.

The mechanism of hearing (in the mammal)

Sound waves pass into the **external ear tube** and set the **ear drum** vibrating. Like the diaphragm in a microphone or loudspeaker, the eardrum has no natural frequency of vibration, that is, it is aperiodic. It responds with forced vibrations to any frequencies within its overall range. Its vibrations are communicated to the three bones in the **middle ear chamber**, and they set the **oval window** vibrating. The oval window is much smaller than the ear drum and it is in contact with fluid on its other surface. The bones therefore not only transmit, but also gear down the amplitude of the vibrations to the oval window. The fluid in the inner ear is relatively incompressible, so the vibrations imparted to it by the oval window are transmitted to the **round window**. The membrane in the **cochlea** upon which the sensory cells stand is narrowest at the base of the cochlea, getting progressively wider towards its apex. It is a membrane whose natural frequency of resonance varies along its length. A stretched wire, when plucked, vibrates at a particular frequency, its pitch, and it cannot be made to vibrate at other than that frequency. The cochlear membrane behaves like a number of stretched strings at different tensions. It may be likened to a miniature piano frame with correctly tensioned strings, embedded within a membrane. When any particular region of the membrane vibrates the sensory cells in that region are distorted and therefore set up nerve impulses in the nerve endings associated with them. The cochlea behaves like a wave analyser, transforming particular sounds into patterns of nerve impulses relayed to the brain in the auditory nerve. The human ear is sensitive to vibrations over a range of from 20 to 20 000 cycles per second. Some animals are sensitive to frequencies higher than this. Below the lower limit vibrations may be *felt* if they are large enough. With increasing age the ability to hear high notes diminishes, due to degeneration of the basal part of the cochlear membrane.

The membranous labyrinth

The cells in the cochlea are stimulated by the mechanical distortion of their hair-like processes, brought about by a vibrating membrane. The upper portion of the labyrinth contains patches of sensory cells which are also stimulated by mechanical distortion, but the mechanism which brings it about is rather different. Groups of cells are found in the **utricle, saccule** and in the **ampullae of the semi-circular canals**. Their hair-like processes, in the semi-circular canals, are embedded in masses of gelatinous material containing granules of calcium carbonate. Movements of the head in space cause displacement of the gelatinous material and therefore stimulation of the sensory cells. The arrangement of three semi-circular canals in the three planes of space means that different combinations of sensory cells are stimulated by movement in different directions. The utricle, and also the saccule, are lined with patches of sensory cells whose filaments are stimulated, in each case, by the weight of a small calcareous structure, the otolith, which is free to move under the action of gravity. The utricle and saccule therefore signal the position of the head with respect to gravity.

Section through the human ear

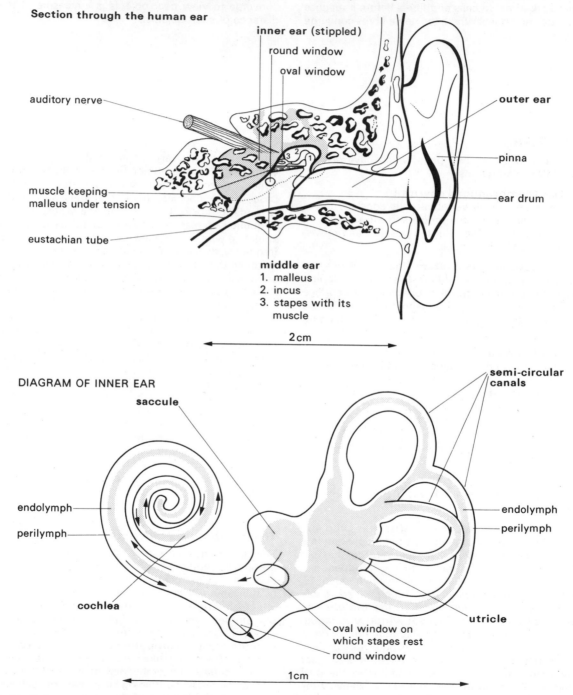

inner ear (stippled)
round window
oval window
auditory nerve
outer ear
pinna
muscle keeping malleus under tension
ear drum
eustachian tube

middle ear
1. malleus
2. incus
3. stapes with its muscle

2 cm

DIAGRAM OF INNER EAR

saccule
semi-circular canals
endolymph
perilymph
endolymph
perilymph
cochlea
utricle
oval window on which stapes rest
round window

1 cm

A transverse section through the cochlea

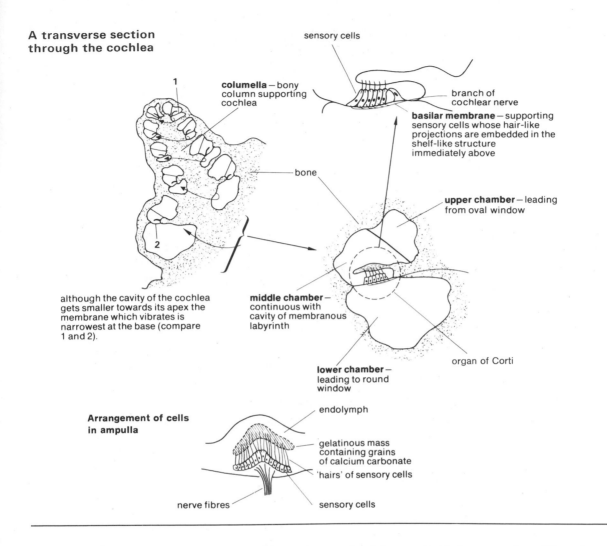

sensory cells

columella — bony column supporting cochlea

branch of cochlear nerve

basilar membrane — supporting sensory cells whose hair-like projections are embedded in the shelf-like structure immediately above

bone

upper chamber — leading from oval window

although the cavity of the cochlea gets smaller towards its apex the membrane which vibrates is narrowest at the base (compare 1 and 2).

middle chamber — continuous with cavity of membranous labyrinth

lower chamber — leading to round window

organ of Corti

Arrangement of cells in ampulla

endolymph

gelatinous mass containing grains of calcium carbonate

'hairs' of sensory cells

nerve fibres

sensory cells

1　Obtain an ox eye from the butcher. Dissect it and identify the different parts.

2　Make a pin-hole in a thin piece of cardboard. Hold the cardboard about an inch from your eye so that you can look through it. Next hold a pin by its pointed end in your other hand and raise it head uppermost between your eye and the cardboard. An upside down, enlarged shadow of the pin will be seen, apparently on the other side of the cardboard. How might this be explained?

3　Cover your left eye and gaze at the cross with the right eye. Hold the page about 22 cm from your eye. What has happened to the circle?

4　Test your ability to distinguish taste. Work in pairs. The individual being tested should be blindfolded and his nose should be clipped. Test with pieces of carrot, potato and apple. Try to distinguish between butter and margarine.

5　Working in pairs, explore different parts of the skin. The individual being tested must be blindfolded. Test ability to distinguish different stimuli applied together with a pair of dividers brought gently against the skin. Compare reactions for different distances between the tips of the dividers. Test sensitivity of lips, tongue, tips of the fingers and small of the back. Contact sensitivity can be explored with a wisp of cotton wool. Hot and cold responses can be investigated with steel knitting needles kept in warm water and in ice.

+

36 PLANT RESPONSES

Plants are able to react to light, gravity, temperature, chemical substances, water and touch. Their responses are of two basic types:

(a) **Tropic responses**: responses involving movement by part of a plant in which the direction of movement is determined by the direction of the stimulus.

(b) **Nastic responses**: movements by the plant, or a part of the plant, in response to a stimulus which is non-directional, e.g. temperature change, or movements which are independent of the direction of the stimulus.

Tropisms are either:

Positive: movement is towards the source of the stimulus, or
Negative: movement is away from the source of the stimulus.

When a plant stem turns towards light it is exhibiting **positive phototropism**: when a stem grows upwards it exhibits **negative geotropism**.

Plant responses may be due to different rates of growth or, less often, to differences in turgidity. Plant responses are slow when compared with animal responses.

THE EFFECT OF LIGHT

The more important effects of light on plants are as follows:

(a) It is necessary for photosynthesis to take place.

(b) Chlorophyll is not formed in the absence of light.

(c) The time at which flowering occurs is determined by the amount and duration of the light falling on the plant.

(d) It affects the rate at which growth occurs.

(e) Stems, leaves and, to a lesser extent, roots react directionally to light. Stems usually grow towards light, roots grow away from light, and leaves arrange themselves in a plane at right angles to the direction of light.

(f) The so-called sleep movements which some plants show may be due to the presence, or absence, of light.

The phototropic response of the stem

Experiment:

Three groups of cress or mustard seeds are germinated under identical conditions. When the seedlings are an inch or two high the three groups are treated as follows (care being taken to see that the seedlings are supplied with adequate air and water):

Group I are placed in complete darkness.

Group II are exposed to continuous light from all sides.

Group III are placed in a box so that light reaches them from one side only.

The seeds are then inspected at intervals for the next two or three days.

The results show that stems grow faster in the dark. Seedlings in group I have grown taller than those in group II. In group III the stems have grown more on the side where there is less light and have therefore turned towards the light. The stem is positively phototropic.

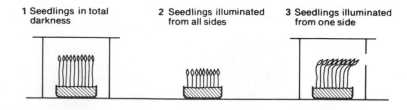

1 Seedlings in total darkness
2 Seedlings illuminated from all sides
3 Seedlings illuminated from one side

THE RESPONSE OF THE SHOOT AND ROOT TO GRAVITY

Take three jam jars, or gas jars, and line with blotting paper. Fill the jars with damp sawdust. Place two or three bean seeds in each jar between the glass and the paper, having previously soaked the beans for some hours. Cover the outside of each jar with black paper.

1

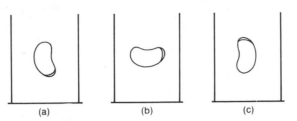

soaked broad beans in jars:
(a) Scar (point of attachment) pointing down
(b) Scar to the side, micropyle uppermost
(c) Scar uppermost

2 AFTER ONE WEEK

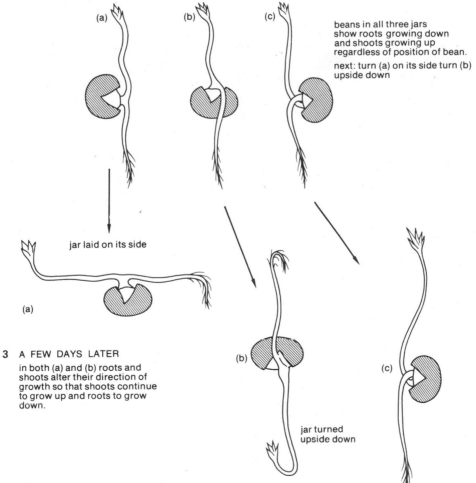

beans in all three jars show roots growing down and shoots growing up regardless of position of bean.

next: turn (a) on its side turn (b) upside down

jar laid on its side

(a)

3 A FEW DAYS LATER

in both (a) and (b) roots and shoots alter their direction of growth so that shoots continue to grow up and roots to grow down.

(b)

(c)

jar turned upside down

Plant responses

Roots are positively geotropic, i.e. they grow towards the centre of the earth. Stems are negatively geotropic. They grow away from the centre of the earth. Leaves arrange themselves at right angles to the direction of gravity. There are some exceptions to this general statement. Some stems (rhizomes, runners, creepers) grow horizontally. Some leaves (Iris, the Grasses) grow vertically. Smaller branch roots do not necessarily grow straight down.

Presence of auxin speeds growth
(see p. 201)

THE EFFECT OF LIGHT ON THE STEM

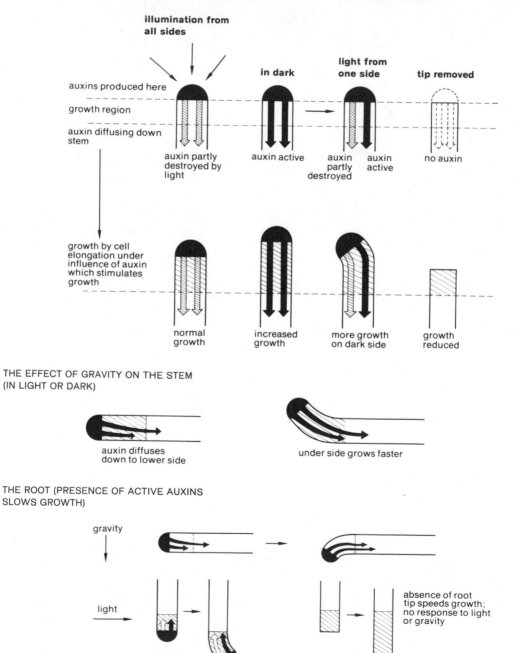

illumination from all sides

in dark

light from one side

tip removed

auxins produced here

growth region

auxin diffusing down stem

auxin partly destroyed by light

auxin active

auxin partly destroyed

auxin active

no auxin

growth by cell elongation under influence of auxin which stimulates growth

normal growth

increased growth

more growth on dark side

growth reduced

THE EFFECT OF GRAVITY ON THE STEM (IN LIGHT OR DARK)

auxin diffuses down to lower side

under side grows faster

THE ROOT (PRESENCE OF ACTIVE AUXINS SLOWS GROWTH)

gravity

light

absence of root tip speeds growth; no response to light or gravity

Explanation of phototropic and geotropic responses in stem and root

Much of the investigation into the mechanisms of plant responses has been carried out on oat seedlings, and these are summarised in the accompanying diagrams. As far as is known the auxins (see p. 201) present in the root are chemically identical with those in the stem tip. The differing responses of stems and roots to auxins are thought to be due to differing sensitivities. When present in sufficient concentration, auxin will inhibit growth in both stems and roots while promoting growth at lower concentrations. It would seem that root growth is slowed by much lower concentration than that required to slow down stem growth.

HYDROTROPISM

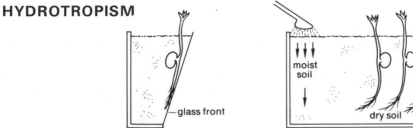

SIDE VIEW   FRONT VIEW

The positive hydrotropism of roots can be demonstrated in several ways. In the glass-fronted box shown above, the roots will grow down the side of the glass and may be kept under observation. If the soil is watered only at one end of the box the roots will grow in that direction. The glass is kept covered except when inspecting the roots. The mechanism of hydrotropism is not known.

TOUCH RESPONSES

There are three main kinds of plants which show touch responses. These are:

(a) Climbing plants Various parts of plants can be adapted for holding fast—modified leaflets, petioles, branch stems, main stems. Contact with a possible support, which may be another plant, causes a speeding up of growth on the opposite side of the stem.

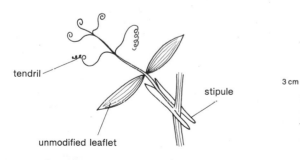

Sweet Pea—climbing plant. Tendrils sensitive to touch, are modified leaflets; one pair of unmodified leaflets act as normal leaves. The whole structure is a compound leaf.

(b) Insectivorous plants These plants are often found in regions where nitrates are in short supply, i.e. regions where the soil is acid and waterlogged. They supplement their nitrogen intake by utilizing animal protein.

Sundew (*Drosera*) (found in Britain).
The tentacles and upper leaf surface of *Drosera* produce a sticky secretion. An insect landing on the leaf is held fast by the liquid. The tentacles fold over to enclose the insect, bending of the tentacles being due to unequal growth on the two sides. The insect is then digested by protein enzymes. *Drosera* is found in acid bogs.

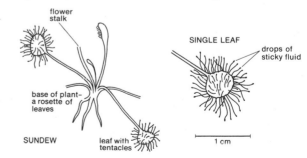

Venus Flytrap is found in Carolina. It acts rather like a mouse trap, the trigger being three hairs found on each half of the upper surface of the leaf blade. The mid rib is thickened and the closing of the trap is brought about by a sudden increase in the water content of cells on its under surface.

187

(c) **Sensitive plants** Some plants show quite spectacular reactions to rough handling. The most well known example is the sensitive mimosa (*Mimosa pudica*). A sharp tap applied to any part of the plant, or shaking it, causes the leaflets to collapse together in pairs and the whole complement of leaves hangs down. This is due to a sudden loss of water from cells on the lower side of the mid-rib of the leaflet and from the lower surface of the base of the petiole.

Of the three kinds of touch response referred to above only the first, the reactions of tendrils to contact, is classified as a tropism. The others are examples of **nastic responses.**

OTHER NASTIC RESPONSES

Some leaves show so-called **sleep movements**. During the daytime they are held outspread in the usual fashion of leaves, but at night they collapse and hang in the vertical position. Many flowers, the daisy for example, close up their petals at night, while those flowers which are pollinated by night-flying insects close up during the day. The mechanisms involved are sometimes due to changes in the water content of cells at the base of the leaves, or petals, and are sometimes brought about by differential growth. The stimulus is either the alternation of light and dark, or the variation in temperature between day and night.

The differences between plant and animal behaviour

It is clearly not possible to compare plant and animal responses in all their aspects. Plants, for example, exhibit only relatively simple responses. They show no patterns of behaviour comparable to instinctive or learned behaviour in animals. It is perhaps worth while to compare the essential features of reflex action with those of a tropism.

(a) Both are adaptive forms of behaviour. Each is brought about by a stimulus, and the response acts in such a way as to maintain, or restore, conditions favourable to survival.

(b) Plants and animals react to the same basic types of stimuli with the single exception of sound.

They differ from each other in that:

(a) Reflex action is usually rapid. Most tropic responses are relatively slow.

(b) Reflex action involves either the contraction of muscles or the secretions of glands. Tropisms are growth responses.

(c) Reflex action involves the stimulation of sense cells which are distinct from the effector organs. No sense cells are involved in a tropism which is limited to a region of active growth.

(d) Reflex action requires a communicating system between receptor and effector. No such structure is present in the plant.

(e) Reflex actions are not necessarily localized, whereas plant responses generally are.

(f) Reflex action is triggered by the stimulus, but its particular form is determined by internal factors. The tropism is a directed response involving movement. The direction of movement depends on the direction of the stimulus.

Leaves of sorrel (Oxalis sp.) showing nastic response in the form of 'sleep movements'.

left Open, daytime position.
right Closed, night position.

37 SUPPORT AND LOCOMOTION

Mammals have to support themselves, maintain their body shape, and move about on dry land without the help of the buoyancy that water gives to aquatic animals. That they are able to do this depends on the proper functioning of three kinds of tissue. These are:

(a) **Bone**, which forms the skeleton, an internal framework of articulated levers.

(b) **Muscle**, which does work by contraction.

(c) **Tendons** and **ligaments**, which attach muscles to bones, and bones to bones respectively.

Bones are mainly *resistant to compression*, although they have a high resistance to bending and stretching forces also.

Ligaments and tendons have great *tensile strength*, like ropes.

Muscles do work by contracting, although they too act like ropes.

When a mechanical tissue is required without the rigidity or hardness of bone, **cartilage** (gristle) is found.

The skeleton

Although the skeleton functions as a whole it is convenient to think of it as consisting of the following parts:

(a) **vertebral column** (backbone), forming the axis of the skeleton. All other parts of the system are attached to it, either directly or indirectly.

(b) the **rib cage** and **sternum**.

(c) the **limbs**, together with their anchorage within the body (**girdles**).

(d) the **skull**, consisting of the cranium or brainbox, the upper and lower jaws, and the bones of·the middle ear.

The vertebral column

The backbone in all vertebrates has two important properties:

(a) It cannot shorten in length. Bone is a dense, hard material with high resistance to compression.

(b) It is flexible. This is because it is made up of a number of separate parts (**vertebrae**) separated by discs of a softer, slightly compressible material (cartilage). Each vertebra can move to a limited extent with respect to its neighbours, and this adds up to a considerable flexibility over the length of the backbone.

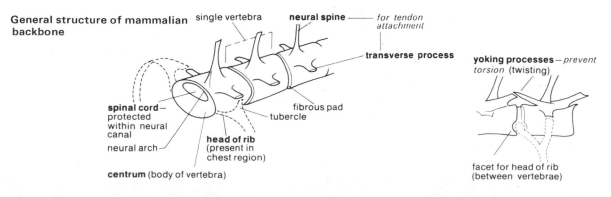

General structure of mammalian backbone

single vertebra

neural spine — *for tendon attachment*

transverse process

yoking processes — *prevent torsion* (twisting)

spinal cord — protected within neural canal

fibrous pad

tubercle

neural arch

head of rib (present in chest region)

centrum (body of vertebra)

facet for head of rib (between vertebrae)

In fishes (p. 62) the backbone is uniform in structure along its length. This is because it is important for locomotion but not for support, which is supplied by water. The backbone bends chiefly in the horizontal plane during movement. In the four-legged mammal the backbone provides for support as well as locomotion. The upthrust by the supporting columns of the legs is concentrated at the ends of the trunk, whereas the weight of the body acts downwards at all points. The backbone of the mammal therefore differs along its length because the different regions do different jobs and are subject to different stresses. The movement of the backbone is mainly in the vertical plane. The differences between the different vertebrae are summarized below:

1 In the neck (7 **cervical** vertebrae). These have very much reduced projections; they are very small and are therefore very flexible as a whole. The first two are specially modified as the **atlas** supporting the skull and providing for nodding movements of the head, and the **axis** which by means of its joint with the atlas allows for rotational head movement.

2 In the chest (12 or 13 **thoracic** vertebrae). These form an integral part of the rib cage, to which the shoulder blade is anchored by means of muscles. The vertebrae are therefore capable of only limited movement. They have long backward sloping spines, attached to which are ten-

dons and ligaments which pass forward towards the head.

3 In the waist (7 **lumbar** vertebrae). These are subject to the greatest stress in both locomotion and support. They have large transverse processes and neural spines sloping forward, away from the direction of stress.

4 In the hip (5 **sacral** vertebrae). These are fused together. They form an integral part of the box girder forming the hip girdle.

5 The tail (**caudal** vertebrae—number varies according to species). These are subject to low levels of loading. There is no spinal cord in the tail. They possess few or no processes, being reduced to simple spools at the end.

The backbone in support

The weight of the body acts downwards at all points, but it is supported at two points only by fore and hindlimbs. If the backbone were a simple beam it would be subject to forces as shown in the diagram. These would tend to make it bend. It is clear that the backbone does not function as a simple beam because it is made up of separate parts, and its role in movement requires that it should be flexible. To function as a simple beam would require that the material between the vertebrae should be as strong as bone. In most mammals the backbone, together with the limb skeletons, muscles and tendons,

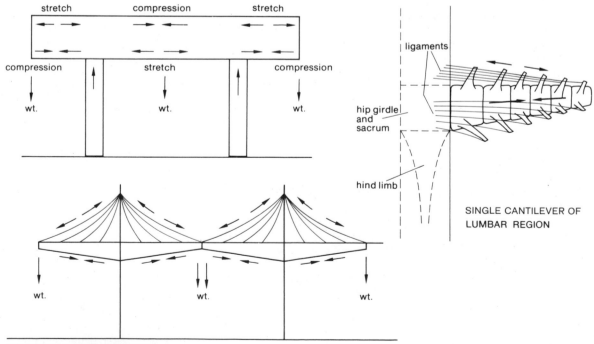

DOUBLE CANTILEVER BRIDGE

SINGLE CANTILEVER OF LUMBAR REGION

Paired cantilever support system in a mammal

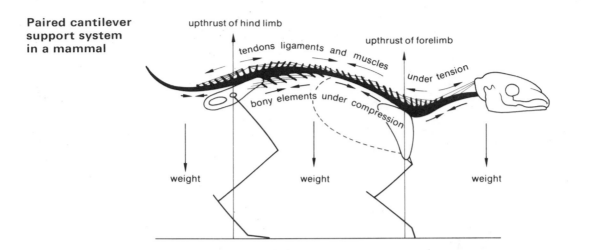

functions as a pair of **cantilever systems** such as are used in bridge design. Some parts, bone, are subject to compression while others, especially ligaments and tendons, are subject to tension.

The rib cage

Ribs in most vertebrates, other than mammals, function rather like the metal rods found in reinforced concrete. They give additional support by stiffening the body wall in the region of the body cavity. They are present in most of the length of the body wall between the girdles, in fishes, for example, but do not impede movement of the body which is in the horizontal plane. In the mammal, where the back-bone moves in the vertical plane, the ribs are restricted to the chest region and are increasingly shorter towards the rear. The last three (floating ribs) are separate from the remainder, which are linked ventrally to the sternum. The ventral ends of the fixed ribs are made of flexible cartilage. These ribs, the series of bony plates forming the sternum and the attached muscles (intercostals), form a box of variable volume which is important in respiration (see p. 128).

The limb skeleton

The skeleton of the leg in mammals is based on the pentadactyl (five finger) plan general to all land vertebrates.

The pentadactyl plan is capable of considerable adaptation. It may vary according to whether it is a forelimb or a hindlimb, and it may vary according to the particular means of locomotion.

Differences between fore and hindlimbs

(example—rabbit) (see p. 192)
The forelimb in the rabbit does most of its work *in resisting forces tending to make it collapse*, i.e. when

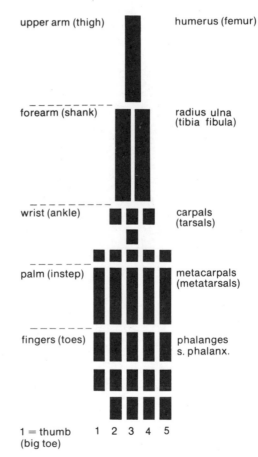

landing after a forward leap, and when digging. *The elbow joint points to the rear* and the mechanical efficiency of the lever system hinged at the elbow is increased by the attachment of the extensor muscles of the upper arm to the extension of the forearm

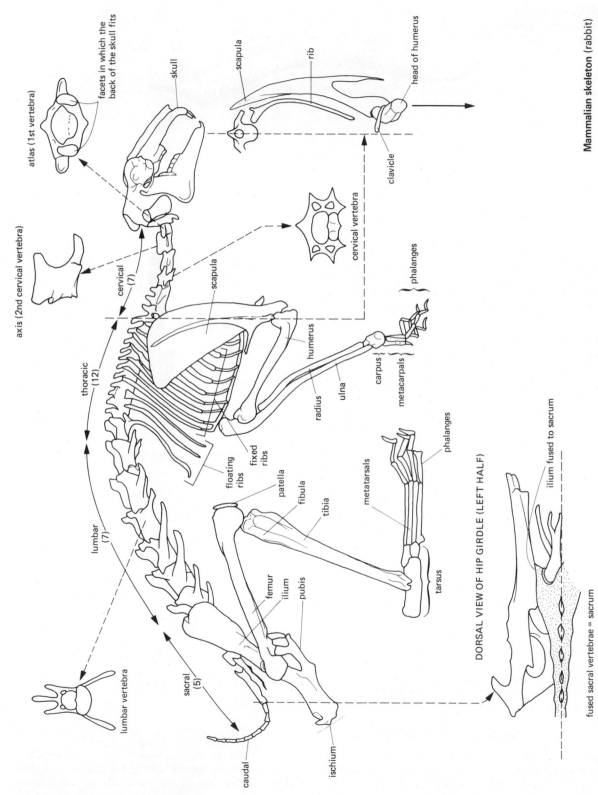

atlas (1st vertebra)

facets in which the back of the skull fits

skull

scapula

rib

head of humerus

clavicle

axis (2nd cervical vertebra)

cervical vertebra

cervical (7)

scapula

thoracic (12)

humerus

radius

ulna

carpus

metacarpals

phalanges

lumbar (7)

floating ribs

fixed ribs

patella

fibula

tibia

metatarsals

phalanges

tarsus

femur

ilium

pubis

sacral (5)

lumbar vertebra

caudal

ischium

DORSAL VIEW OF HIP GIRDLE (LEFT HALF)

ilium fused to sacrum

fused sacral vertebrae = sacrum

Mammalian skeleton (rabbit)

(olecranon process). The two bones of the forearm are bound together, giving increased strength. Some reduction in the number of wrist bones has taken place.

The hindlimb provides the thrust in forward movement. It functions like a flat spring consisting of three levers of approximately equal length. *The knee points forward.* The hindlimb does its effective work *in extension.* Its efficiency as a lever is enhanced by its greater length. The efficiency of the system is increased by the backward extension at the ankle, but at the knee the tendons are carried across the joint above the knee cap. There are only four digits, the ankle bones are reduced in number, and the shank bones are both unequal and partly fused.

The limb girdles

Both fore- and hindlimbs are anchored in the body wall by the limb girdles against which they pivot. It is believed that at some time, a long way back in the evolution of vertebrates, the fore- and hindlimb girdles were much more nearly alike than they are in mammals. In many vertebrates both consist of three bones on each side. They have become modified differently in response to the differing needs of support and locomotion at the two ends of the body.

The **hip girdle** is formed from three bones fused together on either side. They meet in the mid line and are also firmly attached to the backbone. The whole structure forms an extremely efficient **rigid box girder**. It is subject to considerable force as it transmits the thrust of the hindlimbs to the backbone, and since the attachment of hindlimb does not lie in the same line as the backbone, this is a shearing force, hence the girder structure.

The **shoulder girdle** consists of two bones on either side. The **scapula** is a triangular flat bone lying against the rib cage, separate from its fellow on the other side. The two **clavicles** act as simple struts bracing the ends of the scapula against the sternum in the mid line. Clavicles are not specially important and are absent from many mammals. The arrangement of the shoulder girdle is so different from that of the hip because the fore limbs act as shock absorbers rather than as propulsion units. The landing shock is distributed over a wide area and is taken up by the muscles and tendons holding the scapula in place. No connection is possible between the two halves of the girdle because of the rib cage.

The skull

The jaws and teeth have been considered elsewhere (p. 116); the bones of the ear also (p. 181). The skull is a complicated structure in all vertebrates and cannot be considered in detail here. It has two main functions. It provides support for the delicate soft tissues of the brain. These cannot tolerate distortions of any kind if they are to carry out their functions properly. It also serves to protect the organs in the head, including the brain, from mechanical damage. The bones of the skull do not meet completely at birth. In the human baby there is a fontanelle (hole) which persists for about two years. The bones are connected by irregular edges, **sutures**, which are visible in the exposed skull, disappearing with age as the bones fuse completely.

JOINTS OF THE BODY

Swivel (or rotating) joint:

Found between the atlas and axis vertebrae. Involved in shaking of the head.

Ball and socket joint:

Found at the shoulder and at the hip. This type of joint allows for movement in all directions and for limited rotation.

Similar joints, where the surfaces in contact are only slightly curved and are called condylar joints. These are found between the wrist bones and between the skull and the atlas. These are capable of rather more limited movement in all planes.

Hinge joint:

Found at the knee and the elbow and in the finger joints. These permit movement in one plane only.

Immovable joints:

The joints between the bones of the skull.

The composition of bone

Bone is made up of inorganic materials as well as organic substances. The inorganic fraction, which makes up about 70 per cent of the dry weight of bone, consists chiefly of **calcium phosphate** with some magnesium phosphate. The organic fraction consists of **bone cells** together with **protein fibres**. The two components, organic and inorganic, are so completely inter-related that either may be destroyed without altering the shape of the bone. If the bone is subjected to strong heat, to decompose the organic matter, it becomes white and brittle, crumbling easily. If on the other hand a bone is soaked in dilute acid the mineral content is removed and the bone then becomes soft and flexible, because of its meshwork of protein fibres.

The complex structure of bone is a consequence of the fact that the dense ground substance of mineral matter is impermeable to liquids, dissolved matter and oxygen. For bone to remain a living tissue a complex network of spaces must link the bone cells with their blood supply.

Support and locomotion

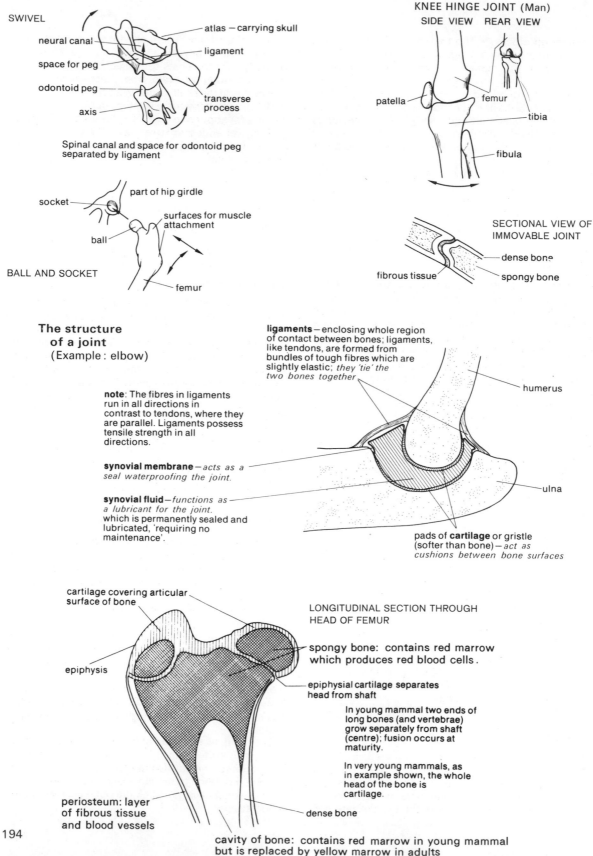

SWIVEL

neural canal

space for peg

odontoid peg

axis

atlas — carrying skull

ligament

transverse process

Spinal canal and space for odontoid peg separated by ligament

KNEE HINGE JOINT (Man)
SIDE VIEW   REAR VIEW

patella

femur

tibia

fibula

socket

ball

BALL AND SOCKET

part of hip girdle

surfaces for muscle attachment

femur

SECTIONAL VIEW OF IMMOVABLE JOINT

dense bone

fibrous tissue

spongy bone

The structure of a joint
(Example : elbow)

ligaments — enclosing whole region of contact between bones; ligaments, like tendons, are formed from bundles of tough fibres which are slightly elastic; *they 'tie' the two bones together*

humerus

ulna

note: The fibres in ligaments run in all directions in contrast to tendons, where they are parallel. Ligaments possess tensile strength in all directions.

synovial membrane — *acts as a seal waterproofing the joint.*

synovial fluid — *functions as a lubricant for the joint.* which is permanently sealed and lubricated, 'requiring no maintenance'.

pads of **cartilage** or gristle (softer than bone) — *act as cushions between bone surfaces*

cartilage covering articular surface of bone

epiphysis

periosteum: layer of fibrous tissue and blood vessels

LONGITUDINAL SECTION THROUGH HEAD OF FEMUR

spongy bone: contains red marrow which produces red blood cells.

epiphysial cartilage separates head from shaft

In young mammal two ends of long bones (and vertebrae) grow separately from shaft (centre); fusion occurs at maturity.

In very young mammals, as in example shown, the whole head of the bone is cartilage.

dense bone

cavity of bone: contains red marrow in young mammal but is replaced by yellow marrow in adults

The structure of bone

1 DENSE BONE

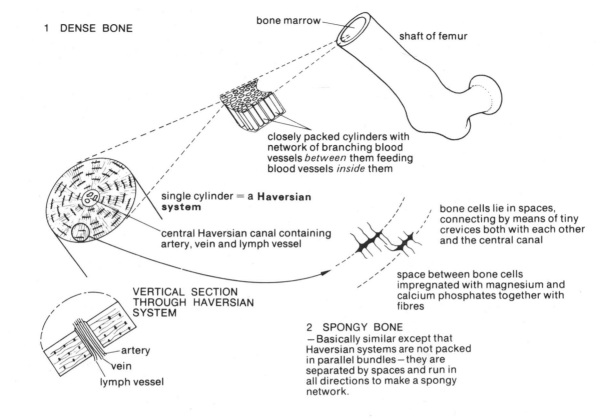

bone marrow

shaft of femur

closely packed cylinders with network of branching blood vessels *between* them feeding blood vessels *inside* them

single cylinder = a **Haversian system**

central Haversian canal containing artery, vein and lymph vessel

bone cells lie in spaces, connecting by means of tiny crevices both with each other and the central canal

space between bone cells impregnated with magnesium and calcium phosphates together with fibres

VERTICAL SECTION THROUGH HAVERSIAN SYSTEM

— artery
— vein
— lymph vessel

2 SPONGY BONE
—Basically similar except that Haversian systems are not packed in parallel bundles—they are separated by spaces and run in all directions to make a spongy network.

The structure of muscle

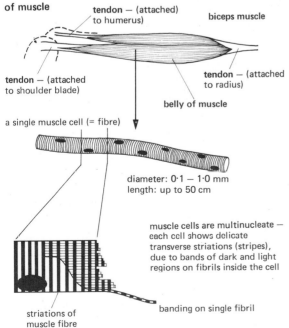

tendon — (attached) to humerus)

biceps muscle

tendon — (attached to shoulder blade)

tendon — (attached to radius)

belly of muscle

a single muscle cell (= fibre)

diameter: 0·1 — 1·0 mm
length: up to 50 cm

muscle cells are multinucleate — each cell shows delicate transverse striations (stripes), due to bands of dark and light regions on fibrils inside the cell

striations of muscle fibre

banding on single fibril

Muscles

A muscle does work by shortening. Since there is no change in volume the muscle diameter increases. The muscles attached to the skeleton are composed of **striped muscle fibres**. They contract only if stimulated by nerve impulses. Such fibres are capable of very powerful contractions but are easily fatigued. In man such muscles are called **voluntary muscles** because they are under conscious control.

Other muscle types

Heart muscle fibres are similar in appearance to skeletal muscle except that the fibres are joined to each other by cross-connections. Heart muscle differs from skeletal muscle in that it contracts automatically without the need for nervous stimulation. Its contractions are slower, less powerful but also less easily fatigued.

Support and locomotion

Smooth muscle fibres are found in the walls of hollow organs, such as the gut and the bladder, attached to hair follicles in the skin, and in the iris diaphragm. They are capable of less powerful but longer-acting contractions and are not under conscious control. They have no striations.

The Action of Muscles

The muscles of the mammalian body are generally organized in **pairs** which have **opposing actions**. If we consider the elbow joint as an example, bending (flexing) of the arm is brought about by the contraction of the muscles lying in front of the upper arm. Straightening (extension) of the arm is the result of contraction of the muscles behind the upper arm. These two sets of muscles are called, respectively,

The action of muscles

'extending' muscles shown in black
'flexing' muscles dotted

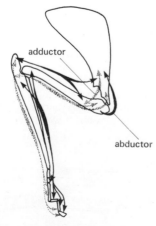

adductor

abductor

adductor

abductor

Structures involved in movement of the elbow joint

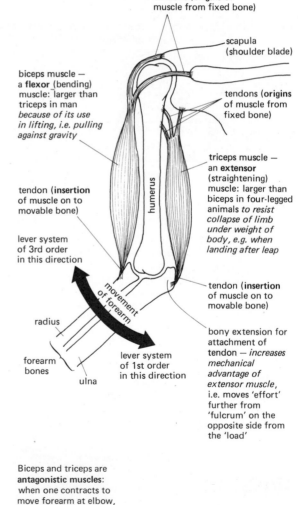

tendons (**origins** of muscle from fixed bone)

scapula (shoulder blade)

biceps muscle — a **flexor** (bending) muscle: larger than triceps in man *because of its use in lifting, i.e. pulling against gravity*

tendons (**origins** of muscle from fixed bone)

triceps muscle — an **extensor** (straightening) muscle: larger than biceps in four-legged animals *to resist collapse of limb under weight of body, e.g. when landing after leap*

tendon (**insertion** of muscle on to movable bone)

humerus

lever system of 3rd order in this direction

movement of forearm

tendon (**insertion** of muscle on to movable bone)

radius

forearm bones

ulna

lever system of 1st order in this direction

bony extension for attachment of tendon — *increases mechanical advantage of extensor muscle*, i.e. moves 'effort' further from 'fulcrum' on the opposite side from the 'load'

Biceps and triceps are **antagonistic muscles**: when one contracts to move forearm at elbow, the other relaxes. When both muscles contract simultaneously, they oppose each other and the elbow is held steady.

flexors and **extensors**. Except when a flexor muscle contracts reflexly in response to an unpleasant stimulus, movement of the limb is the result of the combined action of both **antagonistic muscle sets**. The ability to maintain a limb in any particular position is similarly due to the balanced pull of opposing muscles. The diagram shows in simplified form the arrangement of muscles in the limbs of a rabbit. It is not a drawing of the actual arrangement of the muscles, since the situation is very much more complicated. Bones moving with respect to each other under the force of contracting muscles can be regarded as simple machines, in this case, as **levers**. Muscles can shorten only to a limited extent and it therefore depends on where the point of muscle attachment lies in relation to the joint as to whether the lever amplifies the movement at the point of muscle attachment, or the force it generates. It cannot do both.

Some lever systems of the body

2nd order lever

Operates on the same principle as a wheelbarrow. A small upward force applied to the handles can overcome a much larger force (weight) acting downwards in the barrow. Similarly a relatively small muscular effort is required to raise the body weight. In each case the effort acts through a greater distance than the load moves but in neither case is this important.

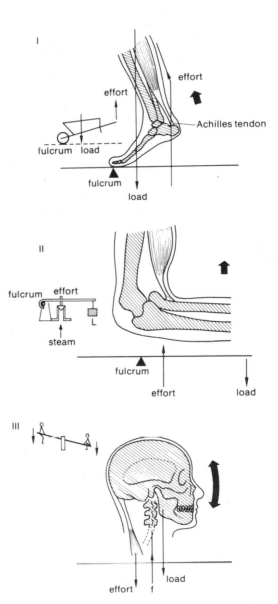

3rd order lever

Operates when lifting a weight on the upturned hand. The effort is applied close to the fulcrum. A small movement of the point of muscle attachment provides a large movement at the hand. A large upward force must be exerted by the muscle to balance a small force, acting down, owing to the weight of an object in the hand. In this case expenditure of muscular energy is less important than freedom of movement.

1st order lever

This is a 'balanced' lever system. The muscle's function is to act as a holding structure, and the force owing to the weight of the head is almost equal to the force exerted by the muscle.

LOCOMOTION

Movement in mammals shows so much variation that it is possible to make only a few general remarks here and these will be confined to movement over the ground using all four limbs.

Walking

Movement of the trunk is not involved, the backbone remaining fairly rigid. Forward thrust is generated by the hind limb. The movement of the whole lever system of the leg is brought about by alternate contraction of abductor and adductor muscles passing between the girdle and the femur (humerus). During the power stroke, when the leg moves backwards in relation to the body, the extensor muscles are contracted. The recovery stroke, when the leg is carried forward, begins with the contraction of the flexor muscles which lift the leg clear of the ground. At the end of this stroke the leg is again fully extended. The legs move in a definite sequence, for example: LEFT FRONT: RIGHT HIND: RIGHT FRONT: LEFT HIND. Only one leg is raised at a time, the remaining three forming a tripod with the centre of gravity of the animal lying within it.

Trotting

This is similar to walking. The rate of movement of the legs is speeded up so that each leg is raised before the one preceding it in the sequence has been lowered. Two legs are therefore raised at any time and the body weight is supported by two legs diagonally opposite.

Galloping

This involves a complete change of sequence. Instead of a diagonal sequence of movement it changes to an alternation between front and hind limbs. In the simplest case, as illustrated by the stoat, fore limbs move together followed by movement of hind limbs together.

Usually both front and both hind limbs are not raised and lowered simultaneously. The sequence may be:

LEFT FRONT: RIGHT FRONT: LEFT HIND: RIGHT HIND. This is called a transverse gallop and is illustrated by the horse.

Or it may be:

LEFT FRONT: RIGHT FRONT: RIGHT HIND: LEFT HIND. This is called a rotatory gallop and is used by the dog.

When galloping at full speed the sequence of actions is speeded up and so telescoped that the fore limbs have left the ground before the hind limbs have made contact. When this is coupled with the whiplash action of the backbone, as in the cheetah, very high speeds indeed may be reached.

Sequence of movement in rotatory gallop of dog

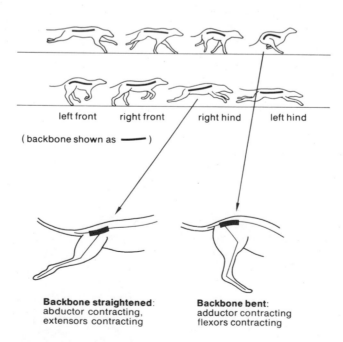

left front   right front   right hind   left hind

(backbone shown as ━━)

Backbone straightened: abductor contracting, extensors contracting

Backbone bent: adductor contracting flexors contracting

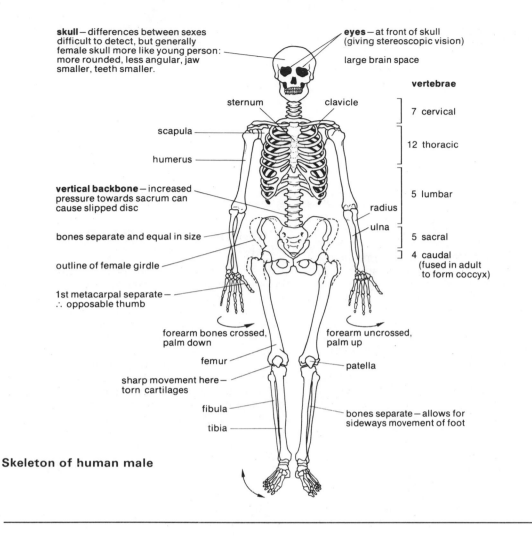

skull — differences between sexes difficult to detect, but generally female skull more like young person: more rounded, less angular, jaw smaller, teeth smaller.

eyes — at front of skull (giving stereoscopic vision)

large brain space

vertebrae

sternum

clavicle

7 cervical

scapula

12 thoracic

humerus

vertical backbone — increased pressure towards sacrum can cause slipped disc

5 lumbar

radius

ulna

bones separate and equal in size

5 sacral

4 caudal (fused in adult to form coccyx)

outline of female girdle

1st metacarpal separate — ∴ opposable thumb

forearm bones crossed, palm down

forearm uncrossed, palm up

femur

patella

sharp movement here — torn cartilages

fibula

bones separate — allows for sideways movement of foot

tibia

Skeleton of human male

---

1 Examine prepared slides of bone under the microscope.

2 Obtain a large bone from your butcher. Very carefully saw it transversely and vertically through the head. Try to identify the different parts. Remove pieces of tendon and lean meat (muscle) attached to the bone. Separate out some of the fibres from each with mounted needles and examine under a microscope.

3 Visit your local museum. Note carefully the differences in the arrangement and proportions of limb bones in mammals which move in different ways.

38 GROWTH

A characteristic feature of living organisms is that they grow. In most cases the cessation of growth is the prelude to death, although the system may have a long 'run-down' time, as it has in man and many other organisms. It can be said that life means growth, for a living system spends its time taking in materials from outside and transforming them into the tissues of its own body, thus increasing in size. In the flowering plant these materials are carbon dioxide, water and a variety of inorganic salts in solution. In animals they are complex carbon compounds directly or indirectly derived from plants. With a few exceptions, the fungi for example, growth is accompanied by an increase in cell number. Man begins as a single cell, the fertilized egg, which is the first of about thirty-two cell generations needed to form the full-grown adult. As growth proceeds the cells undergo changes in form and function to become specialized for particular tasks. All the cells which are descendants of the fertilized egg carry the same hereditary material, or information. The ultimate fate of any particular cell therefore depends upon where it happens to be. The experimental transplanting of cells in non-human embryos from one region to another has demonstrated that this is so, and that how a particular group of cells changes depends on the chemical influences it is subjected to.

Sooner or later, as growth slows and old age, or senescence, sets in, cells lose their capacity to divide. Recent research has shown that the ability of cells to divide ceases after about sixty divisions in man, although why this should be is not known.

GROWTH OF A FLOWERING PLANT

Growth in the flowering plant is restricted to definite regions, called **meristems**. In meristems the cells remain unspecialized and are capable of undergoing continuous division. Elsewhere the cells become transformed into cells with particular functions, and this often leads to the death of the cell. This process is often accompanied by the enlargement of the cell, due to an increase in the size of its vacuole. It is therefore more accurate to say that growth in the flowering plant consists of cell division *and* cell enlargement. Since growth involves cell division and therefore an increase in cell number, the rate at which growth proceeds depends on how much growth has already occurred. Obviously the more cells there are the greater the rate of increase. A typical growth curve for a plant is shown below.

Since flowering plants are immobile, growth in length is necessarily additive. This is most clearly seen in the root. A root has as one of its functions the anchoring of the plant in the soil. As it increases in girth it exerts a pressure on the earth around it, rather in the way that a wallplug anchors a screw in the wall, by outward expansion and pressure against the sides of the hole. A root which increased in

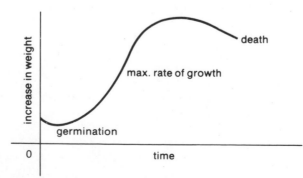

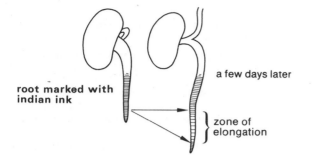

root marked with indian ink

a few days later

zone of elongation

length, uniformly throughout its length, would be moving through the soil, if it were possible for it to move at all, and its ability to resist the forces tending to pull it out of the ground would therefore be reduced. The root-hairs which absorb water and minerals from the soil survive only for a short period. Their survival at all would be impossible if the root were moving relative to the soil.

Apical growth is the only kind possible in the stem of a large tree. The pressure at ground level, due to the weight of timber above, clearly rules out the possibility of active vertical growth in this region. A number of factors have some effect on plant growth :

Growth substances

Growth is under the control of chemical substances called **auxins**. These are produced in the tips of root and shoot. They suppress growth in lateral buds. When the tips of shoots are removed the lateral buds then begin active growth. This is most commonly seen in trees that have been lopped and it is the reason why it is possible to grow dense hedges with privet, or beech. The capacity of lateral or dormant buds to grow when the influence of main shoots is removed also accounts for the great capacity of plants to regenerate tissue when damaged.

Growth of stem in a flowering plant

tissues carried upward as stem below increases in length

Apical meristem — *region of active cell division; cells small, thin-walled, closely packed; nuclei showing all stages of division; no vacuoles visible; dense cytoplasm.*

Region of growth of stem — *in length*

cells actively elongating - vacuoles appear and increase in size, new material added to cell wall; cells begin to specialise, except in cambium

cambial strand

vascular bundle

cambium extends between bundles

xylem

phloem

cambial cylinder

pith

cortex

cambial cylinder

Region of secondary thickening — *division within cambium cuts off new xylem cells towards the centre of the stem and new phloem towards the outside—as cambium moves outwards, cell division planes along the radii increase the number of cambial cells to keep pace with increase in girth*

remains of original vascular bundle   **1st 2nd 3rd and 4th year's growth of xylem**   **phloem**

Light

All plants grow faster in the dark. This is because the auxins are partly inactivated by light. Permanent darkness causes **etiolation**, i.e. growth of spindly, yellow stems.

Temperature

Rate of growth, other factors remaining constant, varies according to temperature. In temperate zones growth effectively ceases in winter. Apart from the effects of temperature, growth is unlimited.

Other physical factors

The characteristic shape of a plant may be modified by many factors, for example:

Trees growing on clifftops are compact, and contoured according to the direction of the prevailing wind.

Trees growing in streets often have extremely extensive rooting systems, since water is less likely to be present below road surfaces.

GROWTH IN A MAMMAL

Growth in the mammal differs from plant growth in four important ways:

(a) There are no special regions where growth takes place. Growth occurs throughout the tissues, i.e. it is intercalary.

(b) Growth involves cell division but not usually increase in cell size.

(c) Growth is limited. This is related to the fact that mammals have a fixed shape. This is a necessary consequence of the fact that mammals are mobile organisms.

(d) Growth is independent of temperature.

Mammals resemble plants in that:

(a) Growth is under the control of chemical substances.

(b) The rate and extent of growth depends in some measure on the balance and quantity of food.

(c) The growth curve for a mammal resembles the same general form as that for a flowering plant.

Although growth in mammals is intercalary, nervous tissue is exceptional in that no division, and therefore no increase in the number of nerve cells, takes place after birth.

Although the rate of growth depends upon availability of food in the form of a balanced diet, evidence tends to suggest that well fed mammals not only grow faster and bigger and mature earlier than mammals maintained at just above starvation level, but they die earlier also. Increase in size also means a change in shape. Different parts of the body grow at different rates and there is a change in form which is related to absolute size (see p. 205).

Although mammals show no overall increase in size when they are adult, growth processes still continue. New blood cells, for example, continue to be made throughout life.

Three basic sorts of change are involved in the process of ageing.

(a) Muscle cells as well as nerve cells no longer divide. The total number diminishes as a proportion fail.

(b) Supporting tissues change. Elastic tissue, in the skin, the lens, the walls of arteries, becomes replaced by fibrous tissue with resulting wrinkles, long sight and high blood pressure. The protein collagen, making up one third of the body protein, acts as a binding material in the body and it becomes progressively stiffer with age throughout the body as a whole.

(c) The cells which continue to divide in adult life progressively lose their ability to do this. The body therefore shrinks and loses weight, although this may be offset, in human beings,

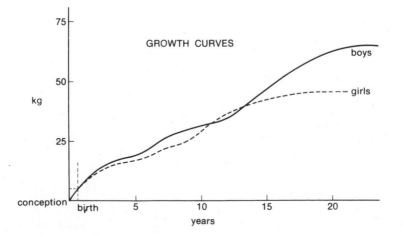

by the tendency to eat more food than is actually required, thus increasing the body's fat reserves.

GROWTH IN OTHER ORGANISMS

In **single-celled organisms** such as bacteria, or *Amoeba,* the notion of growth as applied to multicellular organisms cannot strictly refer to individuals. Cells derived by asexual reproduction from a single cell form a colony of individuals genetically alike. Such a colony is called a clone and in some respects the growth of a clone can be compared with the growth of a multicellular organism. It is tempting to push the comparison further, especially with organisms which have a sexual phase also. Nevertheless it is important to remember that ideas about ageing which apply to multicellular organisms cannot apply to organisms, like bacteria, which have no sexual process of reproduction.

In **Mucor**, which is not divided into cells, growth means the continuous lengthening of filaments, but this is presumably a modified form of cell division since the number of nuclei in the cytoplasm increases. In more complex plants, such as **seaweeds**, **mosses** and **ferns**, growth takes place as the result of the activity of an apical meristem. Unlike that of the flowering plant, the apical meristem of these plants consists of a single cell which undergoes continuous division.

As far as other animals are concerned, the rate and extent of growth depends on availability of food. Except in birds, which resemble mammals in their growth patterns, the rate of growth depends also on temperature. In almost all cases growth takes place throughout the body. With the exception of birds and insects, growth is unlimited in that there is no definite upper size limit. Growth in **insects** and other **arthropods** occurs stepwise. Size remains constant until a moult occurs. This is followed by a rapid growth phase before the new cuticle hardens.

Tapeworms (see p. 28) show a growth pattern which is in some ways like that of plants. The region immediately behind the scolex acts rather like a meristem.

Animals, unlike plants, often have a different form and habit when young. Such a young animal is called a **larva**. Before becoming adult the larva undergoes a change of form, or **metamorphosis**. Such an organism generally belongs to one or other of four groups. It may be:

(a) sedentary, like a barnacle, or many of the colonial coelenterates.

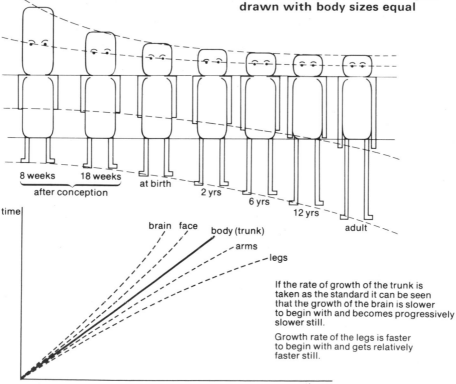

Human beings at different ages, drawn with body sizes equal

8 weeks   18 weeks   at birth   2 yrs   6 yrs   12 yrs   adult

after conception

time

brain face body (trunk) arms legs

If the rate of growth of the trunk is taken as the standard it can be seen that the growth of the brain is slower to begin with and becomes progressively slower still.

Growth rate of the legs is faster to begin with and gets relatively faster still.

growth rate

(b) an internal parasite, in which case the larva may show special adaptations which help it to withstand exposure to the hostile outside world (see p. 30).

In **(a)** and **(b)** the larva is the dispersal phase.

(c) a member of one of the more advanced insect orders (bee, wasp, butterfly, moth or two-winged fly) (see p. 48). These have highly specialized ways of life, particularly with respect to nutrition and locomotion, and such specialized ways are not accessible to the young.

(d) an amphibian (see p. 65). Such animals are only partly adapted for dry land life, and the young must live in water until their structure has become sufficiently complex to live on dry land.

Allometric growth

Animals which have a definite fixed shape, such as arthropods and vertebrates, show changes in proportion as they increase in size. This is because different parts of the body grow at different rates. Furthermore, the rate at which any part grows alters according to size of the body as a whole. Allometric growth can be illustrated by reference to human beings. Some parts of the body show a slower growth rate than the body as a whole while others show a faster growth rate. As a rough generalization, the farther away from the head the faster the growth rate. The brain, and its outgrowth, the eye, grow most slowly. The facial area grows rather more; hence the change in shape from a baby's face to that of an adult. Arms grow rather faster than the head. A child can put its arm over the top of its head and touch the ear on the opposite side at about the age of five years but not before. Legs grow faster and form proportionately more of the total height in tall people than in short ones. Allometric growth can best be explained by drawings of a collection of individuals of different sizes.

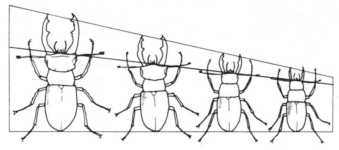

Differential growth in stag beetles (male)

Adult beetles, like all insects, do not grow — the size of any beetle is related to its growth *before* it becomes a pupa. The diagram shows how the proportions of the insect vary with its absolute size.

Diplodocus: amphibious dinosaur from late triassic period— about 170 000 000 years ago

An example of proportion changes with absolute increase in size.

Length — up to 30 m.
Weight — 25–35,000 kg
Brain weight — 0·5 kg.

As animals get progressively larger their heads *in relation to* their bodies get smaller.

39 SIZE

CELLS

A living cell is continuously taking in dissolved substances from its environment and returning to the environment the waste products of its own activities. The transport of materials, in either direction, is the result of physical diffusion, and takes place across the cell surface. The rate of diffusion *per unit area* depends on various factors—for example, the size of the molecules involved, and the difference in their concentrations on the two sides. The total amount entering or leaving the cell in a given time will depend on the total surface area of the cell. Now although the ability of a cell to meet its needs depends on its surface area, the extent of those needs depends on its volume, and it so happens that these two, volume and surface area, are not related in a direct way for any particular shape. For a given shape, volume is proportional to the cube of the linear dimensions while surface area is proportional to their square, when expressed in the same units. Consider a cube with sides of length 1 mm.

Its volume $= 1^3 = 1$ mm^3
Its surface area $= 6 \times 1^2 = 6$ mm^2
Thus there are 6 mm^2 *of surface per mm^3*

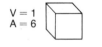

If the dimensions are doubled
Volume $= 2^3 = 8$ mm^3
Surface area $= 6 \times 2^2 = 24$ mm^2
In this case there are 3 mm^2 *of surface per mm^3*

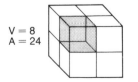

The surface area to volume ratio can only be increased by altering the shape:

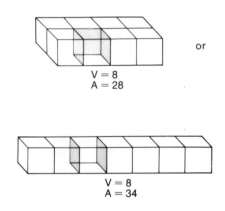

For any given *volume* the surface area can be as large as one likes, by making the solid figure flatter or longer, but for a cell of given *shape* there is a limit to its size, which is determined by the ratio of its surface area to volume.

Spherical cells have the smallest surface area in relation to volume and are therefore much more limited in size than other shapes. Nerve cells, for example, may be several feet in length because their surface area is large in relation to their volume. The limitations on cell size are not only a matter of surface area. The activity of a cell is a function of its nucleus, and in most cells the ratio of nuclear volume to cell volume is about the order 1:7. When the growing cell reaches this ratio it either divides or dies. Xylem cells (see p. 55), for example, enlarge by vacuolation and live only for a short time. Phloem cells, which also enlarge, divide into two to form a sieve tube which has no nucleus and a much smaller companion cell.

205

MULTICELLULAR ORGANISMS: ANIMALS

As long as animals remain simply organized, increase in size must be closely linked with increase in surface area. Hydra is a simple sac-like organism, and no cell is more than a very short distance from the surface of the animal. This applies equally to the cells lining the enteron, since they are in contact with the outside world through the mouth. The largest coelenterate is a jellyfish which may be as much as two metres across, but its surface area in relation to its volume of living tissue is very large. Tapeworms illustrate another way in which large size may be attained without specialization. When an animal has a ribbon-like shape there is no theoretical limit to its size, since all parts of the body are close to an external surface.

However, coelenterates are mostly small, and tapeworms live in a rather specialized environment where the body is supported by the host. Further increase in size requires that two other conditions should be satisfied: (a) the absorptive surfaces should increase in area without increase in volume, and (b) there should be an effective alternative to diffusion between such surfaces and the rest of the tissues. All large animals have specialized regions for gaseous exchange to take place. The lungs of a man, for example, take up a small proportion of his volume but their total internal surface area is equivalent to the area of a tennis court (see p. 125). Similarly, the lining of the gut has its surface area increased by the presence of villi (see p. 108). The circulating blood ensures the rapid, efficient transport of dissolved food materials and the respiratory gases to and from the tissues, i.e. it reduces the distances over which diffusion must take place. The insects are a special case because, although they possess a blood system, they rely on simple gaseous diffusion along a system of air tubes for the transport of oxygen to the tissues. Their size is therefore limited by the distances over which diffusion is a working possibility.

Further size increase in land animals is possible only if a rigid supporting framework is present. This also has limitations. The weight of an animal is proportional to its volume. The strength of its legs is proportional to their area of cross section (doubling its length does not make a leg stronger). Doubling all the linear dimensions therefore increases the weight by a factor of *eight* but the strength of the legs only *four* times. With overall increase in size the legs must therefore be proportionately thicker.

In warm-blooded animals the relation between volume and surface area puts a final limit to size, even in marine forms where support is not a problem. The amount of heat produced is proportional to volume. Its loss is dependent upon surface area. The largest mammals are limited to seas near the poles where the temperature gradient across the surface of the animal is consistently at a maximum. At the other end of the scale, minimum size in mammals is determined by the need to *conserve* heat, and the smallest mammals must spend a very large part of their lives eating just to replace heat lost from the surface of the body.

MULTICELLULAR PLANTS

The limitations that apply to animals are largely a consequence of the fact that they move about to seek their food. Plants are not subject to this limitation. They too must possess a surface area appropriate to their needs i.e. to carry out gaseous exchange and to absorb water and minerals from the soil. Even in such a large plant as an oak tree no living part of it is ever more than a very short distance from the external environment. The living tissue forms a thin film extending over the surface of the dead supporting wood (see p. 162). Organization is arranged along the vertical axis of the plant. Downward movement of materials presents no special problems and the movement of water up the tree is not limited by height, within the size range of any trees that exist at the present time. Ultimate size in plants is most likely to be limited by inbuilt ageing mechanisms already referred to (p. 202). There appears to be no physical limitation to their size.

40 FORM

Animals and plants show such a variety of form that it is not possible to make more than a few general remarks here. Their basic patterns are fixed by their basic needs. Within this broad framework the actual form of the organism is determined by its way of life and the particular local conditions under which it lives.

ANIMALS

These mostly display one or other of two basic kinds of symmetry. Some are organized, like the coelenterates, according to a plan of radial symmetry (see p. 26). The majority, however, actively move about in search of food. These are **bilaterally symmetrical**. This means that the animals can be divided into two halves each of which is a mirror image of the other. Since they move, they necessarily have a front (or **anterior**) end. Here the sense organs are concentrated, together with the beginning of the food canal and any structures associated with the taking in of food. Animals live in a world subject to gravity. The upper surface (called **dorsal**) is therefore different from the lower surface (called **ventral**). The more complex animals (e.g. earthworms, insects, vertebrates) are segmented. This means that their bodies are divided along their length into sections which are more or less alike. The co-ordinated interaction of these segments provides the basis for much more effective movement than is usually possible in unsegmented animals.

Increase in size, and life on land, both involve the development of hard parts to support the soft tissues, either as an external skeleton (insects) or as an internal framework (vertebrates). The presence of a skeleton makes possible the development of specialized appendages which can be used for a variety of purposes, including locomotion. From each basic pattern a great deal of adaptation is possible. Some indication of the variety of mammals has already been given (p. 79), and their adaptations to different diets have also been referred to (p. 109). Even a detailed specification such as that of a mammal can be subject to a great deal of modification, according to habitat and climate, as well as diet.

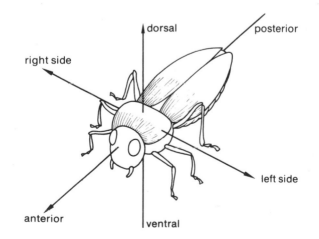

PLANTS

Like mammals or insects, flowering plants are a remarkably uniform group, although perhaps this may not be at once apparent. The flowering plant lives in two worlds, that of the soil and that of the atmosphere above it. The soil supplies its needs for anchorage, water and mineral salts, but the light needed for photosynthesis (see p. 93), and the carbon dioxide, must be obtained above ground. These differing needs from very different environments impose a division of the plant into shoot and root. They also impose a vertical organization on the plant. The need for light requires that the aerial parts must possess a large surface area. The absorption of materials by diffusion or osmosis also demands a large surface, though not necessarily a similar one. These two needs are met by a system of branching. Finally the need for support of aerial plants means that the plant is basically symmetrical about its vertical axis.

These aspects of plant form are the results of the method of nutrition of the flowering plant and the fact that it is immobile. Variation from this basic structure depends on a number of factors. For example, the features of the environment may play an important part. These may be:

1 Climatic conditions (temperature, rainfall, amount of light).
2 Physical conditions (drainage, slope, height above sea level, wind).
3 Soil (composition and physical properties).
4 The presence of other living organisms.

These differing environmental features are reflected in the adaptations which flowering plants show to different ways of life. They may be parasitic, or insectivorous. They may be climbers or creepers. They may live totally submerged in freshwater or in deserts, at high altitudes on mountains or in salt marshes. All these and many more have some bearing on the particular adaptations of form that a flowering plant may show.

ADAPTATION - STEMS

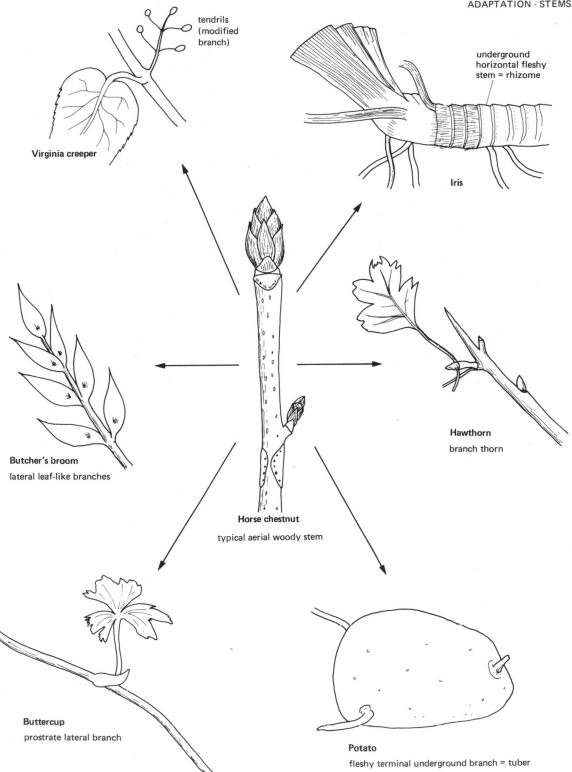

tendrils
(modified
branch)

Virginia creeper

underground
horizontal fleshy
stem = rhizome

Iris

Butcher's broom
lateral leaf-like branches

Horse chestnut

typical aerial woody stem

Hawthorn
branch thorn

Buttercup
prostrate lateral branch

Potato
fleshy terminal underground branch = tuber

209

Form

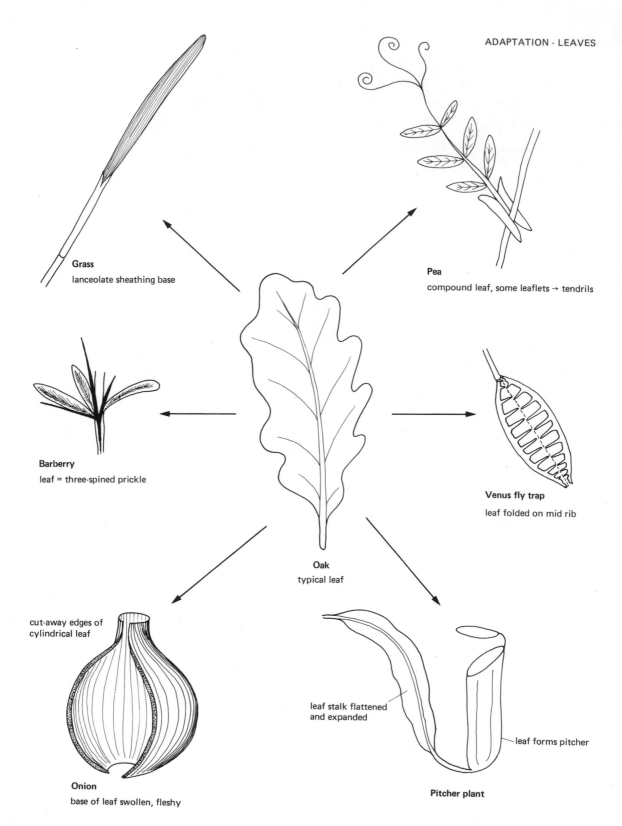

Grass
lanceolate sheathing base

Pea
compound leaf, some leaflets → tendrils

Barberry
leaf = three-spined prickle

Oak
typical leaf

Venus fly trap
leaf folded on mid rib

cut-away edges of
cylindrical leaf

Onion
base of leaf swollen, fleshy

leaf stalk flattened
and expanded

leaf forms pitcher

Pitcher plant

210

ADAPTATION IN MAMMALS
(not to scale)

Sea (dolphin)

stream-lined: no hair: thick layer of
fat in dermis (blubber): no hind limb:
no external ear: horizontal tail fluke:
dorsal fin

Cold (Polar bear)

large size (up to 750 kg): long, thick
white hair: thick layer of fat under skin

Soil (mole)

eyes small: no external ears: front limbs
broad, powerful, shovel-like: dense
short fur to exclude soil particles

Desert (camel)

broad flat feet: hump — food store —
also protects against sun's heat:
matted hair prevents excessive
evaporation of water: double eye lashes

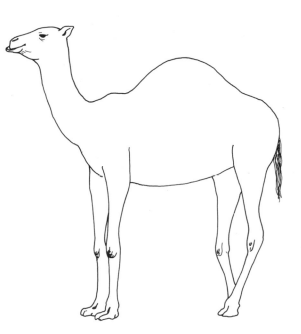

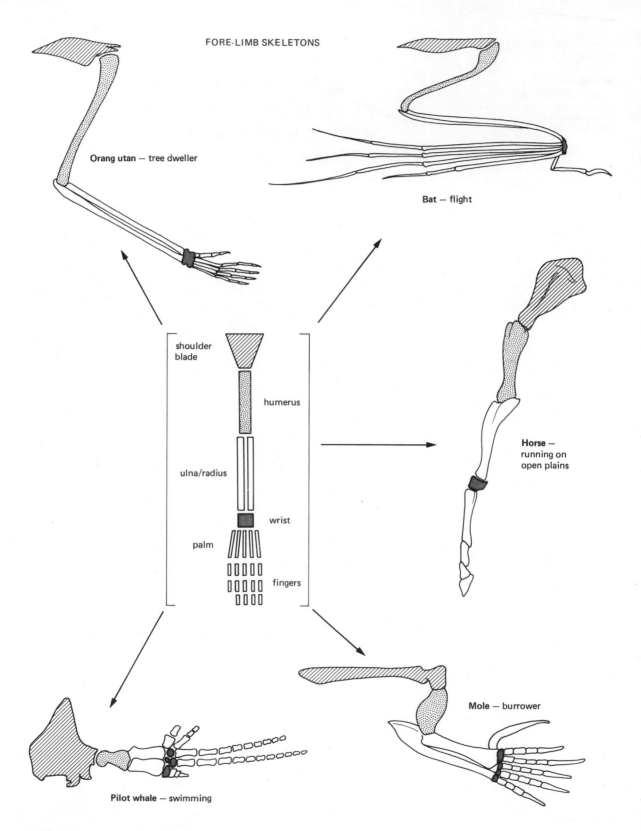

FORE-LIMB SKELETONS

Orang utan — tree dweller

Bat — flight

shoulder
blade

humerus

ulna/radius

wrist

palm

fingers

Horse —
running on
open plains

Pilot whale — swimming

Mole — burrower

One of the most important distinguishing features of living organisms is their ability to reproduce their own kind. Although there are many different ways in which reproduction may take place they all involve either an asexual or a sexual process.

Asexual reproduction

Each individual formed as a result of asexual reproduction has only **one parent**. All the offspring of such a parent are alike. They share the same hereditary characteristics both with each other and with the parent. The cells of the parent divide just as they do in the course of normal growth, but at some stage there is a physical separation of the products of cell division, which then give rise to new organisms. The pattern of asexual reproduction varies according to the type of organism involved

1 One-celled organisms, e.g. *Amoeba* (see p. 5) reproduce either by **binary fission** (the division of one cell into two equal daughter cells which then separate) or by **fragmentation** of the nucleus and cytoplasm into a number of **spores** each of which can grow into a new organism.

2 Some multicellular animals, e.g. hydra (see p. 25), reproduce by **budding**. Cell division leads to the growth of new individuals which appear as buds on the body of the parent. These later separate and become independent.

3 Many organisms, e.g. fungi (p. 15), mosses (p. 33), and ferns (p. 36) produce **spores**, single cells enclosed in a resistant outer wall. These arise as a result of cell division in specialized regions of the plant body, e.g. the sporangia of fungi, the capsules of mosses and underneath the leaves in ferns. Under favourable conditions each spore is capable of giving rise to a new plant.

4 In the flowering plants asexual reproduction occurs when various organs of the plant body become detached and continue to grow. These may be parts of stem or root, leaves or bud.

5 Complex animals are not normally capable of asexual reproduction. A few arthropods can reproduce by **parthenogenesis**. This refers to the development of eggs without fertilization

and is a modified form of sexual reproduction. Male bees, for example, normally develop from unfertilized eggs. Aphids produce a number of generations during the summer months all of which are females which have developed from unfertilized eggs.

Asexual reproduction can be regarded as an extension of the normal growth process, and a method of bringing about a rapid increase in population under conditions which are favourable to growth.

Sexual reproduction

This differs from asexual reproduction in a number of ways.

1 It does not necessarily mean an increase in numbers.

2 It involves in every case the **fusion** of two cells, usually from two different members of the species.

3 Each fusion product contains hereditary material which is a combination of that contributed by the fusing cells and is therefore different from either. The individuals arising as a result of sexual reproduction are different in many respects both from each other and from the parents. Sexual reproduction is therefore a source of **variety** in population.

4 In many organisms sexual reproduction appears to be an adaptation to survival under adverse conditions. The fusion product is often a resistant, dormant stage, germinating only under favourable conditions. Examples are the zygospore of *Spirogyra* and the seeds of annual flowering plants.

5 Sexual reproduction is a property displayed by all living organisms except bacteria, some fungi and a few other single-celled organisms, such as *Amoeba*.

The cells which fuse are called **gametes**. In simple unspecialized organisms the gametes are often alike in appearance and behaviour. Sometimes the whole living part of the animal or plant becomes transformed into one or more gametes (see *Spiro-*

gyra p. 12). In such cases the formation of gametes necessarily means the disappearance of the parent. In multicellular organisms in which there is some division of labour (see p. 85) among the body cells, certain parts of the body are specialized for the production of gametes. These are of two kinds:

(a) an egg cell (ovum): immobile, containing food reserves, and

(b) a sperm cell (male gamete). This is much smaller, mobile, and contains no food reserves. In the lower plants and in animals, the sperm cell possesses its own means of locomotion. In the flowering plant it depends on external agents for its transport.

In some cases both types of gamete are produced within the body of the same individual organism (earthworm, tapeworm, some flowering plants). They are said to be **hermaphrodite**. Where this is the case mechanisms usually exist to prevent fusion of gametes from the same organism. In other cases, (arthropods, vertebrates, and some flowering plants) the sexes are separate. The different types of gamete are produced by different types of individual. The egg cell is produced by *females* and the sperm cell by *males*. When a male gamete fuses with an egg cell the egg is said to be **fertilized**. Fertilization often takes place externally in the case of animals or plants which live permanently in water. In land animals and plants fertilization takes place within the body of the female where the sexes are separate, or within the female structures in hermaphrodites.

The development of gametes

Cell division (mitosis) as it usually occurs has already been described (p. 86). In cells which are going to give rise to gametes, increase in number takes place in the same way but the final development of gametes is preceded by a cell division which is markedly different.

The reason for this is as follows. All body cells contain chromosomes in their nuclei. They are, in effect, a set of chemical instructions determining how the cell will behave or develop in a given situation. This hereditary material is organized as particles, or **genes**, arranged in a single row along the length of each chromosome. The number of chromosomes is constant for every species of living organism and is an even number, since each type of chromosome is represented twice. Each male and female gamete contributes equally to the complete set of chromosomes. The chromosome number in each gamete is therefore half that of the fertilized ovum (**zygote**). It is also half the number found in any typical body cell, since body cells are formed as the result of division of the fertilized egg.

It follows that at some stage during the production of gametes there must be a special type of cell division in which the normal body cell chromosome number is reduced to a half. This kind of division is called **meiosis**. It differs from mitosis in that the chromosomes take up position on the spindle in pairs. Members of each pair then separate into the two daughter cells which result from the division.

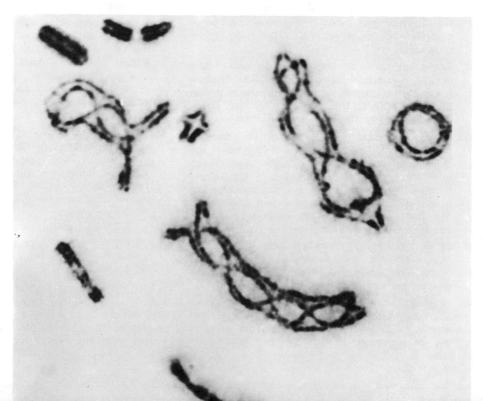

Patterns of chromosomes during meiosis. The chromosome threads are linked together, and when they later separate, parts of them may become interchanged; this produces variation

Reproduction in flowering plants

VEGETATIVE PROPAGATION

Large numbers of flowering plants are able to reproduce asexually. Man often exploits this property of plants for his own benefit. Potatoes, for example, are never grown from seed. The potato is the swollen terminal part of a lateral underground stem arising in the axil of a scale leaf. Each potato is capable of giving rise to a new plant in the following year. Many spring flowers, such as crocuses and snowdrops, are spread entirely by asexual means, since they flower before the insect pollinators are active. They spread by the separation of new bulbs and corms, which are detached branches. The long drooping stems of weeping willow, or *Forsythia*, produce adventitious roots where they touch the ground, and are then capable of independent existence. Creeping buttercup produces runners, which are long branches lying on the ground. Buds and adventitious roots are produced at intervals, and these grow into new plants. By such means plants can spread and colonize large areas in a short time without the need for sexual reproduction. Some plants can be spread artificially. Varieties of apple are propagated by means of grafts.

To summarize:

(a) Vegetative reproduction is a means of rapid spread under favourable conditions, without the need for seed production and without dependence upon external agents.

(b) It is a means whereby plants may be grown from year to year with the same genetic inheritance, so that particular varieties may be maintained, for example, strawberries, apples, pears, potatoes.

(c) It has a limitation in that since all such plants are genetically alike they are all susceptible to the same disease organisms. A disease that attacks one plant attacks them all. (A most famous example was the famine caused by failure of the potato crop, due to attack by blight, in Ireland in 1845 (see p. 16).

ORGANS OF VEGETATIVE PROPAGATION

(Structures not produced by sexual reproduction, capable of surviving independently of the rest of the plant).

Organ	Description	Examples
Rhizome	Underground, horizontal, branching stem, always swollen with food reserves.	Iris, solomon's seal, couch-grass
Stem tuber	Swollen tip of underground, lateral stem. Arises below ground in the axil of a scale leaf.	Potato, artichoke
Corm	Short, vertical, underground stem. Swollen with reserve food. Only two or three years' growth present at any time.	Crocus, gladiolus
Bulb	Underground disc-like stem bearing close-set leaves with fleshy bases containing stored food.	Daffodil, tulip, onion
Suckers	Underground, lateral branches. Their ends turn up and produce buds.	Mint, pear
Runner	Lateral branches arising close to the ground. They grow rapidly along the ground, producing buds and adventitious roots at intervals. These become separate plants.	Strawberry, creeping buttercup
Offsets	Short stout runners terminated by a single bud.	Houseleek
Stolon	When a weak stem falls over and touches the ground, its tip swells, it develops adventitious roots and further growth is continued by a lateral bud.	Blackberry
Root tubers	Swollen fibrous roots each capable of developing into a new plant.	Dahlia
Bulbils	Fleshy axillary buds.	Lily, Lesser celandine
Leaf buds	Bud detaches from leaf border and grows into new plant.	*Bryophyllum* (a leaf succulent. Buds grow around edges of leaves)

DAHLIA – root tuber

stem base

adventitious roots– *become swollen with food reserves; when separated, each gives rise to new plant developed from bud at tip.*

stored food inulin

IRIS – **rhizome** – underground horizontal main stem – *branches separate as older parts rot away*

lateral branch (formed from lateral bud)

vein scars

leaf scars

swollen stem

internode

node

developing **lateral bud**

adventitious roots (roots arising direct from stem)

aerial shoot *develops from terminal bud.*

aerial shoot – (foliage leaves with sheathing base)

rhizome *contains food stored as complex carbohydrate*

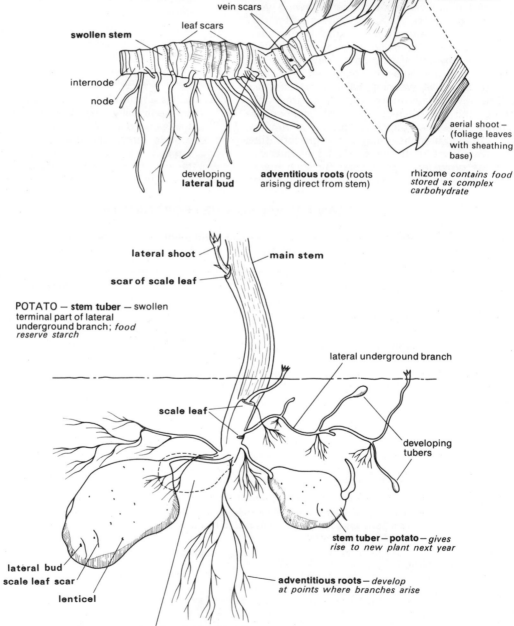

lateral shoot

scar of scale leaf

main stem

POTATO – **stem tuber** – swollen terminal part of lateral underground branch; *food reserve starch*

lateral underground branch

scale leaf

developing tubers

lateral bud
scale leaf scar

lenticel

remains of old potato tuber

stem tuber – potato – *gives rise to new plant next year*

adventitious roots – *develop at points where branches arise*

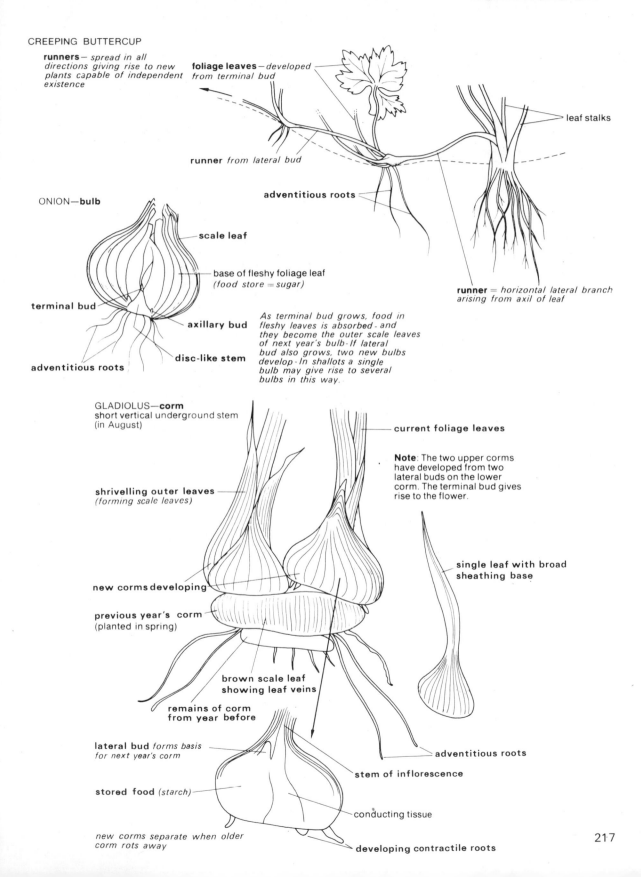

CREEPING BUTTERCUP

runners— *spread in all directions giving rise to new plants capable of independent existence*

foliage leaves—*developed from terminal bud*

leaf stalks

runner *from lateral bud*

adventitious roots

runner = *horizontal lateral branch arising from axil of leaf*

ONION—bulb

scale leaf

base of fleshy foliage leaf *(food store = sugar)*

terminal bud

axillary bud

adventitious roots

disc-like stem

As terminal bud grows, food in fleshy leaves is absorbed - and they become the outer scale leaves of next year's bulb - If lateral bud also grows, two new bulbs develop - In shallots a single bulb may give rise to several bulbs in this way.

GLADIOLUS—corm
short vertical underground stem
(in August)

current foliage leaves

Note: The two upper corms have developed from two lateral buds on the lower corm. The terminal bud gives rise to the flower.

shrivelling outer leaves
(forming scale leaves)

single leaf with broad sheathing base

new corms developing

previous year's corm
(planted in spring)

brown scale leaf showing leaf veins

remains of corm from year before

lateral bud *forms basis for next year's corm*

adventitious roots

stem of inflorescence

stored food *(starch)*

conducting tissue

new corms separate when older corm rots away

developing contractile roots

217

Sexual reproduction—
the structure of the flower

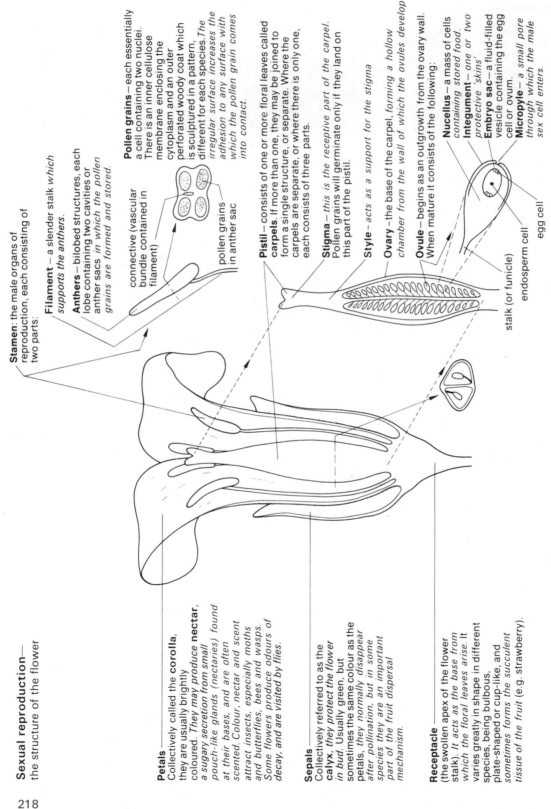

Stamen: the male organs of reproduction, each consisting of two parts:

Filament – a slender stalk *which supports the anthers.*

Anthers – bilobed structures, each lobe containing two cavities or anther sacs *in which the pollen grains are formed and stored.*

connective (vascular bundle contained in filament)

pollen grains in anther sac

Pollen grains – each essentially a cell containing two nuclei. There is an inner cellulose membrane enclosing the cytoplasm and an outer perforated woody coat which is sculptured in a pattern, different for each species. *The irregular surface increases the adhesion to any surface with which the pollen grain comes into contact.*

Pistil – consists of one or more floral leaves called **carpels**. If more than one, they may be joined to form a single structure, or separate. Where the carpels are separate, or where there is only one, each consists of three parts.

Stigma – *this is the receptive part of the carpel.* Pollen grains will germinate only if they land on this part of the pistil.

Style – *acts as a support for the stigma*

Ovary – the base of the carpel, *forming a hollow chamber from the wall of which the ovules develop*

Ovule – begins as an outgrowth from the ovary wall. When mature it consists of the following:

Nucellus – a mass of cells containing stored food.

Integument – *one or two protective 'skins'*

Embryo sac – a fluid-filled vesicle containing the egg cell or ovum.

Micropyle – *a small pore through which the male sex cell enters.*

stalk (or funicle)

endosperm cell

egg cell

Petals
Collectively called the **corolla**, they are usually brightly coloured. *They may produce nectar, a sugary secretion from small pouch-like glands (nectaries) found at their bases, and are often scented. Colour, nectar and scent attract insects, especially moths and butterflies, bees and wasps. Some flowers produce odours of decay, and are visited by flies.*

Sepals
Collectively referred to as the **calyx**, *they protect the flower in bud.* Usually green, but sometimes the same colour as the petals, *they normally disappear after pollination, but in some species they are an important part of the fruit dispersal mechanism.*

Receptacle
(the swollen apex of the flower stalk). *It acts as the base from which the floral leaves arise. It varies greatly in shape in different species, being bulbous, plate-shaped or cup-like, and sometimes forms the succulent tissue of the fruit (e.g. strawberry).*

SEXUAL REPRODUCTION

Flowering plants, like mammals, are highly successful land organisms. Less advanced plants, such as the mosses and ferns, depend upon external water in which the male sex cells swim in order to reach the egg cell. Mammals, because they are mobile, are able to mate, and the sperms are transferred to the eggs in an internally fluid medium, called semen (see p. 233). Flowering plants also retain the egg cell within the tissues of the parent and are independent of external water. Since they are immobile they depend upon external agencies for the transmission of the male sex cell, contained within a resistant pollen grain. The reproductive structures are the flowers. The following points about flowers should be specially noted :

1 They are temporary structures, often produced in very large numbers.

2 Each may be regarded as a specially modified shoot, consisting of special leaves.

3 Each consists of a stem generally bearing four types of floral leaf. In order, from the outside there are sepals, petals, stamens and carpels. Flowers are usually hermaphrodite. Sometimes flowers lack either carpels or stamens and are therefore male or female. These single-sex flowers may occur in the same plant, as in hazel, or in separate plants, as in holly.

Pollination

Pollination is the transfer of pollen grains from the ripe anther to the stigma. If this occurs within the same flower or between different flowers on the same plant, it is called self-pollination ; if it occurs between flowers from different plants of the same species it is referred to as cross-pollination. Garden peas are normally self-pollinated. Apple blossom produces no fruit unless pollen from another tree is deposited on the stigma. Dandelion can be self-pollinated if cross-pollination does not occur.

When the anthers are ripe they dehisce, that is, they split down one side and the pollen grains are exposed. These are brushed on to the body of an insect visitor and later rubbed off onto a stigma, or else they are blown by the wind. Flowers differ greatly, depending on whether they are insect or wind pollinated.

Wind pollinated flowers

Small, inconspicuous and usually gathered together into inflorescences.

Petals and sepals much reduced in size, or green, or entirely absent.

Anthers large and pendulous, i.e. the filaments are long, carrying them clear of the flower.

Anthers are usually versatile, i.e. they are attached only at their mid points so that they shake about in the slightest breeze.

Large quantities of smooth, dry pollen are produced.

Stigmas are large, branched and feathery, suspended clear of the flower by long styles.

The inflorescence is carried on a long stalk high above the leaves, as in grasses, or appears before the leaves, as in many trees.

Neither scent nor nectar is produced.

(Most British trees and grasses are wind pollinated : many trees possess single sex flowers, for example, hazel and poplar.)

Insect pollinated flowers

Sometimes solitary or in the form of an inflorescence.

Conspicuous, with brightly coloured petals, or gathered together in an inflorescence if relatively inconspicuous.

Produce nectar and scent.

Produce less pollen (except in 'pollen' flowers, which produce no nectar).

Pollen is rough and sticky.

Flower structure often shows complex adaptation, often to a particular kind of insect to increase the effectiveness of pollination.

Since most plants are cross-pollinated, and therefore cross-fertilized, they often exhibit structural features which either prevent self-pollination, or make cross-pollination more probable. The stamens may have short filaments so that the anthers are below the level of the stigma. They may shed their pollen outwards away from the stigma. The stigmas and anthers may ripen at different times in the same flower. The stamens and stigmas may be separated into different flowers. The growth of pollen grains, when they land on the stigma in the same flower, may be slowed down or inhibited altogether.

Fertilization

After successful pollination the following events take place :

The pollen begins to germinate under the stimulus of the sugary secretion produced by the stigma.

The inner membrane, together with its contents, puts out a tube, the pollen tube, which grows out through one of the pores in the outer wall of the pollen grain.

The two pollen grain nuclei pass into the pollen tube.

Types of fruit

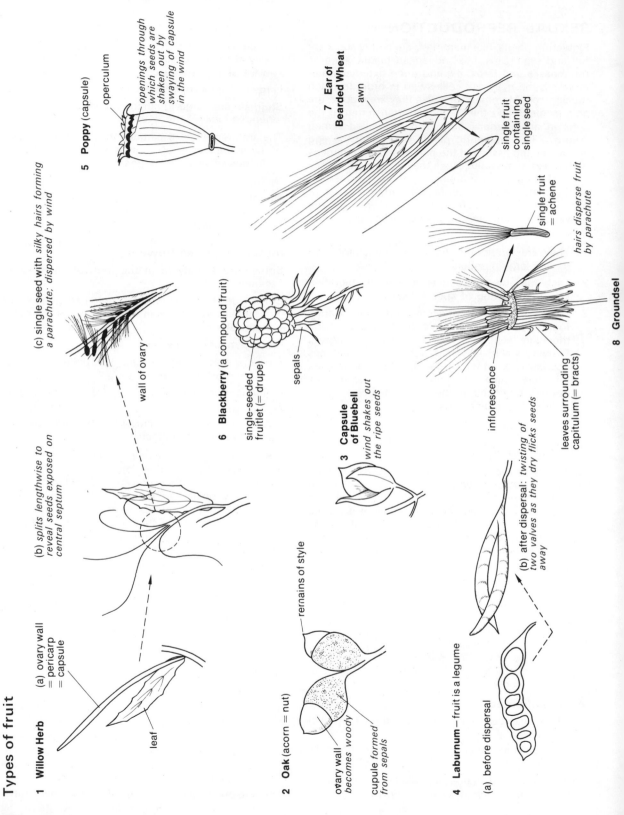

1 Willow Herb

(a) ovary wall = pericarp = capsule

leaf

(b) *splits lengthwise to reveal seeds exposed on central septum*

(c) single seed with silky hairs forming *a parachute; dispersed by wind*

wall of ovary

5 Poppy (capsule)

operculum

openings through which seeds are shaken out by swaying of capsule in the wind

7 Ear of Bearded Wheat

awn

single fruit containing single seed

2 Oak (acorn = nut)

remains of style

ovary wall *becomes woody*

cupule *formed from sepals*

6 Blackberry (a compound fruit)

single-seeded fruitlet (= drupe)

sepals

3 Capsule of Bluebell

wind shakes out the ripe seeds

single fruit = achene

hairs disperse fruit by parachute

inflorescence

leaves surrounding capitulum (= bracts)

8 Groundsel

4 Laburnum – fruit is a legume

(a) before dispersal

(b) *after dispersal: twisting of two valves as they dry flicks seeds away*

The pollen tube is positively chemotropic and negatively aerotropic, i.e. it is attracted by the chemical secretions of the stigma and it grows away from the air.

It grows down the style and into the ovary wall. Penetrating the ovary wall, it enters the micropyle of the ovule.

One of the pollen grain nuclei disappears; as the pollen tube nucleus, its task is finished. The other divides into two.

After entry of the pollen tube into the ovule, one nucleus fuses with the nucleus of the egg cell, the other fuses with the nucleus of the endosperm cell.

The formation of the fruit and seed

After fertilization has occurred, most of the flower parts wither away, since their purpose has been served. Traces of some of them may still persist after the fruit has formed; for example, in apples the remains of the petals and sepals may be seen at the point opposite the stalk. In oranges they may be seen at the point of attachment. In some plants, some of the flower parts may persist, with a different function, connected with dispersal. The style may form a hooked spine in animal-dispersed fruits such as *Avens*. The pappus forms a parachute in dandelion (see p. 224). The receptacle sometimes swells and becomes succulent as in apple. In all plants the main changes after fertilization involve the ovary and its contents. These can be summarized:

Ovary: Its wall is now called the **pericarp**. It increases in size, becoming:

leathery, as in buttercup

woody, as in oak or hazel

dry and brittle, as in most legumes

or fleshy as in succulent fruits, with differentiation into two or more distinct layers (see p. 222).

Ovule: This becomes the **seed**. Its integument becomes the **testa**. The micropyle, which serves for the entry of the pollen tube, persists and serves for the entry of water in the germinating seed. The fertilized egg cell develops into the embryo plant. The endosperm cell also undergoes division, forming a tissue called the endosperm. This grows by digesting the tissue of the nucellus which disappears. In non-endospermic seeds the endosperm is digested and absorbed by the fleshy cotyledons of the embryo.

The transformed ovary with its contents becomes the **fruit**.

The dispersal of fruits and seeds

The function of the fruit is to make possible the dispersal of the seed. It may do this by the active expulsion of the seeds by some kind of mechanical means, or it may be dispersed itself together with the contained seeds. The following is a summary of the main ways in which dispersal is achieved.

1 Wind dispersal

The fruit (more rarely the seed itself) has a large surface area in relation to its volume, therefore offering proportionately greater resistance to the air as it falls towards the ground. This delays its rate of descent and the greater the wind speed, the farther away it lands from the parent plant. Large surface area is achieved in several ways.

(a) *The development of wings*

Examples: Elm, sycamore, ash—the wing is an outgrowth of the pericarp.

Lime—the wing is a persistent bract.

(b) *Parachutes*

Examples: Dandelion, groundsel, ragwort—formed from pappus.

Clematis—formed from persistent style.

Willow herb—the seed develops hairy outgrowths which form the parachute.

(c) *Small size*

This does not apply to fruits, but some seeds are small enough to be dispersed by the wind without the development of special aids. Example: Orchids.

2 Censer mechanisms

The fruit sometimes opens, or develops openings which allow the seeds to be shaken out when the wind blows the fruit about on its stalk. Sometimes the apertures are closed except in dry conditions.

Examples: Poppy, bluebell.

3 Water dispersal

Not common. Generally found in plants which live in or near water. The seed or fruit has a spongy or fibrous layer in which air is trapped, causing the seed or fruit to float.

Examples: Seed covered with spongy 'aril'. This floats for a short period while being carried by current and eventually sinks.

Examples: water lily, alder.

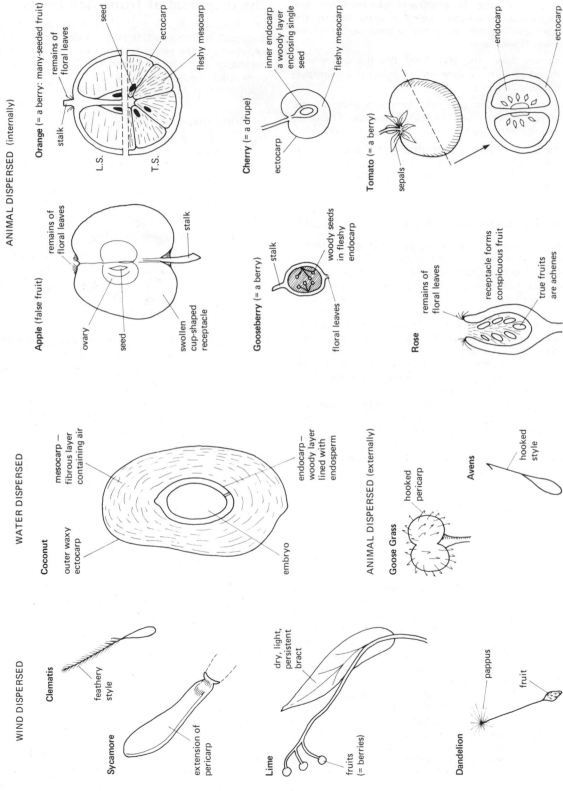

Dispersal of fruits

WIND DISPERSED

Clematis — feathery style

Sycamore — extension of pericarp

Lime — dry, light, persistent bract; fruits (= berries)

Dandelion — pappus; fruit

WATER DISPERSED

Coconut — mesocarp — fibrous layer containing air; outer waxy ectocarp; endocarp — woody layer lined with endosperm; embryo

ANIMAL DISPERSED (externally)

Goose Grass — hooked pericarp

Avens — hooked style

ANIMAL DISPERSED (internally)

Orange (= a berry: many-seeded fruit) — seed; remains of floral leaves; ectocarp; fleshy mesocarp; stalk; L.S.; T.S.

Cherry (= a drupe) — inner endocarp a woody layer enclosing single seed; fleshy mesocarp; ectocarp

Tomato (= a berry) — endocarp; ectocarp; sepals

Apple (false fruit) — remains of floral leaves; stalk; ovary; seed; swollen cup-shaped receptacle

Gooseberry (= a berry) — stalk; woody seeds in fleshy endocarp; floral leaves

Rose — remains of floral leaves; receptacle forms conspicuous fruit; true fruits are achenes

Fruit possessing a fibrous layer containing air and an outer resistant layer. Carried by ocean currents and tides. Example : coconut.

4 Animal dispersal

(a) *The fruit or seed develops hooks which catch in the fur of animals.*

Examples: Goose-grass. The pericarp develops hooks.

Wood avens. The hook is a persistent style

Burdock. The whole inflorescence is dispersed by persistent hooked bracts.

(b) *The fruit is eaten by an animal (succulent fruits).* Brightly coloured. Fleshy tissue which is edible and sweet tasting. Either the inner wall of the fruit, or the testa of the seed is woody, to prevent the digestion of the seed. In the 'false' fruits, the succulent tissue devel-ops from a different part of the flower, such as the receptacle (see strawberry, apple, etc.)

Examples : Cherry, plum (drupes). Single-seeded fruits, the inner wall forming a stone containing the seed.

Blackberry, raspberry. Each flower gives rise to a collection of tiny drupes.

Gooseberries, currants, grapes, tomatoes, bananaś, oranges, cucumbers. These are all 'berries'. There is no woody inner wall to the fruit. There are several seeds each having a woody testa.

Strawberry. The receptacle is the succulent part. The fruits are woody achenes on its surface.

Apple, pear. The succulent tissue is the receptacle. The ovary is the 'core' containing seeds with a woody testa (pips).

Creeping buttercup (*Ranunculus repens*)— a dicotyledon

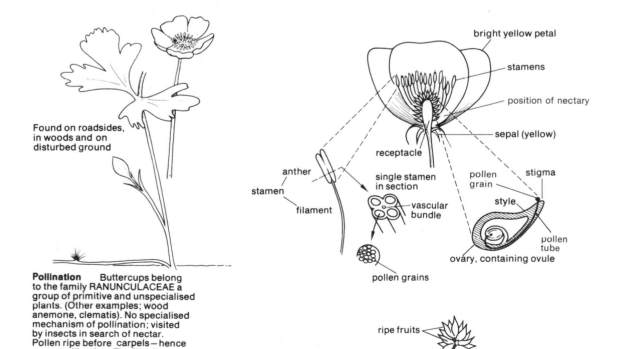

Found on roadsides, in woods and on disturbed ground

Pollination Buttercups belong to the family RANUNCULACEAE a group of primitive and unspecialised plants. (Other examples; wood anemone, clematis). No specialised mechanism of pollination; visited by insects in search of nectar. Pollen ripe before carpels—hence cross-pollination. The floral leaves die away, and wall of ovary becomes tough and leathery. Dispersal mechanism unspecialised—possibly wind dispersal.

5   Mechanical dispersal

The seeds are dispersed from the fruit, which remains attached to the plant. Usually caused by unequal drying of the different layers of the pericarp, which therefore acts as a catapult.

Examples:

> All legumes—pea, laburnum, broom, gorse.
>
> Some capsules—violet, pansy.

6   Casual dispersal

Many seeds are dispersed accidentally. They may be picked up in particles of mud on the feet of birds and mammals and deposited some distance from the place where they were first dispersed from the parent plant. Others may be dispersed by animals which use fruits or seeds as a winter store of food, for example nuts collected by squirrels.

VARIATION IN FLOWER STRUCTURE

Flowering plants are classified mainly on the basis of their flower structure into a number of families. The following five flowers have been selected as representing five basic patterns in flower structure. These are:

1   Buttercup—perfect flower, symmetrical about a central axis.

2   Lupin—imperfect flower, symmetrical about one plane only.

3   Dandelion—composite flowers (miniature flowers or florets) gathered into an inflorescence which mimics in some ways the structure of a solitary flower. (Typical of the Compositae.)

4   Iris—monocotyledon flower, no differentiation into sepals and petals, floral leaves in threes or multiples of three; often elegant, complex flowers.

5   Rye Grass—wind pollinated flower.

Dandelion *(Taraxacum officinale)* - a dicotyledon

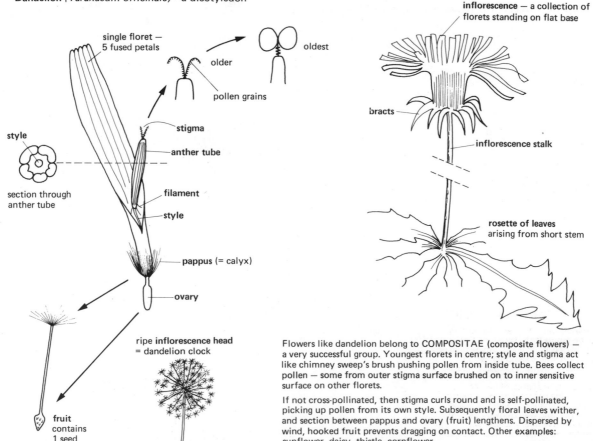

Flowers like dandelion belong to COMPOSITAE (composite flowers) — a very successful group. Youngest florets in centre; style and stigma act like chimney sweep's brush pushing pollen from inside tube. Bees collect pollen — some from outer stigma surface brushed on to inner sensitive surface on other florets.

If not cross-pollinated, then stigma curls round and is self-pollinated, picking up pollen from its own style. Subsequently floral leaves wither, and section between pappus and ovary (fruit) lengthens. Dispersed by wind, hooked fruit prevents dragging on contact. Other examples: sunflower, daisy, thistle, cornflower.

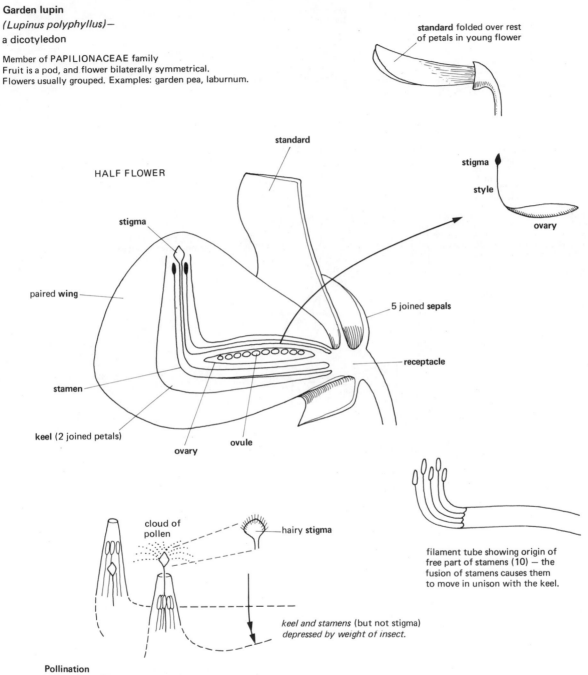

Garden lupin

(Lupinus polyphyllus)—

a dicotyledon

Member of PAPILIONACEAE family
Fruit is a pod, and flower bilaterally symmetrical.
Flowers usually grouped. Examples: garden pea, laburnum.

standard folded over rest
of petals in young flower

HALF FLOWER

standard

stigma

style

ovary

stigma

paired **wing**

5 joined **sepals**

receptacle

stamen

keel (2 joined petals)

ovary

ovule

cloud of
pollen

hairy **stigma**

filament tube showing origin of
free part of stamens (10) — the
fusion of stamens causes them
to move in unison with the keel.

keel and stamens (but not stigma)
depressed by weight of insect.

Pollination
Hairy stigma, hinged separately from
keel and stamens, acts like chimney
sweep's brush driving out pollen in a
shower against bee's undersurface.
Bee visits flower in search of nectar
at base of petals.

After fertilisation
Floral leaves disappear and ovary becomes
a fruit (legume). This splits along one side
as it dries, twists, and flicks out seeds
(explosive mechanism).

Reproduction

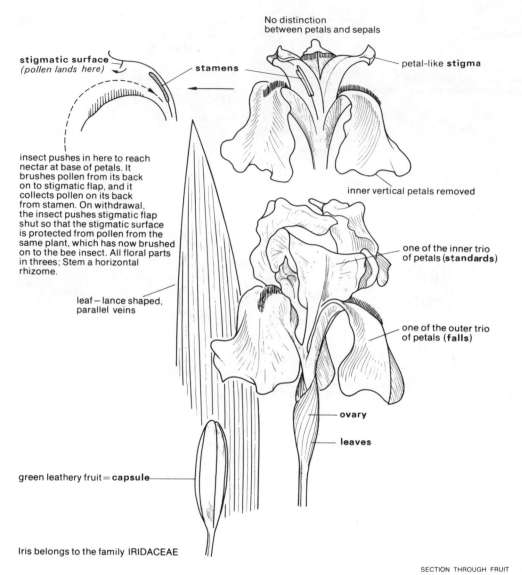

No distinction between petals and sepals

petal-like **stigma**

stamens

stigmatic surface
(pollen lands here)

insect pushes in here to reach nectar at base of petals. It brushes pollen from its back on to stigmatic flap, and it collects pollen on its back from stamen. On withdrawal, the insect pushes stigmatic flap shut so that the stigmatic surface is protected from pollen from the same plant, which has now brushed on to the bee insect. All floral parts in threes; Stem a horizontal rhizome.

inner vertical petals removed

leaf — lance shaped, parallel veins

one of the inner trio of petals (**standards**)

one of the outer trio of petals (**falls**)

ovary

leaves

green leathery fruit = **capsule**

Iris belongs to the family IRIDACEAE

Iris(*Iris sibirica*) — a monocotyledon, insect pollinated

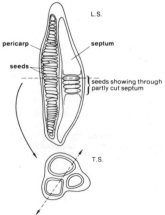

SECTION THROUGH FRUIT

L.S.

pericarp

septum

seeds

seeds showing through partly cut septum

T.S.

Perennial rye grass

(*Lolium perenne*)—a monocotyledon, wind pollinated

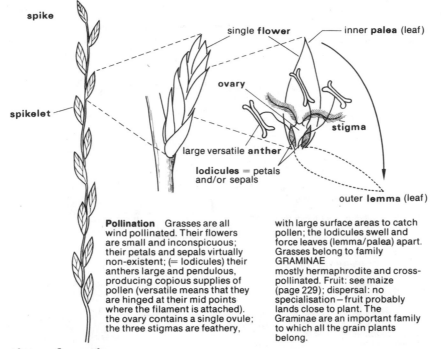

spike

spikelet

single **flower**

inner **palea** (leaf)

ovary

stigma

large versatile **anther**

lodicules = petals and/or sepals

outer **lemma** (leaf)

Pollination Grasses are all wind pollinated. Their flowers are small and inconspicuous; their petals and sepals virtually non-existent; (= lodicules) their anthers large and pendulous, producing copious supplies of pollen (versatile means that they are hinged at their mid points where the filament is attached). the ovary contains a single ovule; the three stigmas are feathery,

with large surface areas to catch pollen; the lodicules swell and force leaves (lemma/palea) apart. Grasses belong to family GRAMINAE mostly hermaphrodite and cross-pollinated. Fruit: see maize (page 229); dispersal: no specialisation—fruit probably lands close to plant. The Graminae are an important family to which all the grain plants belong.

Germination of seeds

Seeds do not normally germinate as soon as they are formed except, perhaps, in ephemerals, which are plants which produce several generations in one summer and are therefore plants with a very short life span. Seeds are a survival stage, and are resistant to cold and to desiccation. It is therefore necessary that they should not germinate before favourable conditions exist. Germination thus follows a period of dormancy, and sometimes subjection to low temperatures is essential if germination is to occur.

The conditions under which germination normally occurs are the same conditions under which enzymes work. Growth must be preceded by the breakdown of insoluble food reserves, and this is carried out by enzymes. Provided that water is present and the temperature is above a necessary minimum, germination will start, but no visible external changes will be apparent unless oxygen is also present.

Experiment to demonstrate the conditions necessary for germination

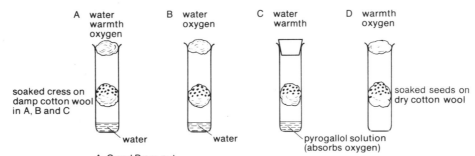

A water warmth oxygen

B water oxygen

C water warmth

D warmth oxygen

soaked cress on damp cotton wool in A, B and C

soaked seeds on dry cotton wool

water

water

pyrogallol solution (absorbs oxygen)

A, C and D are put in a warm place, B is placed outside if wintertime or in a refrigerator. Test tube A acts as a control for each of the others and is the only one in which germination occurs.

Non-endospermic seeds

food stored in swollen seed leaves (cotyledons)

Broad bean
Food reserve: starch and proteins. Cotyledons remain below ground (= **hypogeal**).

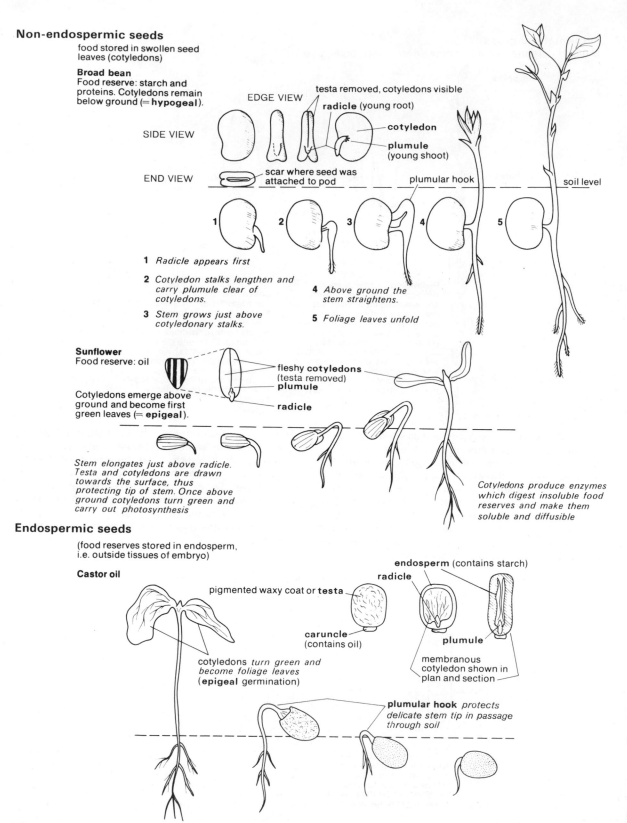

EDGE VIEW

testa removed, cotyledons visible

radicle (young root)

SIDE VIEW

cotyledon

plumule (young shoot)

END VIEW

scar where seed was attached to pod

plumular hook

soil level

1   2   3   4   5

1 *Radicle appears first*

2 *Cotyledon stalks lengthen and carry plumule clear of cotyledons.*

3 *Stem grows just above cotyledonary stalks.*

4 *Above ground the stem straightens.*

5 *Foliage leaves unfold*

Sunflower
Food reserve: oil

fleshy **cotyledons** (testa removed)

plumule

Cotyledons emerge above ground and become first green leaves (= **epigeal**).

radicle

Stem elongates just above radicle. Testa and cotyledons are drawn towards the surface, thus protecting tip of stem. Once above ground cotyledons turn green and carry out photosynthesis

Cotyledons produce enzymes which digest insoluble food reserves and make them soluble and diffusible

Endospermic seeds

(food reserves stored in endosperm, i.e. outside tissues of embryo)

Castor oil

endosperm (contains starch)

pigmented waxy coat or **testa**

radicle

caruncle (contains oil)

plumule

membranous cotyledon shown in plan and section

cotyledons *turn green and become foliage leaves* (**epigeal** germination)

plumular hook *protects delicate stem tip in passage through soil*

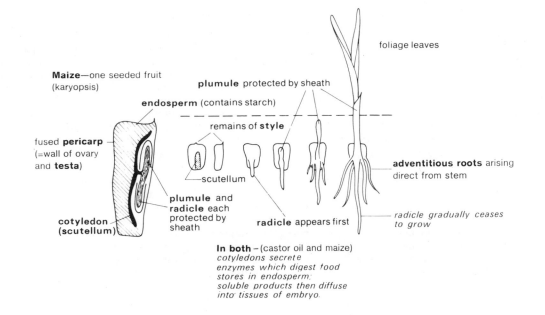

Maize—one seeded fruit
(karyopsis)

plumule protected by sheath

foliage leaves

endosperm (contains starch)

remains of **style**

fused **pericarp**
(=wall of ovary
and **testa**)

scutellum

adventitious roots arising
direct from stem

plumule and
radicle each
protected by
sheath

cotyledon
(scutellum)

radicle appears first

*radicle gradually ceases
to grow*

In both – (castor oil and maize)
*cotyledons secrete
enzymes which digest food
stores in endosperm;
soluble products then diffuse
into tissues of embryo.*

Mammalian reproduction

As land vertebrates, mammals show many features in their mode of reproduction which are adaptations to life on dry land. Aquatic vertebrates, such as the fishes, mostly shed eggs and sperm into the water, where fertilization takes place. Amphibians, like the frog (see p. 67), must return to water for the purpose of reproduction. In both cases the egg is supplied with yolk which acts as a food store for the developing embryo until such time as it is able to fend for itself. The mortality rate is high and so also is the production of fertilized eggs.

Reptiles reproduce entirely on land. Fertilization of the egg is internal. It is laid encased in a hard shell and contains a large quantity of yolk. The embryo growing inside is enclosed within fluid-filled membranes so that it is effectively growing inside its own private pond and it will hatch at a much later stage in its development than the frog. In fact, the period in the frog's life up to the time of metamorphosis corresponds to the development of the reptile up to the time of hatching. Hence the much greater supply of yolk. The situation is very similar in birds, the main differences being that the egg is kept warm by the parents so that it develops at constant temperature, and the parents also feed the fledgling, which is still relatively helpless when hatched.

In mammals the egg is **fertilized internally** and probably more effectively than in the reptiles or birds. The male possesses a **penis** which is introduced into the **vagina** of the female during mating. The sperm cells are therefore always contained within a fluid medium from the time they leave the testis, where they are produced, until they meet the eggs in the upper part of the **oviduct**. The oviduct in the mammal is specially modified in its lower region. Part is modified to form a **uterus**. The fertilized egg is not released from the body, neither is it particularly yolky. Instead it attaches itself to the wall of the uterus and develops inside the mother's body. The last part of the oviduct, the only unpaired structure, the vagina, receives the penis of the male during mating and acts as the birth canal when the young mammal is ready to be born.

When the sexes are separate, as they are in mammals, mechanisms must exist to bring about the meeting between egg and sperm cell. In most mammals the key role is played by the females. Males are generally able to produce sperm and to mate at any time of the year, or at least, at any time during the **breeding season**. Most mammalian species breed once or perhaps twice a year. The eggs are shed into the oviduct and the female is said to be **in heat**. If she is isolated from the male at this time she very soon reverts to a sexually inactive state which persists until the next time that **ovulation** occurs, that is, until the next breeding season. In normal circumstances the male is attracted to the female, often by her scent, which is given off by the secretions of special glands. Sometimes **mating** is preceded by an elaborate form of **courtship behaviour**. Whether fertilization occurs or not the female ceases to be in heat after a few days and reverts to the sexually inactive state.

Some mammals, necessarily small ones, are able to breed all the year round, or through a considerable part of the year, producing a succession of litters. The female comes into heat at regular intervals. If mating occurs and she becomes pregnant she will next be in heat at or about the time that she finishes suckling her young.

In the rabbit the **oestrus cycle** lasts seven days. The female is in heat every seventh day. although she is receptive to the male to a lesser degree at other times. Ovulation does not normally occur however in the absence of the male, and is usually induced by the act of mating. Sometimes the mere sight of the male is enough to cause ovulation. The female produces litters at monthly intervals from January onwards. Breeding ceases in the autumn because at this time the males become sexually inactive. They start to moult, the testes are retracted into the body cavity, and they become temporarily sterile.

Development

Development of the egg begins only after a sperm cell has entered it, i.e. after fertilization has occurred. The entry of a single sperm cell triggers off changes in the egg which prevent the entry of other sperm. Although the sperm look alike there are, in fact, two kinds, and the sex of the new individual is determined by which type of sperm enters the egg. By the time the egg has reached the uterus it has already divided several times and consists of a solid ball of cells. It sinks into the thickened wall of the uterus and the **embryo**, as it is now called, continues to undergo rapid cell division. The **implanting** of the embryo in the wall of the uterus is communicated, in a way at present unknown, to the ovary, where the **corpus luteum** (see below), continues to enlarge and to produce the pregnancy hormone which halts the production of further eggs. The cells of the embryo become rearranged to mark out the main regions of the body and then they begin to change into blood cells, nerve cells, muscle

Female reproductive system (rabbit)

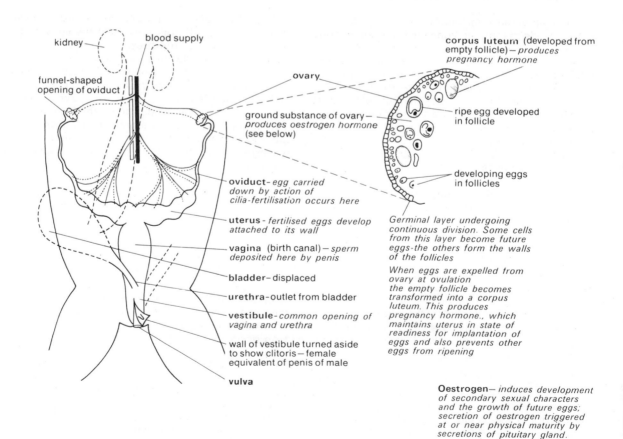

kidney

blood supply

funnel-shaped opening of oviduct

ovary

corpus luteum (developed from empty follicle) — *produces pregnancy hormone*

ripe egg developed in follicle

ground substance of ovary — *produces oestrogen hormone (see below)*

developing eggs in follicles

oviduct — *egg carried down by action of cilia-fertilisation occurs here*

uterus - *fertilised eggs develop attached to its wall*

vagina (birth canal) — *sperm deposited here by penis*

bladder — *displaced*

urethra — *outlet from bladder*

vestibule — *common opening of vagina and urethra*

wall of vestibule turned aside to show clitoris — *female equivalent of penis of male*

vulva

Germinal layer undergoing continuous division. Some cells from this layer become future eggs-the others form the walls of the follicles

When eggs are expelled from ovary at ovulation the empty follicle becomes transformed into a corpus luteum. This produces pregnancy hormone., which maintains uterus in state of readiness for implantation of eggs and also prevents other eggs from ripening

Oestrogen — *induces development of secondary sexual characters and the growth of future eggs; secretion of oestrogen triggered at or near physical maturity by secretions of pituitary gland.*

cells and so on. At a very early stage the main organ systems are recognizable, and the greater part of uterine development is taken up by growth in size.

The embryo is at first dependent upon the fluid in the uterus for its nutrition but soon it develops a structure, the **placenta**, which is in part an outgrowth of its own gut. This forms close connections with the uterine wall, and develops blood vessels which lie close to the uterine blood vessels; exchange of respiratory gases, dissolved food materials and excretory products occurs between the embryo blood stream and the mother's circulation. When the embryo, or **foetus** as it is usually called in the later stages, reaches full term, slow rhythmic contractions of the muscular uterine wall, rather like the peristaltic action of the intestine, push the embryo along the uterus. Once the foetus passes

into the vagina, it is rapidly expelled from the body, shortly followed by the placenta, still connected to the newly born infant by the **umbilical cord**. The mother bites the cord, through which blood soon ceases to flow, she eats the placenta and washes her babies. The **mammary glands** have been developing during pregnancy, and suckling soon begins. The period from fertilization to birth, in the rabbit takes about twenty-eight days. If however the mother is exposed to stress, for example, overcrowding or shortage of food, reabsorption of the embryos can occur, even as late as fourteen days after fertilization. Within a day or so after the birth of the litter mating may occur again, and the female may become pregnant once more. Rabbits are unusual in that the female becomes sexually active so soon after the birth of her litter.

Male reproductive system (rabbit)

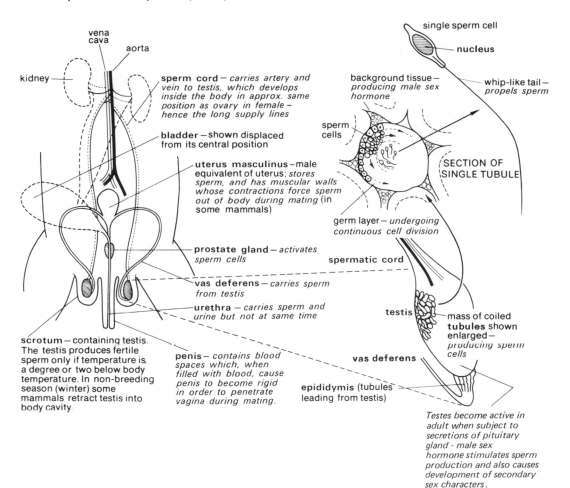

vena cava

aorta

kidney

sperm cord — *carries artery and vein to testis, which develops inside the body in approx. same position as ovary in female — hence the long supply lines*

bladder — shown displaced from its central position

uterus masculinus — *male equivalent of uterus; stores sperm, and has muscular walls whose contractions force sperm out of body during mating (in some mammals)*

prostate gland — *activates sperm cells*

vas deferens — *carries sperm from testis*

urethra — *carries sperm and urine but not at same time*

scrotum — containing testis. The testis produces fertile sperm only if temperature is a degree or two below body temperature. In non-breeding season (winter) some mammals retract testis into body cavity.

penis — *contains blood spaces which, when filled with blood, cause penis to become rigid in order to penetrate vagina during mating.*

single sperm cell

nucleus

background tissue — *producing male sex hormone*

sperm cells

whip-like tail — *propels sperm*

SECTION OF SINGLE TUBULE

germ layer — *undergoing continuous cell division*

spermatic cord

testis

mass of coiled tubules shown enlarged — *producing sperm cells*

vas deferens

epididymis (tubules leading from testis)

Testes become active in adult when subject to secretions of pituitary gland - male sex hormone stimulates sperm production and also causes development of secondary sex characters.

MAMMALIAN REPRODUCTIVE CYCLE—FEMALE

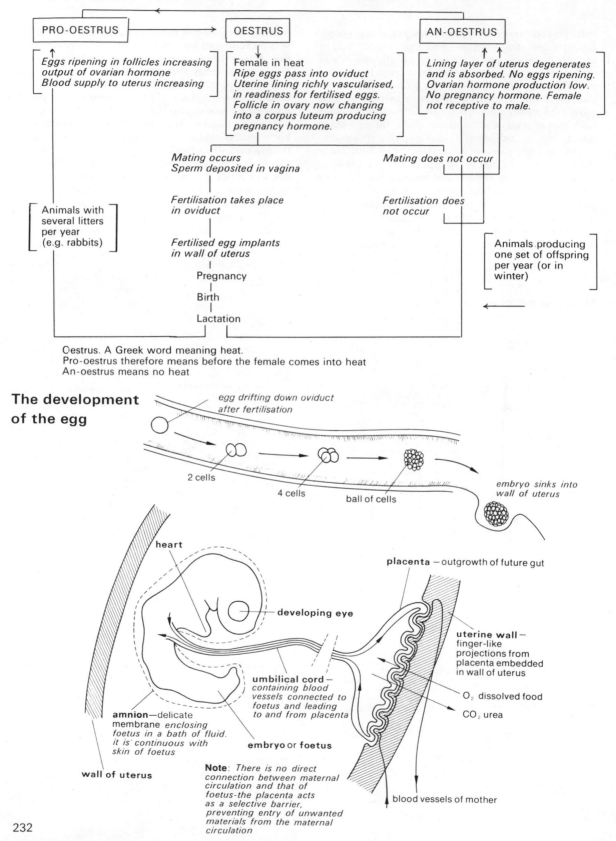

| PRO-OESTRUS | → | OESTRUS | | AN-OESTRUS |

*Eggs ripening in follicles increasing output of ovarian hormone
Blood supply to uterus increasing*

Female in heat
Ripe eggs pass into oviduct
Uterine lining richly vascularised, in readiness for fertilised eggs.
Follicle in ovary now changing into a corpus luteum producing pregnancy hormone.

Lining layer of uterus degenerates and is absorbed. No eggs ripening. Ovarian hormone production low. No pregnancy hormone. Female not receptive to male.

*Mating occurs
Sperm deposited in vagina*

Mating does not occur

Animals with several litters per year (e.g. rabbits)

Fertilisation takes place in oviduct

Fertilisation does not occur

Fertilised egg implants in wall of uterus

Pregnancy

Birth

Lactation

Animals producing one set of offspring per year (or in winter)

Oestrus. A Greek word meaning heat.
Pro-oestrus therefore means before the female comes into heat
An-oestrus means no heat

The development of the egg

egg drifting down oviduct after fertilisation

2 cells

4 cells

ball of cells

embryo sinks into wall of uterus

heart

placenta — outgrowth of future gut

developing eye

uterine wall—
finger-like projections from placenta embedded in wall of uterus

umbilical cord—
containing blood vessels connected to foetus and leading to and from placenta

O_2 dissolved food

CO_2 urea

amnion—*delicate membrane enclosing foetus in a bath of fluid. it is continuous with skin of foetus*

embryo or **foetus**

wall of uterus

Note: *There is no direct connection between maternal circulation and that of foetus-the placenta acts as a selective barrier, preventing entry of unwanted materials from the maternal circulation*

blood vessels of mother

232

HUMAN REPRODUCTION

Since man is a mammal his reproductive processes are basically similar to those of other mammals. There are, however, some important differences:

1 The uterus, unlike that of most mammals, is a single structure.

2 Only one egg is shed at ovulation so that only one embryo develops at a time. Multiple births do occur, but rarely. Twins, which are most common, occur once in about eighty births.

3 Possibly because single births are the rule, development within the uterus takes much longer (40 weeks) in relation to size, than in any other mammal.

4 The reproductive cycle in the human female is called the menstrual cycle. It lasts for approximately twenty-eight days, although there may be considerable variation about this figure. In other mammals, except the apes and some monkeys, the thickened lining of the uterus is gradually re-absorbed if fertilization does not take place. In human beings it breaks down suddenly and leaves the body as the menstrual flow. This flow begins about fourteen days after ovulation and may last from three to five days. It consists of a mixture of blood and cell debris. The menstrual flow, or **menstruation**, as it is often called, takes its name from the Latin word *menstruus*, meaning monthly. It begins, irregularly at first, at about the age of thirteen, and recurs at about twenty-eight day intervals until the individual is about fifty years old, being interrupted only by pregnancy. It used to be thought that menstruation was due to the influence of the moon, which also has a twenty-eight-day cycle. Primitive peoples still regard the whole process with awe and often treat a woman as an evil, unclean thing to be avoided at all costs when she is menstruating.

5 There is nothing in the behaviour of human females corresponding to the sexual behaviour of other female mammals at the time of ovulation, that is, at oestrus. This is a very short time when the non-human mammalian female is not only receptive but may actually seek out the male. At other times she actively drives off any male who tries to approach her sexually. In the human female there is no period of heat; in fact there are no obvious indications that ovulation has occurred. On the other hand there is nothing corresponding to an-oestrus in human beings. There is no point in the female's reproductive period when she is unable to engage in mating behaviour. This is not to say that she does but merely that she can. After all, there are many other factors to take into account in human

behaviour besides the purely physiological ones. Man is a social animal and, as far as studies of many primitive human cultures can tell, generally monogamous (marriage between one man and one woman). The different pattern of reproduction in human beings is therefore important as a kind of social cement. Indeed, organized society would have been impossible under other circumstances because, apart from creating the basis for continuing bonds between individual men and women, it creates, more important still, the basis for a stable, secure environment for children, who take longer to grow to maturity and independence than any other species of young mammal.

Puberty and adolescence

For the first dozen years of life boys and girls are very much alike. The ovaries in girls and the testes in boys remain undeveloped, as do all other parts of the reproductive system. Dress children alike, trim their hair in the same style and length, and they are almost indistinguishable in appearance, if not in behaviour. The beginning of puberty is marked by the development of the secondary sexual characters. The pituitary gland at the base of the brain begins to secrete hormones which stimulate the ovaries, or testes, to produce germ cells, and also their respective sex hormones. In girls this stage is marked by the commencement of menstruation. The breasts begin to develop and fat deposits are laid down under the skin to form the typical adult female shape. Hair grows under the arms and in the pubic region. In boys corresponding changes occur, although somewhat later than in girls. The body becomes more muscular. The Adam's apple enlarges and the voice deepens. Hair grows in the armpit and in the pubic region. It may appear on the chest and the face need shaving. The penis and testes enlarge and occasional discharges of semen (the fluid containing the sperm cells) occur during sleep. Both boys and girls begin to show a heightened awareness of and interest in the opposite sex.

Puberty, therefore, refers to all those physical changes which transform a child into an individual capable of sexual reproduction. There is, however, more to the business of becoming a man or woman than a capacity to reproduce. Adolescence generally refers to the period during which a person changes from an essentially childish outlook on life to the mature and independent outlook of a typical adult. Adolescence begins when puberty begins but it continues usually into the late teens. It is a transition stage when the individual is neither child nor adult. Adolescents are often said, by their parents, to be difficult. Parents are often regarded by their adolescent children with resentment, for all sorts of reasons. There is room for understanding on both sides, but more particularly by the adults. It is a time when young people are particularly sensitive and unsure of themselves, and this is reflected in a readiness to blush, a tendency to spots and a great concern with their personal appearance and other people's reaction to it. They are especially sensitive to the opposite sex.

COMPARISON OF SEXUAL REPRODUCTION IN FLOWERING PLANTS AND MAMMALS

The fact that we refer to sexual reproduction in flowering plants and in mammals indicates that there are certain features of the process which are common to both. Nevertheless, since they differ so fundamentally there are various ways in which the basic processes of sexual reproduction differ also. Some of these are summarized below:

Similarities

1   In both kinds of organism sexual reproduction involves the fusion of unlike gametes.

2   In both, fertilization occurs within the tissues of the parent, in the ovule of the flower, in the oviduct of the female mammal.

3   The embryo begins its development internally in each case.

4   After separation from the parent each has its needs catered for. The embryo plant is dispersed with a supply of food and the young mammal is fed by maternal secretions

5   In each case the offspring differs from its parent.

Differences

1   Mammals are dioecious, that is, the sexes are always separate. Plants are usually hermaphrodite, although some species are dioecious.

2   The reproductive organs of plants are the flowers, which are temporary structures renewed annually. Mammals are born with an immature set of sex organs which function when the animal is mature and remain functional for most or all of its life.

3   The male sex cells in the plant are transferred passively, by wind or insect, and enclosed in a tough resistant case, the pollen grain. Mammals are mobile and the male sex cells are transferred during the act of mating.

4   Fertilization in the plant is aided by the active growth of the pollen tube, since the male gamete is inactive. The male sperm cell actively swims in search of the ovum.

5   After fertilization the mammalian egg begins a continuous process of development at constant temperature. This continuous development continues after birth, being dependent only upon availability of food. The embryo in the flowering plant undergoes an initial period of development, which then ceases and is followed by a dormant period. Growth subsequently depends on the availability of water and suitable temperature.

6   Mammals are dispersed by the fact that they are mobile. Flowering plants depend on external factors for their dispersal, i.e. the wind, animals, water, or the mechanisms built into the fruit.

7   In flowering plants the production of offspring is very great but the survival rate is correspondingly low. In mammals few offspring are produced but their chances of survival are very much higher.

---

1   Collect examples of as many flowers as possible. Try to classify or group them on the basis of their structure. Suggest in each case how pollination might occur.

2   Lupins can be grown quite easily from seed. When the flowers begin to appear, cover a number of them, before they ripen, with small muslin bags to prevent insects from reaching them. As the flowers mature, operate the pollen dispersal mechanism in half your covered specimens. Remove the muslin bag, press down the keel with the tip of a pencil then replace the muslin. Try to answer these questions. Are lupins normally cross-pollinated? Can they be self-pollinated? Is the visit of an insect an essential feature of pollination?

3   Place some pollen in a dilute sugar solution and examine at intervals for signs of growth. Add a small piece of stigma from the same species of flower to the sugar solution and observe the behaviour of the pollen tubes.

4   Collect as many fruits as possible and classify them according to their means of dispersal.

5   Examine prepared sections of ovary and testis under the microscope. If possible study prepared mounted preparations of male and female reproductive systems.

6   If possible, carry out a study of the life history of a small mammal such as can be kept in the biology laboratory, e.g. hamster, mouse, rat or guinea pig.

The study of inheritance is called Genetics (from Gk. *genesis*, creation, generation). It seeks to provide the answers to two questions. Why do living things resemble each other? Why are they different? All members of a given species, for example, man, are distinctly different from members of another species. All are sufficiently similar as to leave no room for doubt that they are members of the same group, yet at the same time each and every individual is distinguishable from every other in the group. There are, of course, degrees of difference. It is easy to distinguish between a European and a Chinese. It is difficult to distinguish between identical twins. Although in the last resort each human being is unique, for instance no two sets of fingerprints are ever identical, all share characteristics to a varying extent with each other. For example:

1   All human beings belong to one or other of two groups according to sex. These are mutually exclusive groups existing in approximately equal numbers.

2   All belong to one or other of four groups according to blood type. In Britain, for example, 40 per cent of the population belong to group A, 40 per cent to group O, 15 per cent to group AB and 5 per cent to group B. Frequencies differ in different parts of the world.

3   It is not easy to sort people on the basis of skin, hair and eye colour. Eye colour might be the basis for perhaps a dozen different groups. Skin colour varies according to latitude.

4   When such a feature as height (or weight) is considered it is no longer possible to fit individuals into a limited number of groups, because this is a characteristic which varies continuously. If a thousand men, all of the same age, are taken at random from, say, the south-east of England, and their heights measured, the results can be plotted on a graph. The curve has a characteristic shape. It is technically known as a *normal* curve with most individuals clustered around the average for the group as a whole. If women are measured instead of men the same shape curve is obtained but displaced towards the lower end of the scale because women are on average shorter than men.

As it is with human beings, so it is with other organisms. Primroses fall into two groups according to whether they are pin-eye or thrum. Some members of the buttercup family are variable in the number of petals they have. A random collection of leaves from a tree will show a normal distribution of length.

The situation is slightly different for species which (a) do not reproduce sexually, such as the bacteria, or (b) reproduce sexually but are closely inbred. Some flowering plants for example are normally self-pollinated. In such species variation is much less marked and is confined to narrower limits.

One of the first problems to resolve is what causes variation, in height, eye colour, and so on. Some differences have been shown to be due to environmental factors. The steady increase in the average height of people in Britain over the last seventy years is a consequence of improved diet. On the other hand the differences between Niloti tribesmen, average height over six feet, and pigmies from the tropical rainforest, average height just over four feet, are almost certainly due to inherited factors.

Generally speaking, characteristics which are *qualitative*, such as sex or blood group, in human beings, or height in garden peas (they are either tall or dwarf), are not affected by environmental factors. An individual comes within one or other of a limited number of exclusive groups for each of these characteristics. Attributes such as height in human beings are *quantitative*, and these seem to be affected both by heredity and environment. The children of pigmy parents are never likely to be much more than four and a half feet tall. If they are well fed they are likely to be an inch or two taller than if they are kept on short rations.

QUALITATIVE (UNIT) DIFFERENCES

In order to understand what is involved in the inheritance of qualitative differences in living organisms it will be necessary to examine some fairly simple straightforward cases.

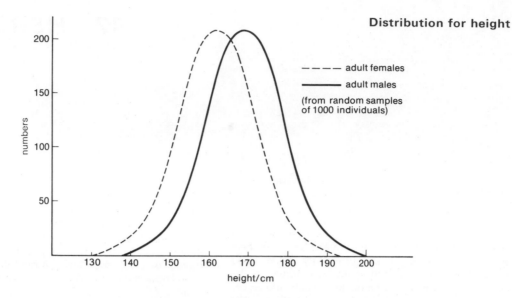

Distribution for height

- - - - - adult females
———— adult males

(from random samples of 1000 individuals)

The first man actually to do this was **Gregor Mendel** (1832-1884) who was abbot at the Augustinian monastery at Brunn, Austria (now Brno, Czechoslovakia). His work was done on the garden pea, which offers a number of advantages for the geneticist. For example, it is normally self-fertilized, although it can be cross-fertilized. It also exhibits a number of clearly defined differences or varieties. Mendel's work was published in an obscure journal in 1866 and it remained undiscovered by the world at large until thirty years later, when its importance was realized.

1   Colour inheritance in the garden pea

Garden peas breed true for flower colour. For example, if peas are planted and the resulting plants produce white flowers, the seeds from those flowers will grow into plants which also produce white flowers. Successive generations continue to produce white flowers and no other colour.

Parent generation

A pure breeding red-flowered plant crossed with a pure breeding white-flowered plant (stamens removed from red flowers, their stigmas dusted with pollen from white flowers—or vice versa).

Seeds from parent generation planted the following year, to produce first filial generation.

These self-pollinate to produce second filial generation consisting of red flowered and white flowered plants in approx. ratio 3:1.

One-third of the red-flowered F.2 breed true. Two thirds behave like F.1 producing both red- and white-flowered plants. White-flowered F.2 breed true.

In this cross the characteristic appearing in the F.1 generation, i.e. red flower, is said to be **dominant**. The characteristic which is suppressed in F.1 but reappears in F.2, namely, white flower, is said to be **recessive**. Pairs of contrasted characteristics which behave in this way are called **alleles**.

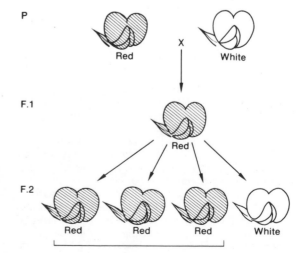

2 Wing form in the fruit fly
(*Drosophila melanogaster*)

Here the sexes are separate. In the F.1 the offspring are allowed to interbreed (brother and sister mating).

(The sexes actually look different but to simplify matters all the specimens in the diagram are drawn alike.)

Most fruit flies have normal wings. There is a variety which has vestigial wings.

The F.1 from such a cross all have normal wings.

F.2 consists of individuals of both kinds in the approx. ratio normal : vestigial = 3 :1

Normal wing is dominant
Vestigial wing is recessive
Normal wing and vestigial are alleles

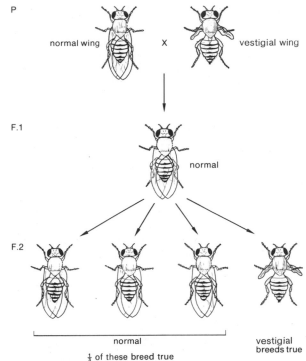

$\frac{1}{3}$ of these breed true
$\frac{2}{3}$ produce vestigial wing offspring

3 Eye colour in man

Human beings exhibit a number of characteristics which are inherited in the same way as the examples already given for the garden pea and the fruit fly. With human beings it is not, of course, possible to carry out experiments. Most of the information is collected by studies of family histories. Ratios cannot be derived from single families because children are not normally produced in large enough numbers. The ratios are obtained by running together the results of studying a large number of families showing the same pattern.

A man with brown eyes and a family history of brown eyes, marries a blue-eyed woman. All their children are brown-eyed. If two such brown-eyed individuals marry, their children will have brown eyes or blue eyes in the approximate ratio 3 :1.

In the parental generation both parents are pure for brown and pure for blue eyes respectively. The F.1 generation are hybrid brown-eyed. Brown is therefore dominant to blue. Two such hybrid brown-eyed individuals produce offspring in the ratio 3 :1 brown to blue.

Each of the examples described above illustrates unit factor inheritance. In each case one of the contrasted characters disappears in the F.1 generation, to reappear in a proportion of the F.2 generation.

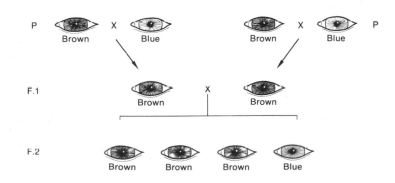

Mendel drew two very important conclusions from his experiments and these are still accepted as valid.

1 Inheritance is always **particulate**. Whatever it is that is transmitted from one generation to the next it must remain separate. Only on this basis can the reappearance of the recessive characteristic be explained.

2 For any pair of contrasted characteristics (alleles), for example, red petal and white petal, only one can be represented in a single gamete. The male or female gamete can contain only one of them. ('Them' in this case refers to whatever it is that causes white or red petal. Mendel called

it a 'germinal factor'.) This is a necessary condition for the reappearance of the recessive characteristic. If it were not separated it could not reappear in the next generation.

The inheritance of red and white petal colour can now be re-stated in symbols. The germinal factor for red will be represented by R and that for white by r. It is conventional to use the same symbol for two alleles and where one is dominant to use a capital letter, the small letter being used for the recessive. Since each individual receives a contribution from each parent it will be represented by two letters.

Parental
generation

Gametes
One kind only in
each case

F.1 a hybrid since its parents are different

Produces two sorts of gamete in equal numbers

Assuming that fusion of gametes is random, all combinations are equally likely

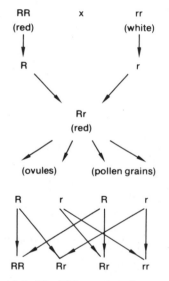

3 red; 1 white; RR is pure breeding, as is rr. Rr produces two sorts of offspring.

The other two examples can be similarly treated. For instance, BB indicates pure brown eyes, Bb hybrid brown eyes and bb pure blue eyes.

So far reference has been made only to the results of breeding experiments. The study of cell behaviour and the significance of events taking place therein was to come much later than the work of Mendel. The events occurring in ordinary dividing cells (mitosis) have already been described (p. 86), as well as those occurring in the special kind of division (meiosis) preceding gamete formation (p. 214). In meiosis the two members of each chromosome pair separate into separate gametes, and as far as can be discovered each pair separates quite independently of every other pair. This, then, is the cell equivalent of Mendel's statement. The **Law of Segregation** can be written as: **For any pair of chromosomes only one can be present**

in a single gamete. The chromosomes are known to be the site of Mendel's germinal factors. Each chromosome carries these germinal units, or **genes**, as they are called, arranged in line order along its length. Since each chromosome is represented twice, each gene is also represented twice. In the garden pea the gene for petal colour is found at a particular position (locus) on a particular chromosome.

From what has been said before it follows that there are two sorts of gene for petal colour. If both are present, as in the F.1 generation, one on each of the paired chromosomes, one, for red colour, completely suppresses the effects of the other. When the gene pair are alike the individual is said to be **homozygous** (Gk. *homo*, same; *zygos*, joining) for the character concerned (RR or rr). When they are different it is said to be **hetero-**

zygous (Gk. *hetero,* mixed, different) (Rr). The fact that all the cells in the plant are derived by normal cell division from the original fusion product of the gametes means that every cell carries the gene pair affecting petal colour, although obviously they show their effects only in the petals. The total complement of chromosomes may carry thousands of gene pairs, affecting all the differing aspects of structure and function. Genes are complex structures which act as sets of chemical instructions, or as miniaturized programmes, controlling the activity of the cell. Every cell carries the same set of instructions. What parts of the programme are implemented depends on where the cell happens to be.

Incomplete dominance

In the examples given so far, where the gene pair is different, one has completely suppressed the effects of the other. Dominance has been complete. This is not always the case. In snapdragons, for example, red-flowered plants crossed with white-flowered plants give very different results from those for the garden pea.

Pure red-flowered plants crossed with pure white-flowered plants. All F.1 have pink flowers.

F.2 consists of $\frac{1}{4}$ red flowered, $\frac{1}{2}$ pink flowered, $\frac{1}{4}$ white flowered.

Multiple alleles

In all the examples so far described, one pair of contrasting characteristics have been involved, that is, two alleles. While no pair of chromosomes can carry more than two alleles at any time it is sometimes the case that the two may be members of a 'family' of alleles. For example, the normal fruit fly has red eyes. There are individuals with eye colour ranging from pale red to white. There are in fact about a dozen alleles for eye colour, any two of which may be present in one individual. Red eye is dominant to all the others, but crosses between any of the other types produce individuals with intermediate eye colour. Human blood groups are determined by a group of three alleles. The factor for Group A is dominant to the factor for Group O. That for Group B is also dominant to Group O. The factors for Groups A and B do not show dominance with respect to each other. The individual inheriting these factors one from each parent is Group AB.

1 If a person with a family history of group A only marries a person of Group O all their children are Group A. These are hybrid for group A. If two such individuals marry they produce children in the ratio 3:1 group A: group O.

2 A similar situation applies to Group B.

3 If pure Group A marries Group B their offspring are AB and the F.2 generation will be as shown.

4 If hybrid A marries hybrid B, all four groups occur in next generation.

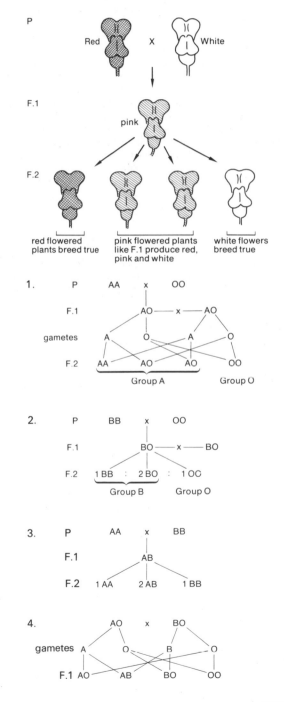

Sex determination

In mammals the determination of sex is dependent upon one pair of chromosomes. In the human female there are twenty-three complete sets of chromosomes. One pair is known as the X chromosomes (X is the twenty-fourth letter of the alphabet and the X chromosomes were so designated when man was thought to have twenty-four pairs of chromosomes). When the number of chromosomes is halved during gamete formation each gamete contains one X chromosome, plus twenty-two others.

In males there are also twenty-three pairs of chromosomes in each body cell. Instead of two X chromosomes, however, there is a single X together with a much smaller Y chromosome which carries no genes, as far as is known. When male gametes are formed, therefore, the X and Y chromosomes separate into separate cells and there are two kinds of sperm. One kind consists of an X chromosome and twenty-two others. The other kind has a Y chromosome and twenty-two others. The sex of the resulting individual depends on which kind of sperm fuses with the egg.

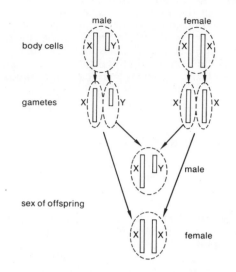

Sex-linked characteristics

The X chromosomes carry genetic factors, just as do other chromosomes. A recessive gene, which must be present in both X chromosomes in order to produce visible effects in females, will produce those effects in males when represented once on the single X chromosome. As a result, sex-linked characteristics show a much higher incidence in males than in females. Three such features are **colour blindness**, **early baldness** and **haemophilia** (blood fails to clot). In the following example

X stands for a normal chromosome carrying the normal gene and X_h indicates a chromosome carrying the recessive gene for haemophilia.

For a woman to be a haemophiliac she must inherit the gene from both parents. Her father will be haemophiliac and probably her maternal grandfather also. In the normal way a sex-linked characteristic is transmitted from father to daughter, who does not show it but transmits it to her sons.

A normal woman marries a man with haemophilia

gametes

Their sons are normal. Their daughters carry the gene for haemophilia on one X derived from father.

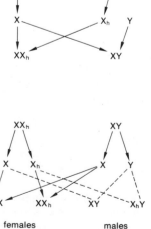

If such a daughter marries a normal man . . . . . .

her gametes are of two kinds.

Half her sons will be haemophiliac. Half her daughters will carry the gene for haemophilia but, like the mother, will not show it.

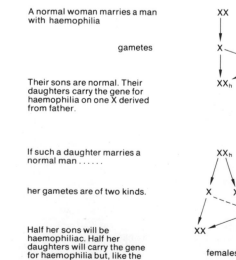

Multi-factor inheritance

Most of the characteristics which organisms show are not of the kind which makes it possible to place them in a limited number of clearly defined groups. Height in human beings has already been referred to as an example of such a quantitative feature. At first sight it might seem that skin pigmentation is a qualitative feature. Africans have dark skins, north Europeans have fair skins. When two such individuals marry their children all have a light brown skin which is intermediate in colour between that of the two parents. The pattern seems like that exhibited by snapdragons, until the results of crosses between such types of children are analysed. When this is done it turns out that while most children are similar to their parents, that is, light brown, they may be both darker or lighter than their parents. They form a continuous series covering the whole range from the dark pigmentation of one grandparent to the fair skin of the other. In fact if a graph is drawn on the basis of intensity of pigment it turns out to be similar to the graph for height. This can be explained only on the assumption that there are several pairs of genes, on different chromosomes, each pair producing intermediate effects like those for petal colour on the snapdragon, and each being distributed to the gametes independently of every other pair.

It is not known how many genes affect skin colour in human beings, except that it is probably many more than four. The same is true of height. As far as height is concerned, its distribution in Britain (see graph p. 236) suggests that most individuals are hybrid for most gene pairs, and that the genetic factors show a random distribution throughout the population. This is the case where height has no significance in survival. On the other hand, skin colour has been important for survival, in the past if not at the present time. It is reasonable to suppose that individuals in tropical countries have a higher proportion of genes for dark skin, while those peoples native to high latitudes lack these genes. The genes for skin pigmentation therefore represent a reservoir of genes. The actual combinations present in each racial group depend on the intensity of ultra-violet radiation to which the group has been subjected and the action of natural selection (see p. 246).

Summary

The structure and function of every living organism is determined by chemical particles called genes. These are arranged in linear order on the chromosomes, the number of which is fixed for every species. Chromosomes are paired and therefore so are the genes. Each pair of genes may be alike, or different, and it is drawn from a family of genes, alleles, peculiar to that part of the chromosome. Each and every cell in the organism contains the same set of genes which are, in effect, a chemical programme determining how the cell shall develop and function. Some characteristics are determined by one gene pair (single factor inheritance). Others are the result of interaction of many pairs of genes (multi-factor inheritance). Still other gene pairs exist which manifest their effects on a number of different characteristics. The sorting of genes occurs in random fashion. It occurs when gametes are formed and when they meet. The particular combinations which are most successful tend to spread throughout a population.

Section III
LIFE AND THE ENVIRONMENT

43 EVOLUTION

The theory of **Organic Evolution** asserts that, since the origin of life on earth, living organisms have continuously undergone change. As a result, modern plant and animal species are very different from those which lived in earlier ages. The theory also states that all living organisms are related, by descent, to organisms of the past. This means that, when changes occur, they are passed from one generation to the next, i.e. they are inherited.

This theory of Evolution directly contrasts with the belief of **Special Creation**—that every type of plant and animal was specially created once and for all by a supernatural being (as the Bible reports God to have done during the few days described in the first chapter of the Book of Genesis). With Special Creation, it is quite clear that living organisms have never changed since their original creation. This view of the creation of life prevailed in Christian Europe until well into the nineteenth century.

Evidence supporting the theory of evolution has been available to man at least since the Greeks discovered fossils several centuries before the birth of Christ. However, it is only comparatively recently (since the mid-nineteenth century) that the evidence has been put forward as a convincing biological theory.

The Greeks found shells of marine molluscs in mountains many miles from the sea. Herodotus, in the fourth century B.C. found fossil shells in Egypt, and concluded that the land there must once have formed part of the sea-bed.

In the thirteenth century, the Dominican monk Albert the Great collected fossils while he was lecturer at Paris University. He realized that they provided a record of life from the past but, because of his strict religious training, he did not accept that the fossils could provide evidence for an evolutionary process. In the centuries following, other eminent men unearthed fossils and considered their significance. Leonardo da Vinci found shells of cockles in the rocks of Lombardy. The current explanation was that, during the Flood, these animals had moved to the Lombardy Plains from the sea. Leonardo thought it very unlikely that an animal as slow-moving as a cockle would travel over several hundred kilometres during the one hundred and fifty days of the Flood.

Various fanciful theories were put forward to explain the unearthing of fossils of those species which do not exist today. They were said to be evidence of God's failure in creation, rejected by him. Even as late as the nineteenth century, it was suggested that fossils were the work of the devil, designed to confuse Man.

The French biologist Cuvier (1769–1832) realized that the fossil record showed a *succession* of different animal populations. He devised a **Theory of Catastrophism** to account for the extinction of earlier forms of life and the apparent creation of new species. Cuvier suggested that the earth had been the scene of a series of great catastrophes. After each catastrophe, the few surviving living organisms reproduced themselves to give populations of organisms which flourished until the next catastrophe, and so on. To explain the appearance of new fossil species, Cuvier suggested that they were not really new at all, but that they would be found elsewhere in the world in rocks of greater antiquity as yet unexplored by geologists.

In spite of a wealth of evidence (see later) supporting the assertion that living organisms have evolved, the ways in which they have done so, i.e. the *mechanism* of evolution, have long been debated. Two of the most influential explanations are now briefly described:

Lamarck (1744–1829)

Lamarck was virtually the first to suggest a definite mechanism for evolution. He believed that organisms are directly affected by their environment. If, for some reason, the environment changes, certain features possessed by an organism might prove more 'useful', i.e. give the organism a better chance of survival. Alternatively, other features might prove of less use. Lamarck, in his 'law of use and disuse', suggested that the former would be developed and the latter would degenerate. As a result of geological change, the land surface might become swampy; according to Lamarck's theory, an animal might spread its toes to support itself on the softer ground, and during the course of its life might develop longer toes or even webbed feet. The organisms inheriting these changes would constitute a new species.

Unfortunately for Lamarck's theory, it is difficult to imagine how a living organism could acquire new characteristics in this way. Also, available evidence suggests that characteristics developed as a result of environmental pressures are not inherited: such characters would somehow have to affect the gametes.

Darwin (1809–1882)

Charles Darwin was an acute observer of plant and animal life. Between 1831 and 1835 he travelled to South America as the naturalist of a surveying expedition in HMS *Beagle*; he collected specimens and made extensive notes. His observations prompted him to produce a theory of evolution in the *Origin of Species*, published in 1859. There are two essential features of Darwin's theory:

(a) that individual members of any species show certain differences from each other, i.e. they show variation (see also p. 235).
(b) that although organisms are enormously prolific, the number of individuals in any species remains remarkably constant.

Clearly, most offspring die at a quite early age, i.e. before they can reproduce themselves, and Darwin concluded that there is a 'struggle for existence' in which only the fittest organisms survive. Evolution would be brought about by **Natural Selection**. Many organisms will die by chance without reproducing, but nevertheless variations will be significant. Organisms with disadvantageous variations will be less likely to survive. Those with advantageous variations will tend to predominate in a population if these variations are passed from parent to offspring. In this way, a population might become better adapted to its environment and in time might constitute a new species.

The essential features of natural selection are still accepted by present-day biologists, though the theory has been modified in detail. The type of variations which Darwin had in mind are only minor variations and probably they are not significant in natural selection. The main source of *substantial* variation in living organisms is the result of alteration to the genes in chromosomes (see p. 86). These alterations are called **mutations** (see p. 247); they are inheritable. The nature of mutations was unknown in Darwin's time.

Most mutations are disadvantageous because they appear accidentally. The small proportion which is not disadvantageous tends to spread through the population simply because the individuals which carry them are more likely to survive and reproduce. Other mutations appear to be neither advantageous nor disadvantageous, though they may later become so should the organism's environment change.

Evidence supporting the theory of evolution

1 Perhaps the most satisfactory evidence that evolution has taken place comes from **palaeontology**, i.e. the study of fossils, the remains of living organisms from past eras. This is because it is the only evidence which is built on a strictly historical basis. It provides the only direct evidence that change has occurred. There are now quite accurate methods for dating rocks. In general, animals 'fossilise' better than do plants, because the hard tissues (skeletons) of animals are more durable. 'Fossils' covers a variety of types of remains, from whole organisms (e.g. the mammoths of Siberia preserved in ice) to imprints in rock (e.g. the outlines of leaves and footprints of animals). Whole skeletons, or just small fragments, have been found. Whole insects have been fossilized, preserved in resin. Organisms have been petrified, their tissues replaced by mineral matter. Sometimes living organisms have formed a mould; their bodies have become buried and when later decomposed, a space is left that could be filled by another material, so producing a cast. Fossils are not uncommon, and students should try to find them themselves. In Great Britain, limestone and chalk hills are particularly good sources.

Gaps are obvious, but the fossil record is remarkably complete for certain species. The evolution of the horse is often quoted, but there are other vertebrate and even invertebrate examples, such as the evolution seen in the shells of sea urchins. In all these cases, the quite gradual changes which are a feature of the theory of evolution are visible in a strict time sequence from one organism to the next.

Vertebrates first appear in the fossil record as fish-like animals in the Ordovician epoch, 360 million years ago. Reptiles were dominant in the Mesozoic era and then declined. Mammals first appear in the early part of the Mesozoic. Flowering plants begin to appear at a similar time.

2 Further evidence to support evolution comes from a study of the structure and functioning of present-day organisms, particularly of vertebrates. The structure of the brain in all vertebrates is so similar that coincidence is most unlikely. Similarly, even though amphibians, reptiles, birds and mammals use their limbs in many different ways for many different functions, they are all based on the same five-digit plan.

The blood proteins of mammals that, for other reasons, are believed to be closely related are more alike than are the blood proteins of apparently unrelated mammals.

3 The distribution of plants and animals over the earth's surface impressed Charles Darwin on his journeys. Marsupial mammals are those that produce their young at a comparatively early stage; development is completed in an abdominal pouch. This method of reproduction seems less efficient than that of placental mammals (see p. 229): where marsupials and placentals are in competition with each other, the marsupials seldom survive. Today, the majority of marsupial species are found in Australia, an island with no native placentals. The most reasonable interpretation of this, which agrees very well with the geological evidence, is that Australia was once attached to a large continental land-mass. Actually, it is thought that Australia was once attached to a land-mass over to the west, on the far side of what is now the Indian Ocean. What happened next is explained by the theory of Continental Drift. Apparently, this land-mass then separated into several smaller masses: Australia, India and Antarctica drifted away, leaving Africa in roughly its present position. The link between Australia and its parent land-mass severed before the evolution of placental mammals. Placentals later evolved on the main land-mass, but were unable to spread to Australia because of the ocean barrier. On the mainland, placentals competed with and led to the decline of the marsupials. Australian marsupials were able to survive because they were protected from competition by their isolation.

There are many other examples of organisms that are not evenly distributed throughout the world. When taken in conjunction with geological data, like the example of the mammals described above, they provide very persuasive evidence of evolution.

4   An increasing weight of evidence for evolution is being obtained from genetics (see p. 235). Man's use of selective breeding techniques results in improved varieties of beef cattle, cereal crops, etc., and these improvements are inherited. Alterations in the Chromosome composition of plants at such centres of botanical research as Kew Gardens have produced new species. Mutations can be induced artificially by bombarding organisms with X-rays, and also by treating them with certain chemicals.

Furthermore, there is ample evidence to suggest that natural selection operates continuously. It is well known that in a population of insects individuals undergo mutations which make them resistant to pesticides; new pesticides have to be developed to kill these mutant forms, otherwise they reproduce and spread throughout the population. Because insects breed at a fast rate mutations appear rather frequently; consequently further mutations may soon appear, resistant to newly-developed pesticides.

The **peppered moth** provides an interesting example of the effect of natural selection. The 'normal' peppered moth has pale wings, mottled with a characteristic peppery pattern. When this moth settles on the lichen-covered trunks of trees it blends with the background and becomes almost invisible. Another variety of the moth has a black body and dark wings. Since the Industrial Revolution the proportion of light to dark moths has decreased in industrial areas, sometimes to the virtual exclusion of the light-winged forms. The reason seems to be that the moths are preyed upon by birds such as robins, which hunt them by sight. Atmospheric pollution from heavy industry has killed the lichen and blackened the bark of trees nearby. In these conditions the mottled moths are no longer camouflaged, whereas the dark-winged forms are at an advantage—the mottled forms are seen and eaten with greater frequency. This proportionate increase in dark-winged moths has occurred only in industrial areas. Elsewhere, e.g. in the highlands of Scotland and in SW England, the trees are still lichen-covered and the mottled moth predominates, presumably because moths of the dark variety are more conspicuous.

1   Many museums have good displays concerned with the geological and biological aspects of evolution and these should be visited.

2   Collect fossils and try to identify them.

3   Read further about the history of biological science, and in particular the work of Lamarck and Darwin. You should be able to discover more information that seems to support the Theory of Evolution.

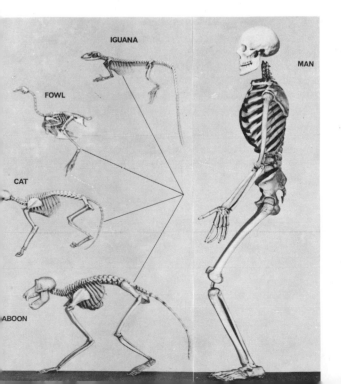

Evolution of the vertebrate skeleton, illustrated by a reptile, bird and three different mammals

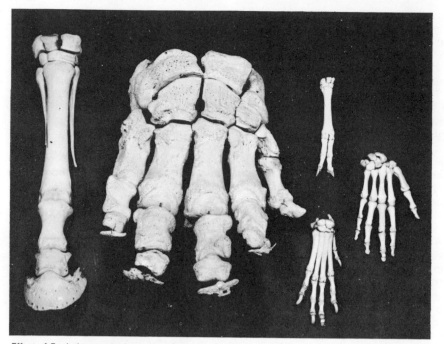

Effect of Evolution on the basic five-digit (pentadactyl) limb of five different mammals. Left to right—feet of horse, elephant, sheep above dog, man

below—Effect of selective breeding on the skulls of dogs; each is derived from the original wolf skull but shows its own characteristics

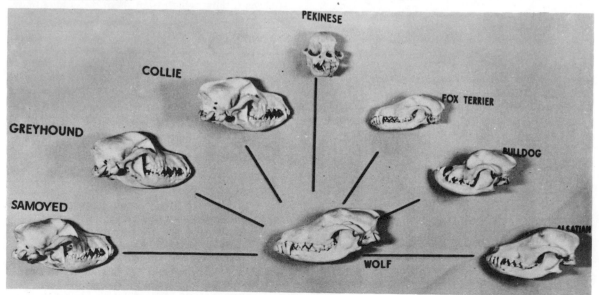

PEKINESE

COLLIE

FOX TERRIER

GREYHOUND

BULLDOG

SAMOYED

WOLF

ALSATIAN

Formation

Soil has been formed from parent rock over many millions of years by a very slow process called weathering. The more important weathering agents are :

1   Frost

Rainwater may become trapped in small crevices on the rock face ; it freezes, expands and produces a lateral pressure which will cause the splintering of the adjacent rock.

2   Temperature changes

Rock expands when heated and contracts when cooled ; hot sun causes the surface layers to expand more than the layers deeper down ; unequal stresses are produced which will cause the rock surface to fragment.

3   Water

(a)   **Mechanical effect**. Continual movement of water, either in liquid or in glacial form, against a rock surface has an abrasive effect and will cause small pieces of rock to be worn away and smoothed.

(b)   **Chemical effect**. Carbon dioxide dissolved in the water forms carbonic acid ; although weak, carbonic acid reacts with the rock surface, particularly with limestone rock, and steadily wears it away.

4   Wind

Air blowing across a rock surface has an abrasive effect especially where, as in desert regions, the air is charged with minute rock particles.

5   Living organisms

Biological weathering, such as the erosion caused by the growth of plant roots, takes place continuously.

Soils do not always lie over their own parent rocks. In Britain movement of wind, water and glaciers (during the ice ages) has moved soils away from their places of origin and deposited them elsewhere. Such soils are called alluvial or sedimentary soils.

Soil profiles

Every soil has a 'profile' which can be seen in vertical section. Soil profiles are made naturally when, for example, coastal cliffs are eroded by wave action or when there has been a subsidence of underlying rock. Quarrying is an activity of man which produces artificial soil profiles, and soil profiles are excavated for scientific investigation. The constituents of the upper layers of a profile in a sedimentary soil will, by definition, seldom bear resemblance to the underlying parent rock. In other soils the parentage of the soil may be clearly seen in its profile. Each profile has three main layers (referred to as 'horizons') :

A soil profile

A horizon

B horizon

C horizon

true (or **top**) **soil** — *on average 20 to 30cm deep seldom more than 1 metre; contains relatively small, fully-weathered particles; generally dark-coloured because of its organic content (humus) and therefore can support the growth of plants.*

sub-soil — *contains relatively large, partly weathered particles and no humus; therefore does not support plant growth*

parent rock

249

The subsoil represents a stage of weathering part-way between the original parent rock and the fully-formed soil. At some time in this process life must have evolved: hardy plants must have colonized the developing soil. The most likely colonizers are bacteria, lichens and certain types of alga and moss. When these died their remains became incorporated into the soil as humus, to feed future generations of perhaps less hardy plants. This process of humus-formation has been continuing ever since.

Soil components and their importance to plant life

The following components are essential, i.e. they must be present if a soil is to maintain its fertility:

1 **Rock particles** of different sizes
2 **Water**
3 **Mineral salts**
4 **Air**
5 **Humus**
6 **Earthworms**
7 **Micro-organisms**

1 ROCK PARTICLES

These form the basis of the soil. The main components are resistant oxides of silicon and aluminium together with complex alumino-silicates, though other elements may be present, e.g. metallic elements such as iron, potassium and magnesium. Calcium carbonate is a common rock, formed from the compacted outer skeletons of microscopic organisms which lived in the seas of previous eras.

The nature of rock particles profoundly affects the properties of the soil that contains them; this in turn determines the species of plants and animals which will live there. Similarly, the nature of the rock particles depends on the parent rock from which they were weathered. Hard rocks such as sandstone and gravels weather to give soils with predominantly large particles: soft rocks such as shales produce small particles. Rock particles are classified rather arbitrarily into four main categories according to their size:

Type of Particle	Diameter (mm)
Coarse sand	2 to 0.2
Fine sand	0.2 to 0.02
Silt	0.02 to 0.002
Clay	0.002 and smaller

Thus coarse sand has the largest, and clay the smallest, particle size. Note that although the measurements are given in diameters even the smallest rock particles have irregular shapes. Particles larger than 2 mm diameter are classed as gravel or stones. Because of their extreme sizes, soils composed exclusively of coarse sand or of clay have extreme properties, and are not fertile. All fertile soils are **loams**, i.e. they are various mixtures of different sorts of particle. The following shows the approximate proportions of particle types in three different loams:

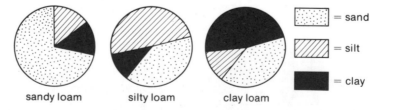

sandy loam   silty loam   clay loam

= sand
= silt
= clay

Soils where sand particles predominate tend to be light and easy to work but do not give good support to the roots of plants. Soils with predominantly clay particles are heavy and sticky and do not favour root penetration.

A rough analysis of any given soil can be made by allowing its components to settle out (sediment) in water. The largest particles, being the heaviest, fall most quickly, whereas the smallest are often too light and remain in suspension. The following sequence of experiments, many of them quantitative, illustrates important aspects of soil; they have the advantage that the student can perform them at almost any time in the year.

A: Experiment to analyse a soil sample by a process of sedimentation

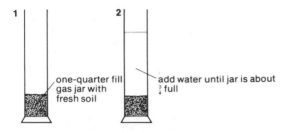

one-quarter fill gas jar with fresh soil

add water until jar is about ¾ full

3 Cover open end of jar, invert several times and then immediately stand jar vertically.

4 Leave undisturbed for a few days. Results are summarized in the diagrams below.

Results  an ordinary soil sample

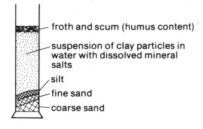

- froth and scum (humus content)
- suspension of clay particles in water with dissolved mineral salts
- silt
- fine sand
- coarse sand

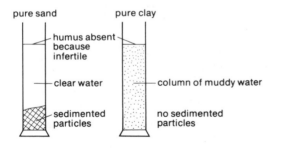

pure sand | pure clay

- humus absent because infertile
- clear water
- sedimented particles
- column of muddy water
- no sedimented particles

Although appearances are frequently deceptive, the colour of a soil may give an indication of its composition and fertility. The red soils of East Devon are quite characteristic, as are the yellow limestone soils of the Cotswolds. Iron compounds in well-aerated soils impart a red-brown colour because the iron is mainly in the oxidized (ferric) state; in water-logged and therefore infertile soils, the iron is mainly in the ferrous state and appears grey.

An important feature of a good loam is that it is **friable**, i.e. it has a crumbly texture. This is due to the aggregation of different sizes of rock particle into **soil crumbs**. Probably no other single feature is more important in promoting soil fertility. It is believed that crumb formation is brought about by earthworms when they egest casts (see p. 39), and by gummy secretions of certain soil bacteria and fungi; also plant roots, particularly of grasses, have

an effect. Clay particles can be made to adhere together and form large aggregates by addition of lime—this process is called flocculation.

B: Experiment to demonstrate the flocculation of clay soil

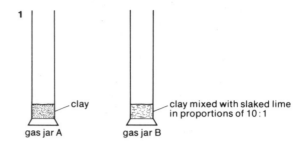

- clay
- clay mixed with slaked lime in proportions of 10:1

gas jar A | gas jar B

2 Three-quarter-fill each jar with water.

3 Cover open ends and shake both jars vigorously. Stand both jars vertically and undisturbed, and observe for up to 30 minutes.

Results are shown in diagrams A and B.

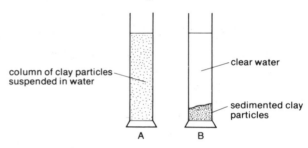

- column of clay particles suspended in water
- clear water
- sedimented clay particles

A | B

This experiment shows the effect of lime in causing clay particles to aggregate into crumbs large enough to settle out under gravity.

C: Experiment to determine the percentage of rock particles in a soil sample

1 Weigh porcelain crucible.

2 Re-weigh crucible filled with fresh soil to obtain weight of soil.

3

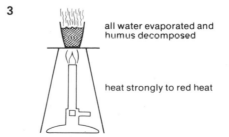

- all water evaporated and humus decomposed
- heat strongly to red heat

4 Cool in a desiccator.

5 Weigh crucible + soil again to obtain new weight of soil.

Result

$$\frac{\text{Weight of soil after heating} \times 100}{\text{Weight of soil before heating}} = \quad \%$$

Importance of Rock Particles

1 These provide anchorage for the plant roots which penetrate amongst them.

2 Their nature and size affect the properties of the whole soil.

2 WATER

The total water content varies considerably with external conditions, but on average is about 10-20 per cent by weight. In an unwaterlogged soil, water is found as a thin film held by force of attraction around each soil particle. As the particles are of irregular shapes and different sizes, a small pocket of air is trapped between the water films. From the way in which it behaves, it is possible to divide the film into two layers:

(a) **available water** (capillary water), which can be absorbed by plant roots.

(b) **non-available water** (hygroscopic water), which is held so strongly by the soil particles that roots cannot exert sufficient suction to remove and absorb it. Since non-available water is left when available water has been absorbed, this means that plants may die through lack of water though the soil contains water.

Diagram showing water in soil
(see also figure on p. 134)

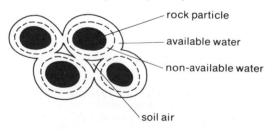

- rock particle
- available water
- non-available water
- soil air

When soil is saturated, for example, during heavy rain, the air spaces are filled with water. Afterwards water (gravitational water) drains away from the soil, drawing air from the atmosphere above into the air spaces.

D: Experiment to determine the percentage of available water in a soil sample

1 Make a large tray with paper and paper-clips. (There is no need to weigh this tray because its weight should be negligible in relation to the weight of soil.)

2 Fill paper tray with fresh soil and weigh to obtain weight of soil.

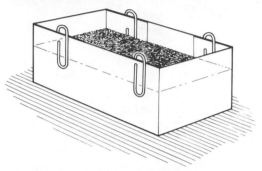

3 Stand tray in laboratory at room temperature.

4 Re-weigh tray at suitable intervals (e.g. every three days) until there is no further loss of weight. This weight loss is due to the evaporation of available water.

Result

$$\frac{\text{Loss in weight after evaporation} \times 100}{\text{Wt. of tray} + \text{soil before evaporation}} = \quad \%$$

(This experiment can be performed using a large evaporating basin of known weight instead of the paper tray.)

E: Experiment to determine the percentage of non-available water in a soil sample

1 Weigh an evaporating basin.

2 Re-weigh basin filled with soil from previous experiment (i.e. with available water removed), to obtain weight of soil.

3 Place basin and soil in a drying oven at 95-99°C (or alternatively on a water bath maintained at the same temperature).

4 Re-weigh basin at suitable intervals (e.g. every three days) until there is no further loss in weight. (On each occasion cool basin first in a desiccator and return to oven after weighing.) The weight loss is due to the evaporation of non-available water at this temperature.

Result

$$\frac{\text{Loss in weight after drying} \times 100}{\text{Weight of soil before drying}} = \quad \%$$

F: Experiment to determine the percentage of total water in a soil sample

1 Weigh an evaporating basin.

2 Re-weigh basin filled with *fresh soil* (i.e. containing both available and non-available water).

Then—proceed with 3 and 4 as in Experiment E above.

Result: $\dfrac{\text{Loss in weight after drying} \times 100}{\text{Weight of soil before drying}} = \quad \%$

Soil which has been completely dried from either Experiment E or Experiment F may be kept in a desiccator for use in Experiment L.

Importance of water

Water is essential for all forms of life (see p. 94), including the macro- and micro-organisms of soil. Flowering plants absorb water from the soil films into their root hairs. It has three main functions:

1 Mineral salts are absorbed from the soil dissolved in water. Both salts (in the xylem) and substances made as a result of photosynthesis (in the phloem) are distributed inside the plant in solution.

2 The hydrostatic pressure of water acts as a partial plant skeleton by maintaining the turgidity of plant cells (see p. 137).

3 Small quantities of water are necessary for photosynthesis (see p. 98).

3 MINERAL SALTS

Soil normally contains less than 0.2 per cent by weight of soluble mineral salts. They are dissolved in the water films surrounding the rock particles. They originate largely from the decomposition of plant and animal remains and from the waste products of these organisms, though some salts are derived from the weathering of rock particles.

G: Experiment to demonstrate the presence of carbonate in a soil sample

Result

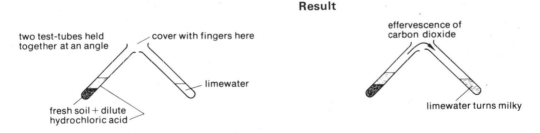

two test-tubes held together at an angle — cover with fingers here — limewater

fresh soil + dilute hydrochloric acid

effervescence of carbon dioxide

limewater turns milky

H: Experiments to demonstrate the presence of other mineral salts in soil

1

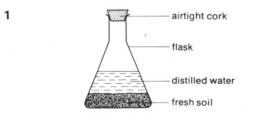

airtight cork

flask

distilled water

fresh soil

3

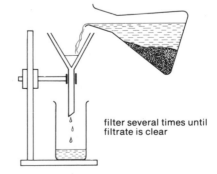

filter several times until filtrate is clear

2 Vigorously shake soil and water together.

4 Reduce volume of filtrate considerably by boiling water away.

5 Test a small portion of concentrated filtrate in a different test-tube for each of the following:

(a) Sulphates
Add some barium chloride solution.
Result: a white precipitate.

(b) Nitrates
Add some 0.5 per cent solution of dipheny-lamine in concentrated sulphuric acid
Result: blue coloration.

(c) Chlorides
Add some silver nitrate solution
Result: a white precipitate.

(d) Phosphates
Add some dilute nitric acid and ammonium molybdate solution
Result: a canary yellow precipitate.

(e) Iron
Add potassium ferrocyanide solution
Result: blue coloration.

The presence of elements such as sodium, potassium and calcium can be demonstrated by completely evaporating the filtrate and then carrying out flame tests on the residual solid.

Importance of mineral salts

1 Many salts absorbed by the roots are essential for healthy plant growth (see p. 100).

2 Salts affect the degree of acidity and alkalinity (pH) of soil, and this in turn affects the organisms which will live there. According to the pH of soil, certain plant nutrients may be insoluble and there-fore unavailable to plants.

The pH of a soil sample can be determined by testing a small quantity of soil filtrate (see Experiment H, stages 1, 2, 3) with a few drops of soil indicator. The colour of the tested filtrate can be compared with the soil indicator chart to obtain an approximate pH value. Increased accuracy can be achieved by testing further portions of the soil filtrate with other more specific indicator solutions.

4 SOIL AIR

Soil air has essentially the same composition as ordinary atmospheric air, although there is slightly less oxygen and more carbon dioxide. In an un-waterlogged soil, air fills the pockets between the constituent rock particles. The size of the particles directly affects the size of the air pockets—larger particles such as sandy particles fit less tightly to-gether and have larger air spaces.

I : Experiment to determine the air content of a soil sample

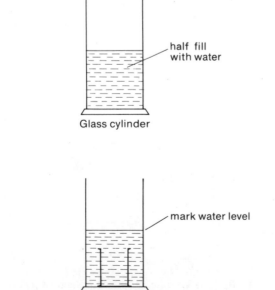

1

half fill with water

Glass cylinder

2

mark water level

Fully immerse suitable tin

3

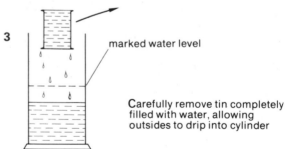

marked water level

Carefully remove tin completely filled with water, allowing outsides to drip into cylinder

4 Empty tin and punch several holes in its base.

5 Drive tin with open end downwards into soil (avoid stony soil).

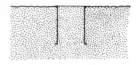

6 Dig out tin completely filled with soil.

7 Cut off surplus soil flush with tin surface at open end (e.g. with a ruler) and remove soil adhering to outer sides of tin.

8

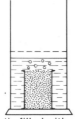

Immerse tin filled with soil
inside cylinder (as in 2); scoop
soil out of tin with glass rod
to ensure complete displacement
of soil air by water

9

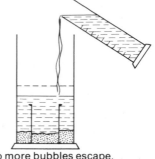

When no more bubbles escape,
add water from measuring cylinder
until original water-level mark
is reached

Result: Volume of water added from measuring cylinder = volume of air in soil sample. (This experiment can be repeated with different soil materials, e.g. equal volumes of sand and clay can be compared to show that clay has a smaller air content.)

Importance of soil air

1 The oxygen contained in soil air is used for the aerobic respiration of soil organisms, including the roots of plants. Lack of oxygen hinders both root growth and the production of mineral salts from humus, and favours the destructive activities of denitrifying bacteria (see 'nitrogen cycle').

2 The size of the soil air spaces (which in turn is dependent on the size of the rock particles) affects:

(a) drainage of the soil. The larger the air spaces the more rapid the drainage—sandy soils are drier than clay.

(b) capillarity of the soil. **Capillarity** is the upward movement of fluid in narrow tubes; it is caused by attraction between the fluid molecules and the walls of the tubes. In very narrow tubes where there is more wall area in proportion to volume of fluid, the force of attraction is greater, and fluid rises higher. The air spaces in soil act as irregular tubes, allowing water to rise. The smaller air spaces of clay soils provide very narrow tubes, so that capillarity is greater here than in sandy soils.

Diagram showing air spaces in soil

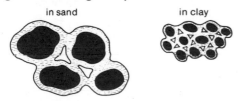

In certain circumstances capillarity may be important, as it brings water from the reservoirs deep in the soil up towards the surface. Without it plant roots might wilt much more quickly on hot dry days, because water evaporated from the soil surface would not be replaced.

J: Experiment to compare the drainage (permeability) of three different soil materials

1 Assemble apparatus pictured below: three funnels plugged with exactly similar wads of cotton wool; each funnel half-filled with an exactly similar volume of different dry soil material packed equally tightly, with stones and large lumps removed, and held vertically over its own glass beaker.

2 Fill three measuring cylinders each with 50 ml of water.

3 Pour each volume of water simultaneously into a different filter-funnel.

4 Leave for a suitable time interval, e.g. when active dripping has ceased.

Results are summarized in the diagram below.

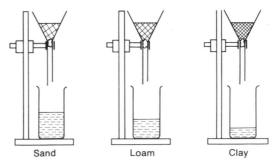

(Quantitative results may be obtained by returning the contents of each beaker into a separate measuring cylinder).

K: Experiment to compare the capillarity of three different soil materials

1 Assemble the apparatus pictured below: three glass tubes open at both ends, clamped vertically side-by-side into a glass trough, each tube plugged with cotton wool at lower end and almost filled with an exactly similar volume of different dry soil material packed equally tightly, with stones and large lumps removed.

2 Fill glass trough almost completely with water.

3 Observe upward movement of water in tubes for several minutes after immersion and at regular intervals (e.g. daily) until no further rise of water is visible.

Results are summarized in the diagram below.

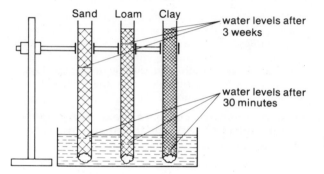

In this experiment water passes very quickly into the sand to fill the large air spaces. The final water level is highest in clay because of its greater capillarity.

(Cress seeds can be sprinkled on to the surface of the soil columns. Those on the clay should germinate before those on the loam; those on the sand probably will never germinate.)

5 HUMUS

Humus is a rather imprecise term. It refers to non-living organic materials, i.e. the dead remains of plants and animals and also their waste products. These materials are in varying stages of alteration by soil organisms. Humus is found predominantly at or near the soil surface; this is because organic materials from most plants and animals fall and accumulate here initially after their death or when these organisms produce waste. As a result of the activities of different animals, particularly earthworms (see p. 39) but also man, humus becomes incorporated into and mixed with the body of the soil.

L: Experiment to determine the percentage of humus in a soil sample

1 Weigh porcelain crucible.

2 Re-weigh almost completely filled with soil *from which all water has been removed* (see Experiments E and F) and which has been kept in a desiccator, to obtain weight of dry soil.

3

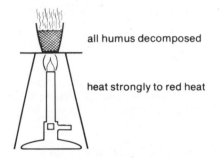

all humus decomposed

heat strongly to red heat

4 Cool in a desiccator.

5 Weigh crucible + soil again, to obtain weight loss owing to decomposition of humus.

Result: $\dfrac{\text{Loss in weight after heating} \times 100}{\text{Initial weight of dry soil}} = \quad \%$

Importance of humus

1 Humus is broken down (e.g. in the nitrogen cycle) to soluble foods in the form of mineral salts, for future generations of plants.

2 Humus improves the texture of soil:

(a) in heavy wet soils, humus separates the soil particles into larger aggregates (soil crumbs see p. 251), so improving aeration.

(b) in light dry soils, humus absorbs water and prevents rapid drainage; both water and dissolved mineral salts are retained in the soil.

3 The humus layer near the soil surface absorbs and retains water like a sponge; this prevents excessive evaporation from the underlying soil during dry weather.

4 Humus provides food for essential soil organisms such as earthworms.

6 EARTHWORMS

The activities of earthworms make them important maintainers of soil fertility; they are considered on p. 39.

7 MICRO-ORGANISMS

Soil contains a variety of bacteria, fungi and other microscopic organisms (e.g. protozoa and round- worms). One gram of soil may contain 100 million micro-organisms.

M: Experiment to demonstrate the presence of micro-organisms in soil

1

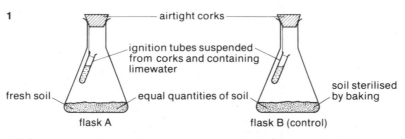

flask A   flask B (control)

2 Leave for several days

Result: Limewater in A turns milky; limewater in B does not change.

This experiment demonstrates the production of carbon dioxide in flask A but not in B. The single difference between these two flasks is that soil in B has been baked. It is reasonable to infer that baking has destroyed organisms which are normally present in soil but which are invisible to the unaided eye. In flask A these living micro-organisms have produced carbon dioxide from their respiration.

Alternative experiment:

1

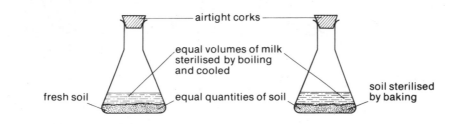

2 Inspect flasks at daily intervals.

Result: Milk in A turns sour before that in B.

From this experiment it is reasonable to infer that the metabolic activities of living micro-organisms in the fresh soil have been responsible for the prema- ture souring of the milk. Although the micro-organ- isms in flask B have been destroyed by baking, the milk in this flask eventually turns sour because of the difficulty of excluding micro-organisms from the air.

Importance of soil micro-organisms

Many different micro-organisms help to maintain the continuity of life in the soil. They are important for two main reasons:

1 They cause the removal by decomposition of the dead remains and waste products of other organisms.

2 They produce basic raw materials for the nutri- tion of future generations of green plants; in turn, green plants provide all other living organ- isms with food.

As a result of these activities micro-organisms derive energy for their own metabolism.

Other soil micro-organisms (e.g. de-nitrifying bacteria) are important for quite opposite reasons; these are harmful in that they destroy humus and its breakdown products, releasing substances that are virtually useless to living organisms.

The element nitrogen is essential to life (it is, for example, a constituent of all proteins). Green plants cannot utilise nitrogen gas but they are able to absorb soluble mineral salts containing nitrogen, e.g. ammonium compounds and nitrates. The **nitrogen cycle** (see diagram), i.e. the way that nitrogen circulates in nature, provides excellent illustration of the activities of soil micro-organisms. The beneficial organisms of the cycle are first **saprophytes** (see p. 110) which putrefy, and then those which **nitrify**. Also beneficial are bacteria which 'fix' or synthesise nitrogen compounds from gaseous nitrogen. Certain of these **nitrogen-fixing** bacteria invade the roots of plants such as clovers, peas and lucernes, causing swellings called root nodules in which the bacteria live symbiotically (see p. 110).

The nitrogen cycle

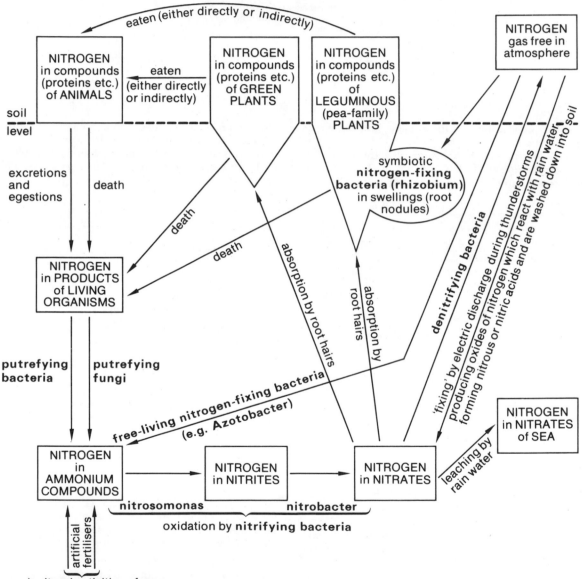

Methods of improving sand and clay soils

The methods listed below are suitable for application on the small scale, e.g. in gardens; elsewhere they might be impracticable or uneconomic.

1 Sand

Add clay and/or silt to make a loam with greater variety of particle sizes; dig in humus (peat, farmyard manure, compost, greencrops, etc.) to promote retention of water and mineral salts; add humus to surface as a 'mulch' to prevent evaporation and to protect from leaching action of heavy rain; hoe to reduce evaporation; add lime, but only if the soil is very acid.

2 Clay

Add sand to make a loam with greater variety of particle sizes; add top-dressing of lime to flocculate clay particles, so producing soil crumbs with improved aeration and drainage (lime also neutralises clay made acid by waterlogging); add humus to promote soil crumb production and so improve aeration and drainage; plough in autumn to promote frost action; roll to increase capillarity and evaporation; build drainage channels.

These improvements to both sand and clay, particularly if applied systematically, should result in increased fertility, especially because they promote the activity of micro-organisms and earthworms.

Generalised comparison between sand and clay soils

	SAND SOILS	CLAY SOILS
1	Light	Heavy
2	Large particles	Small particles
3	Large air spaces	Small air spaces
4	Low capillarity	High capillarity
5	Rapid drainage—water not held by attraction to rock particles. Therefore dry soils	Very slow drainage—water held by attraction to rock particles. Therefore wet soils
6	Mineral salts quickly leached away by rain	Mineral salts attracted by clay particles and retained
7	Low water retentivity	High water retentivity
8	Activity of micro-organisms (and earthworms) low	Micro-organisms active, except where water-logged
9	Very loose consistency —easily eroded by wind	Particles held together by surface tension of water surrounding them —tend to cake when dry—hinders root penetration
10	Low specific heat— show extreme fluctuations of temperature	High specific heat— show more constant temperature though wet clays are rather cold

Agricultural practices

The intensive cultivation of soil poses many problems. Continual removal of crops grown by man for his food, clothing, etc., at the same time removes valuable soil materials. The soil would soon become infertile but for the use of a variety of soil treatment processes.

The following section deals with some agricultural practices which enable man to make the most efficient and economical use of the land in a temperate country such as Britain:

1 Spring Ploughing

This increases the surface area for evaporation. The soil surface slowly crumbles as it dries, giving a good tilth, i.e. a suitable bed for seed germination. More air reaches the soil and this stimulates the micro-organisms to increased nitrate production.

2 Autumn Ploughing

Organic matter (manure, leaves, stubble, etc.) becomes buried and incorporated into the soil more quickly as the upper soil layers are turned over; it decomposes more rapidly. The furrows produced allow winter rainwater to penetrate and prevent running-off; evaporation is decreased.

Frost reaches further into the soil and promotes the weathering of heavy clods to produce a better texture. Also soil pests are killed more effectively.

3 Hoeing and Mulching

Hoeing breaks-up capillary tubes near the surface and reduces water loss from the soil. It allows air into the soil, promotes the activity of micro-organisms and increases nitrate production. It also disturbs the roots of weeds sufficiently to kill them if the soil is subsequently exposed to hot sun.

A mulch is an application of organic material such as straw to the surface. It allows moisture to be retained near the surface layers, checks evaporation, and prevents the leaching effect of heavy rain rushing down through the soil.

4 Rolling and Draining

Rolling pushes the soil crumbs together. This makes more capillary tubes and may bring more water to the surface. It then becomes possible to dry a soil by evaporation and also brings up sufficient water for seed germination.

The laying of drains removes excessive soil water and stimulates additional root growth. It may be used to improve heavy soils.

5 Bare Fallowing

This practice of leaving land uncropped for a season conserves water in the soil and allows increased production of nitrates for succeeding crops. It also allows time for cleaning land.

6 Crop Rotation

Different plants have different nutritional requirements. The requirements of wheat are greater than those of certain other cereals, so wheat can be followed by barley or oats. Alternatively, long-rooted plants may be followed by short-rooted plants so that nutrients are removed from different soil levels. In the succeeding year the field is left fallow or legumes (see 'nitrogen cycle') such as clover are planted to increase the nutrient content of the soil; these crops are subsequently ploughed in. Root crops have lower mineral requirements than do cereals, so that another example of crop rotation is:

WHEAT	—	SUGAR-BEET	—	CLOVER
		or		or
		TURNIPS		BEANS
(1st year)		(2nd year)		(3rd year)

7 Addition of Organic Matter

Organic matter can be added in the form of farmyard manure, or a crop can be ploughed in ('green manuring'). The value of the resulting humus is considered on p. 256.

8 Addition of Artificial Fertilizers

A variety of artificial fertilizers is available for different purposes, e.g. soluble salts such as ammonium sulphate and nitrate, potassium and sodium nitrates and chlorides, 'basic slag' (calcium phosphate combined with calcium carbonate), and 'super-phosphates'. All increase the nutrient content of the soils to which they are applied, but their effects are largely short-term since they do nothing to improve the physical structure of soil.

9 Liming

Addition of lime improves the structure of heavy soils by flocculating the smaller particles (see p. 251); it aids drainage and aeration. Soils become more alkaline, and this promotes the activity of micro-organisms and earthworms. Several plant nutrients which are insoluble in acid soils become soluble and available, though the reverse is equally true. Heavy applications of lime sterilize soil.

---

1 Repeat as many as possible of the soil experiments using different types of soil, in order to obtain comparative results.

2 Examine soil profiles wherever these are revealed, e.g. in disused quarries. Bore down into soil with a hollow cylinder to obtain samples from different depths.

3 When in the country observe agricultural activities at different times of the year and try to find the scientific explanations for these activities.

45 THE BALANCE OF NATURE

A long term view of life on the earth reveals a process of continuous change. For example, the earth during Carboniferous times presented a very different picture from that of today. Nevertheless, there is good reason to believe that the overall situation has not altered materially since then. The species of plants and animals may have altered but the total balance of the life system has not changed. The thin shell at the surface of the earth within which all living things are found is called the **biosphere** and within it are found all the basic materials upon which the maintenance of life depends. These are carbon dioxide and oxygen in the atmosphere and water and mineral salts in the ocean or the soil. The biosphere extends to the limits of light penetration in the sea and to the limits of root penetration on land. It extends into the lower levels of the atmosphere. Nothing is added to nor removed from the total supply of materials, which are continuously being cycled and recycled. The only external factor is the radiant energy of the sun, which is the source of power which maintains the cycle. The biosphere therefore may be regarded as a stable system. It has not always been so. During the very early history of the earth when life first appeared, the atmosphere is thought to have been very different in composition from what it is now, consisting mainly of methane, ammonia, carbon dioxide, hydrogen and water. Its composition has been altered by the presence of living things.

THE CYCLE OF MATERIALS

Carbon (see p. 101): Carbon is the key element in all organic compounds. It may exist as carbon dioxide in the atmosphere, or as part of the tissues of a living organism, or it may be temporarily removed from circulation as the altered remains of a dead organism. Although its rate of circulation, and its distribution, may alter, the total remains relatively constant. Most of it is found in the form of organic compounds, the amount existing as free carbon dioxide being comparatively small. Carbon dioxide forms about 0.03 per cent by volume of the atmosphere. The turnover of carbon dioxide must therefore be a very rapid process. The amount converted into plant tissue per year is probably at least a hundred times as great as the total weight of free carbon dioxide at any time. This is balanced by the process of respiration (see p. 121).

Nitrogen

Nitrogen, like carbon dioxide, exists as a free gas in the atmosphere. Unlike carbon dioxide, there is very much more of it, forming about four-fifths of the total atmosphere. This, however, is mostly unavailable to living organisms. Nitrogen is required for the synthesis of proteins, but since a much smaller proportion is locked into the cycle of living and dead organisms it forms possibly a more stable system than that of carbon dioxide (see p. 258).

Similar cycles exist for the other elements which are of importance to living organisms. Most appear to be fairly abundant outside their role as part of the life cycle—for example, calcium, sulphur, potassium and sodium. Phosphorus presents a different picture. Although it is an important part of plant and animal tissue it is not particularly abundant in the kind of concentration man can exploit. The reservoir of phosphorus is chiefly the rocks of the earth's crust, and a significant transfer of phosphorus from the land to the ocean bed takes place, a transfer which is not balanced by a reverse process.

The energy source

The constant cycling and recycling of materials through living organisms requires a source of energy to maintain it. The ultimate source of energy is the sun. No living organism is able to use light energy directly. It is first transformed into the chemical energy of glucose during the process of photosynthesis. This occurs in the green plant and entails the reduction of carbon dioxide (see p. 95). The energy stored in the form of glucose is released during respiration (see p. 121). Carbon dioxide plays a double role:

1 It acts like a 'solar' battery being charged with energy (the radiant energy from the sun) during

The phosphorus cycle

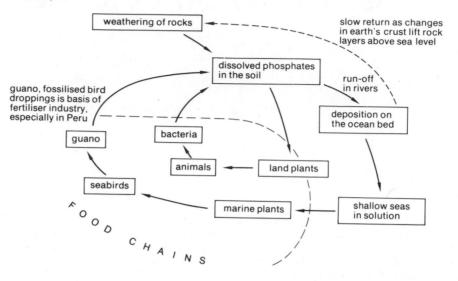

photosynthesis and releasing that energy during respiration.

In an ecosystem the community of organisms as a whole shares the resources of the environment.

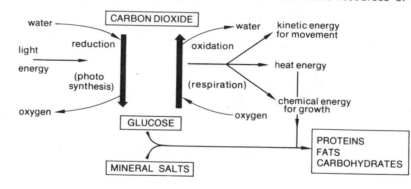

2 It forms the basic material, with water and mineral salts, for the building of plant and animal tissues.

The ways in which carbon and other materials circulate in nature is part of the study called **ecology** (Gk. *oikos*, home; *logos*, knowledge). Ecology forms an enormous field of study part of which is concerned with what are called **ecosystems**.

An ecosystem is a natural unit consisting of all the plants, animals and micro-organisms in an area together with the non-living aspects of the environment. An ecosystem may vary in size from a freshwater pond to an area the size of Greenland. Examples are: a saltwater marsh, a freshwater pond, an oak wood, a coral reef, a rocky shore. Each consists of a balanced interacting system and it is the business of the ecologist to try to understand the many complex forces at work in it.

These are handed on from one kind of organism to another. The sequence starts with green plants, which are **producer** organisms. These are fed upon by herbivores, the **primary consumers**. They, in their turn, are eaten by **secondary consumers**, the carnivores. The nutrient materials are brought back to the beginning of the cycles by the activity of decay organisms. In the simplest case the materials are transferred along what is called a food chain.

Example:

Grass → Cattle → Milk → Man → Micro-organisms
→ Beef ↗

More food materials may pass along a variety of inter-connected routes from the producer organism to the micro-organisms. Such a complex is known as a **food web**.

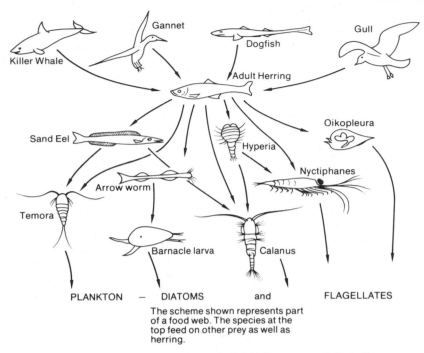

PLANKTON — DIATOMS and FLAGELLATES

The scheme shown represents part of a food web. The species at the top feed on other prey as well as herring.

Autotrophs

'Producer' organisms are called autotrophs, or self-feeders. Autotrophic organisms are either:

(a) Green plants, which use light as an energy source to build up their carbon compounds from carbon dioxide and are also able to synthesize protein (see p. 95). This feeding method is also called **holophytic**.

(b) Some bacteria which gain energy from the oxidation of inorganic substances. Example, *Nitrosomonas*, a soil bacterium, oxidizes ammonia to nitrite (see p. 258).

On land the chief autotrophs are the flowering plants, conifers, ferns and mosses. In the sea, algae,

mainly diatoms and flagellates, are the producer organisms. Except for the seaweeds they are all microscopic and are collectively referred to as **phytoplankton**. The productivity of an ecosystem in terms of the mass of plant tissue it supports depends on a number of factors such as climate, availability of mineral salts, soil structure, drainage and so on.

Heterotrophs

Heterotrophs (other feeders) obtain their carbon in the form of elaborated organic compounds. They are consumer organisms. Those feeding directly on plants are the primary consumers. They are fed upon

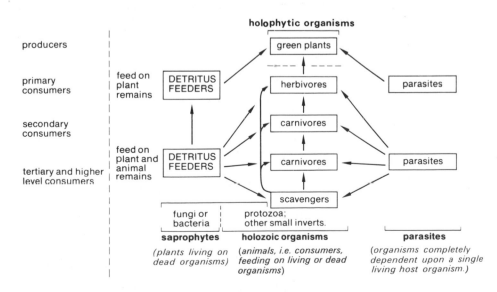

by secondary consumers, and so on down the line.

The above scheme of relationships applies to most ecosystems. In marine habitats the detritus pathway is of far less importance, since the 'phytoplankton' is 'grazed' so effectively that little of it has time to die before it is eaten.

Energy exchange within ecosystems

The balance of nature means that more cannot be got out of a system than is put into it, that there are certain factors that cannot be altered. For example, the sum total of energy falling on the surface of the earth must be regarded as a constant. The efficiency with which plants can take up and transform this energy varies according to climate and availability of water, etc., and we assume that in a stable system nature has utilized the available energy with maximum efficiency. Man can divert the flow of energy through an ecosystem to suit his own purposes. In some instances he can speed up the flow, but in the last resort what he does cannot be worked out like a schedule for profit-making in an industrial firm.

About 500 joules of solar energy fall on the surfaces of land plants per square centimetre of land surface per day. This represents only a small proportion of the total incident solar energy. Some is lost by reflection or by re-radiation. Some is absorbed by the upper atmosphere, and the rest is absorbed and used to raise the land or sea temperature.

Of the light energy that falls on the exposed surfaces of plants some is lost by reflection (see diagram). Another fraction is taken up as the latent heat of vaporisation. Still another fraction is utilized in the process of respiration. What remains is represented by the energy value of the plant tissues. Represented in joules, it amounts to about 1 per cent of the total incident energy falling upon the plant. This is another way of expressing food productivity. It can also be done, for example, by measuring the amount of hay produced in a given field per day or year.

| 15 joules energy in plant tissue | → 1.5 joules in tissues of primary consumer | → 0.15 joules in tissues of secondary consumer |

Energy is lost in: respiration, maintenance of body temperature in warm-blooded animals, movement, metabolism, tissue repair, unabsorbed materials (faeces).

Since there is a drop in available energy it follows that there must be a drop in the amount of living tissue at successive stages. In a natural ecosystem the transfer of energy from producer organism to primary consumer is spread over a number of species, and the speed at which it occurs varies according to species. For example, a ton of meat in the form of rabbit requires about the same amount of plant food to produce it as a ton of beef, but it takes only about a quarter of the time to produce. The controlling factor is the availability of plant food. Eating rabbit meat rather than beef is not the answer to human food problems because the output of a piece of land will be the same whether it is rabbit meat or beef. Put another way, the standing population of rabbits that a piece of land will support is only a quarter the size, in weight, of the population of beef cattle it can support. The turnover rate of animals will be four times as high for the rabbits when compared with the cattle.

Since the energy level falls to 10 per cent at each step in the chain it also follows that the layers in a food web can seldom involve more than perhaps four grades of consumer.

In a natural ecosystem, although considerable variation occurs, the system maintains itself in a state

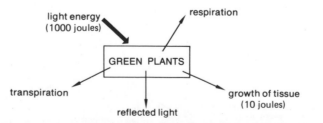

As plants are eaten the energy they represent is transferred to the next level in the food chain or food web. At each step there are energy losses. This figure has been estimated to be about 90 per cent. It means that a piece of land which produces ten tons of grass per year will yield one ton of cattle per year. This energy loss at each step can be represented as shown above.

of long-term equilibrium. The herbivores do not eat all the available food and reduce the habitat to a desert. Carnivores do not eat all the available herbivores so that they then starve to death. In order to get some sort of picture of what happens we need to deal with a number of processes separately, although it must be remembered that this has to be done for convenience and ease of understanding.

Abandoned chalk quarry at Betchworth, Surrey, illustrating plant succession during the colonisation of bare ground

Chalk downland (Chiltern Hills), showing the effect of man in producing a non-climax vegetation—the hillsides are grazed and the plains below cultivated

PLANT SUCCESSION

In a balanced ecosystem, such as a tropical rain forest, we have an example of a stable system which may have taken a very long time to become established and which will stay that way indefinitely. The plant life it supports is described as climax vegetation. It is not possible to provide a documented account of how such an ecosystem comes into being. Nevertheless, a great deal is known about the general way in which it is formed. Soil results from the chemical weathering of bare rock. The first plants to survive on rock are the lichens, and perhaps mosses (see p. 19). Their continuing life cycles transform the rock surface by helping to disintegrate it. Their dead remains contribute to the soil and in so doing create conditions in which other plants can gain a foothold. They in their turn create an environment for still other plants. The process may take a few years or a few millenia until such time as it reaches a stable state, in the example chosen, a tropical rain forest. On a limited scale the process of plant succession can be observed in a variety of situations—for example, old coal tips, disused stretches of road put out of service by road-straightening operations, bomb sites, and so on. Left to its own devices any piece of land will exhibit the process of plant succession.

Ancient Britain was covered by climax vegetation. Most of the lowlands were covered by oak forest. On chalky soils oak would be replaced by beech, on limestone by ash, and on wet lands by alder. The pattern of the countryside as it now is results from man's activities, and since none of it represents climax vegetation it requires work to be done on it. Much of the countryside is obviously non-climax, the meadows and fields and hedges for example. It is not so obvious that heath land or the chalk downs are equally due to human intervention. The chalk downs owe their character to generations of grazing sheep and rabbits. In the absence of these the process of succession begins by the spread of hawthorn scrub. Heath land owes its structure to its use as grazing land, and also to periodic burning. Where these two practices are discontinued heath reverts to woodland. The hedges of the countryside originated as land boundaries, some as late as the eighteenth century.

Factors affecting plant density

The number of plants that are able to grow and maintain themselves in any particular area depends on several factors. Among the physical factors are *light, carbon dioxide concentration* and *temperature*.

An increase in the amount of light falling upon a plant leads to an increase in the rate of photosynthesis, up to a point beyond which there is no increase in the photosynthetic rate. This critical value for light will vary according to species. Increasing the concentration of carbon dioxide will raise this critical value of light. Within limits temperature changes produce similar effects. Within a given region one of these variables will be acting as a **limiting factor**.

The availability of *water* is also an important factor in determining plant growth. Total rainfall in a year is less important than the way in which it is delivered. Much of the rain deposited in light showers, is lost in evaporation. Torrential downpours have a high rate of run-off. Availability of water is a more important factor in tropical regions than light or temperature. In temperate zones light and temperature are the significant factors.

The *composition and structure of the soil* play a very important part in determining how much vegetation a piece of land will support. In a climax vegetation oak woodland allows a much greater turnover of natural resources than does beech. Beech trees have shallow roots. Their canopy of leaves form a mosaic which prevents light from penetrating. The glossy leaves form an impenetrable layer on the woodland floor. As a result, beechwoods possess a very sparse ground vegetation. An oak-wood is much lighter. The roots of the trees are much deeper. There is a rich flora consisting of smaller trees and shrubs together with a carpet of ground vegetation.

Climax vegetation represents a stable situation. In non-climax vegetation there is intense competition between species for survival. This can be observed in a garden which is not looked after effectively. Irises, for example, which maintain themselves along moist river banks, will also grow quite successfully in sandy suburban gardens, provided they are protected from competition. They will spread and form dense woody carpets of rhizomes. If however a few coarse rye grass seeds from the garden next door, or suckers from the Michaelmas daisy patch nearby are not checked, the irises are rapidly overrun and destroyed.

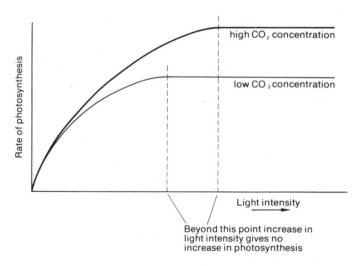

Beyond this point increase in light intensity gives no increase in photosynthesis

POPULATION REGULATION IN ANIMALS

Population regulation in animals is a complex process and ecologists are far from being able to produce a complete account of how it is achieved. Some of the factors which are important are mentioned opposite:

Predator-prey interaction

Herbivorous animals are the food source for various carnivores. The interaction is far from simple but it appears that each acts as a check on the other. The number of herbivores in a given community is kept in check by the activity of the predators. The number of carnivores is restricted by the availability of the prey. Under favourable conditions the number of herbivores tends to rise. This rise is followed after a time lapse by a rise in the number of predators. This slows down the rate of increase in the number of herbivores, a slowing down which is not immediately matched by a drop in the number of predators. The increase in the number of herbivores turns into a decrease, to be followed eventually by a decrease in the number of predators. The populations of the two groups therefore tend to show regular fluctuations slightly out of phase with each other. This pattern is most clearly seen in fairly simple ecosystems. The graph shows the variations in population of Arctic foxes and Arctic hares. The information is based on the numbers of these animals trapped over a period of some years.

Fertility

The fertility of animals appears to bear some relationship to their chances of survival. Rabbits are regarded as particularly prolific breeders. It is sometimes said that were they not preyed upon by so many other species they would rapidly overrun their habitat. While this is obviously true, it is also the case that if rabbits were not able to breed rapidly they would soon be eliminated. In a balanced habitat rabbits breed just sufficiently fast to maintain their numbers. Tapeworms provide another illustration of the proposition that the production of potential offspring is inversely proportional to their prospects of survival (see p. 28).

The swarming habits of the desert locust are perhaps an adaptation to the difficult environment in which it lives. The rise in population which follows rain probably plays an important part in maintaining the species in existence.

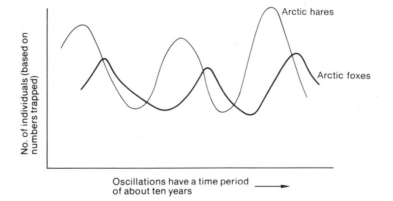

No. of individuals (based on numbers trapped)

Arctic hares

Arctic foxes

Oscillations have a time period of about ten years

Similar fluctuations have been observed in host-parasite relationships. The large white butterfly is parasitized in the caterpillar stage by an ichneumonid fly which lays its eggs in the caterpillar. The grub which hatches behaves as an internal predator and eventually destroys the caterpillar completely. The more large whites there are the more intense the rate of destruction. As a result the large white shows periods of varying population density.

Temperature has an important effect on population levels. In temperate regions where insects may pass the winter either as pupae or as adults, their survival rates vary directly with the severity of the winter. Winter conditions may also affect the numbers of warm-blooded animals, birds for example, which do not hibernate. Although there may not actually be a shortage of food, birds may starve because it is temporarily inaccessible.

Inbuilt mechanisms

While many species of animal are controlled by their predators or parasites there are many species of predatory birds and mammals which are not themselves prey organisms. Evidence suggests that their numbers are regulated by internal mechanisms which are in some way sensitive to the scarcity of food.

Territorial behaviour has been thoroughly studied in many birds, mammals and fish. The male, particularly in the breeding season, exercises proprietary rights over a piece of territory and drives off all other male intruders of the same species. His aggression varies from a maximum at base to a minimum display at the boundary zone. Should his rival from the neighbouring territory be involved, then his aggression increases as he is driven back towards his own territory. Much of this aggression is symbolic and there is seldom any bloodshed. The size of the territory is directly related to the area needed to feed the family and such behaviour patterns spread the species uniformly over the habitat, in so doing acting as a population regulator.

Social behaviour is also thought to play a part in determining population levels. Many birds form colonies in which different status levels exist. There are systems of pecking order. The members at the bottom of the pecking order tend to be squeezed out, or are prevented from mating. In birds which habitually congregate at certain times of the day, for example, starlings, or the grouse population on a heather moor, the behaviour responses of members of the group towards each other appear to determine the overall size of the group, and this varies with abundance or scarcity of food. Limitation in numbers as far as predatory seabirds are concerned operates at their coastal or island nesting sites. The effectiveness of such control mechanisms is illustrated by the accompanying map.

Little is known about the precise mechanisms involved in population regulation of this kind. It appears that when population numbers reach really high densities the members of the community react very much to stress conditions. It is known that among some species, rabbits for example, fertility levels drop very sharply. Females fail to conceive or early failure of pregnancy occurs.

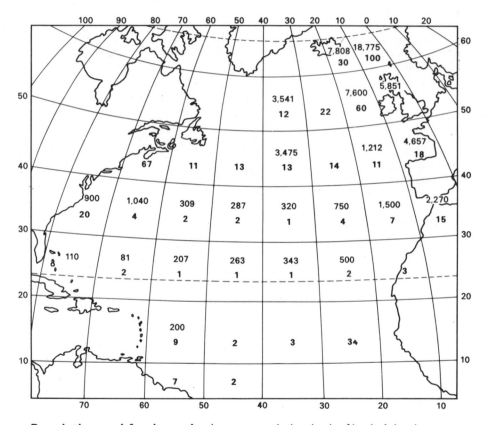

Population and food supply show a correlation in the North Atlantic ocean. The figures in light type give the average plankton count per cubic centimetre of water; the darker figures show the average daily count of ocean birds. (after *V. C. Wynne-Edwards*)

Adapted from *Population Control in Animals* by V. C. Wynne-Edwards. Copyright © 1964 by Scientific American Inc. All Rights Reserved.

46 THE EFFECTS OF MAN ON HIS ENVIRONMENT

Man is part of the ecosystem in which he happens to live and his survival depends on the maintenance and preservation of a stable environment. During the period covered by written history a change has been taking place in man's attitude to his environment. The change has not occurred everywhere, nor always at the same pace. The change has supported the view that man is in some sense a specially privileged animal and that the earth is his inheritance. This is still a relatively new attitude. Primitive communities seem to behave very much as if they are part of a system over which they have no control, which is indeed the case. Hence their beliefs in tribal gods—beliefs in which disease, drought or any other natural disaster is seen as just retribution for past errors.

As a man may survey his garden and feel that it is very much under his control, developing as he wishes, so civilized man has regarded the earth as a whole, particularly since the rise of modern industrial society. Such a view has been encouraged by his success in exploiting the natural resources of the earth and in combating disease, but it assumes that man stands outside the system in some way. Indeed he does, insofar as most civilized men live in urban environments very much of their own creation.

However, the events which have occurred, since industrialization, and especially during the present century, have shown that man is not like a gardener.

In many respects he is much more like a hardy, fertile weed in the garden, overrunning and destroying the rest of its inhabitants. Since his survival depends upon his maintaining a balanced position in the ecosystem it is important that his effects on the rest of the environment should be fully understood. It is not necessary to inspect the other side of the coin. The history of man's achievements, from his own point of view, is already well known.

Although man has always been engaged in the process of altering his environment, until recently his effect upon it was probably no more far-reaching than that of many other successful species. The increase in his influence and capacity to change his environment dates from about two hundred years ago. It has its roots in two causes. These were:

(a) a rise in the population level, and

(b) the transition from an agricultural to an industrial way of life, with its emphasis on consumption of non-replaceable resources.

POPULATION

The graph shows estimated values of past world population figures from about 1600 AD to the present time with an estimated projection forward to 2000 AD. Forward estimates need always to be regarded with some suspicion but at the present time a figure of 6000 million may possibly be on the low side. The graph implies a long slow and steady increase before 1600 AD, but there can be no figures other than rough estimates. It is probable that human populations have fluctuated in the past. The use of primitive tools probably triggered off a sharp rise, as also did the development of an agricultural way of life. At different stages different factors probably played a part in regulating population levels. Pre-stone age man was perhaps under great pressure from the larger carnivores. As human beings came together in large agricultural settlements the effects of parasites and disease-causing organisms would be more pronounced.

The growth curve for population as shown is technically known as an **exponential curve**. A more dramatic phrase refers to it as an 'exploding' curve. It will flatten out at some stage, because of famine if

Man and the environment

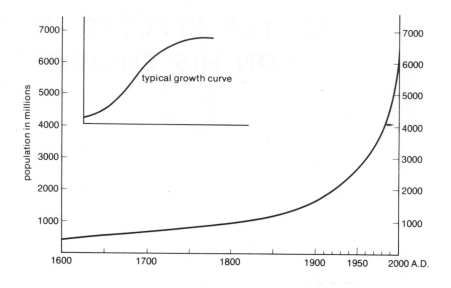

for no other reason, and this is the case for all normal growth curves for individuals as well as populations.

Population regulation has already been considered (p. 266). It is clear that populations are generally stable in the long run, as they must be if the eco-system is to remain in balance. The fecundity (capacity for reproduction) of any species is normally balanced by

(a) the activities of its predators, parasites or disease organisms.

(b) the availability of food.

(c) the operation of internal regulating mechanisms in at least some cases.

If this system of checks and balances is altered in some way—for example, by the elimination of the predators—the number of individuals will necessarily increase until one of the other mechanisms begins to operate as a limiting factor, food shortage, for instance.

Similarly, in the past, human populations have been subject to control:

(a) Disease organisms have played a part in maintaining a high death rate. Occasionally epidemics caused disastrous inroads into whole communities; for example, the Black Death (1348-49) reduced the population of England by a third.

(b) Primitive methods of agriculture meant variable yields; periodic occurrence of famine therefore also played a part in maintaining a high death rate.

(c) Internal mechanisms of population control have taken a variety of forms. The following are just a few examples:

Selective infanticide: the killing of weakling babies by exposure.

Non-selective infanticide: the killing of every second child.

Late age of marriage: in England in 1700 AD the average age at marriage was 29 for men and 28 for women.

There is little evidence to suggest that there has been any significant change in fecundity which might account for a rise in population. Birth rates are subject to fluctuation but, as is clearly shown by advanced societies, populations can continue to rise even though birth rates fall. They do so because the death rate also falls, and is constantly lower than birth rate.

There can be little doubt that the increase in human populations is a consequence of the fall in the death rate. While there can be no argument about the importance of modern techniques of agriculture, improved sanitation and public health, and increasing control over disease organisms in reducing the death rate, they are not the whole story. The rise in population began in England in the mid-eighteenth century. Between 1740 and 1801 the population increased from about 5½ millions to 9 millions. This was before the Industrial Revolution got really under way, so the fall in death rate could not have been due to improved medical services. The evidence suggests that this rise in population was not peculiar to England, that similar rises were apparent even as far away as China. Part of the decline in the death rate is attributed to a sharp reduction in the frequency and magnitude of epidemics. The micro-organism which causes plague has a life cycle which involves two host organisms, the rat flea and the black rat (*Rattus rattus rattus*). The progressive decrease in numbers of the black rat, accompanied by the

270

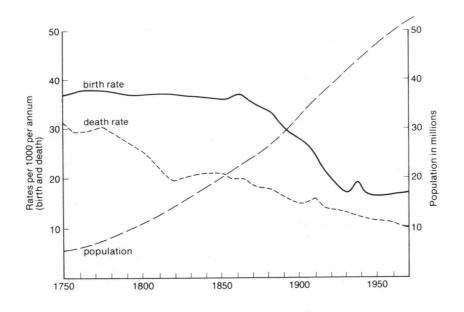

spread of the brown rat (*Rattus rattus norvegicus*), having different habits and different fleas, is thought to be the reason for the decrease in the incidence of plague. Ousting of one species by another is not unknown—for example, the more recent replacement of the red squirrel by the grey squirrel. Micro-organisms are also known to undergo changes which cannot be accounted for. Scarlet fever used to be a particularly virulent disease at the beginning of the present century. It is now so weakened in its effects that it is not always possible to recognize the symptoms of the disease.

The rise in population seems therefore to have begun with a shift in balance within the ecosystem. How much of the continuing rise is due to such causes cannot be known, since it is masked by the general increase in man's ability to counteract the effects of disease organisms by other means. Just how effective these are can be illustrated by reference to one disease (malaria) in one region (Guyana). A campaign mounted by the World Health Organization during the early 1950's resulted in the elimination of the malaria-carrying mosquito. In the decade following the completion of this operation the population rose from 5 millions to 10 millions.

THE GROWTH OF INDUSTRY

The greatest single factor of man's impact on his environment is his increasing numbers. Whatever else he does to the environment its effects must be proportional to his numbers. Throughout historical time man has been altering his environment in a number of profound ways. To some extent this is inevitable. The only human communities which cause negligible change in the environment are hunting tribes and fishermen. No civilization of any kind could evolve without the transition to settled agricultural communities.

The main effects caused by man before the development of highly organized industry can be briefly summarized:

1 Large-scale clearing of forest in Europe and North America.

2 Over-grazing of grasslands, especially in North Africa and the Middle East.

3 Application of unsuitable techniques, or growing unsuitable crops, leading to loss of soil fertility.

4 Large-scale erosion of soil, especially in North America and Australia.

Some of these ill effects on the environment can be remedied, given the incentive. Some of them such as the transformation of North Africa into desert involved slow change over many centuries. The changes which are of immediate concern are the consequences of the change from a predominantly agricultural to an industry-based society, because they operate over a much shorter time-scale and because they are much more far-reaching in their effects. There seem to be at least three important respects in which technical societies are organized which inevitably lead to an upset in the balance of nature. These are:

(a) that they are based on the use of non-renewable raw materials, both as a source of goods and as a source of power.

(b) that they are geared to productivity without serious concern for the fate of the non-usable end-products.

(c) that they are geared to the principle of continuous economic expansion expressed in such terms as 'economic growth' or 'raising the standard of living'.

These issues will be considered again (p. 278) in connection with conservation. Their consequences are summarized below.

NATURAL RESOURCES

These can be divided into two kinds:

Replaceable: timber, food (in the form of food crops and food animals such as cattle and fish), the **raw materials for clothing** (plants such as cotton, leather and wool etc. from animals)

All these materials can theoretically feed back into the cycle either by being burnt or by allowing to decay in the soil. The major problem is to relate their production to the rate of consumption. The use of timber, and the use of sea fish as a food source provide the most serious problems. Modern attitudes to fishing are still essentially the same as those of primitive pastoral or hunting societies. Fishing fleets based on the British Isles have had to travel progressively farther from their home ports to find worth-while fishing grounds as nearer areas have become overfished. This is not really different from the habits of nomadic tribesmen in ancient times who ranged the whole of North Africa in search of pasture for their livestock, eventually reducing large areas to desert.

The timber resources of the earth are estimated to have been reduced by about 50 per cent from their greatest extent since man became a significant factor in nature. Despite an increasing awareness of the need to conserve forests, the rate of consumption of timber still outstrips the rate at which it is being replaced. It is not difficult to see why, in view of the many uses modern man has found for timber. For example, to produce the Sunday issue of the New York Times requires the equivalent of about 0.5 km² of woodland.

Non-replaceable

Modern technological societies are insatiable consumers of a number of raw materials which are not renewable. These are:

Fossil fuels (coal, oil and natural gas).
Metals (in the form of their naturally-occurring ores).
Inorganic fertilizers (particularly phosphates and nitrates deposited as guano by countless generations of seabirds).

Coal, oil and natural gas are sources of fuel required to drive machinery, to provide heat and light, and increasingly to furnish the raw material for the manufacture of innumerable goods, especially plastics. It took approximately 70 million years to form the coal deposits in the earth's crust during the Carboniferous period. If present trends in consumption continue, the twenty-first century will see the final disappearance, not only of workable coal deposits, but also of oil and natural gas.

The difficulty in finding profitable sources of the metallic elements is continuously emphasized by the rising cost of such metals as copper and nickel. The cause of this situation can perhaps best be illustrated by reference to iron ore, which is less likely to act as a limiting factor to the processes of industry than nickel. The ore is mined, perhaps in Sweden or North America, and the pure metal extracted. It may be converted into steel and then built into a motor car body. Some years later, the car now being of little value, it may be abandoned in a field where the steel rusts away to form an iron oxide. This then becomes dispersed in the soil in a non-recoverable form. This is perhaps an extreme picture. Not all cars end up in this way, but quite a large number of empty beer cans and many other end-products of civilized society do.

The need for fertilizers is primarily a need for phosphates. The sources of phosphates in forms which can be used are comparatively limited. They are widespread in nature but not in concentrated form. This is likely to act as a limiting factor in the maintenance of intensive fertile agriculture within the foreseeable future.

LAND

There are a number of ways in which man's impact on the land can be regarded as harmful.

(a) Increasing populations make increasing demands for living space. This necessarily involves the conversion of productive land into non-productive houses and streets. Increasing affluence means more roads and airports, more factories and more dumping grounds. In England and Wales, for example, about 1000 km² of land are lost to agriculture annually.

(b) In industrialized societies farming has tended, like many other activities, to become 'rationalized'. This means increasing the return on capital invested as much as possible. Such improvements have involved the growth of larger farms by the absorption of smaller ones; the elimination of barriers such as hedges to the use of complex, efficient machinery; and the trend towards specialized production instead of mixed farming. This can be regarded as good business practice but it has also meant the extensive use of chemicals to deal with pests of various kinds which thrive better in large areas devoted to single crops. The destruction of trees and hedges has brought problems of soil erosion to large areas no longer protected from the effects of the wind, and in some regions the need to secure a high yield has actually led to a lowering of production due to the excessive use of fertilizers, particularly nitrates.

(c) A failure to appreciate the nature of the forces which operate in natural ecosystems has caused widespread changes in the characteristics of certain regions, where animal species have been introduced from outside. Mention has already been made of the rabbit in Australia. In New Zealand the Scottish red deer and the opossum imported from Tasmania afford other examples. Neither of these species has a natural predator in New Zealand. The deer browses on the leaves of trees within its reach and the opossum, since it can climb, feeds on leaves at the top of the tree as well as upon bark. In a number of remote valleys these animals have spread because of the absence of population checks, and as a result have caused destruction of large areas of forest. This has been followed by large-scale erosion of soil from the slopes of the valleys. The introduction of the prickly pear cactus (*Opuntia inermis*) into Australia from South America at the beginning of the nineteenth century resulted in the infestation of some 200,000 km² of land in Eastern Australia by this weed by 1930.

POLLUTION

Pollution of the environment affects both animals and plants, either by causing a loss in productivity (slower growth and less yield) or by damage to tissues, thus causing illness or disease. In extreme cases it can cause death of plants and animals.

Pollution can occur in the atmosphere, in the soil, in the sea or in freshwater. It can be caused in four main ways:

(a) The production of materials as a result of industrial processes. These may be released into the atmosphere or discharged into streams or rivers.

(b) The widespread use of chemical pesticides and weedkillers.

(c) The production of radioactive materials.

(d) The progressive accumulation of discarded materials.

Air pollution

Pollutants entering the atmosphere produce their effects in one of two ways.

(a) They may enter the tissues of plants and animals directly, either through the stomatal pores or lungs, or at any point on the surface.

(b) They may first be deposited in the soil or in water, then absorbed by plants in solution, and eventually pass into animals by way of the food web.

(a) The main pollutants in this category are produced as a result of **industrial processes**, by **domestic fires**, and by **internal combustion engines**. Their effects tend to be localized, although not invariably so, depending upon such weather factors as wind (direction and force), the extent of vertical air movement, and rainfall. In conditions which severely limit vertical air movement (fog) the effects of pollution are greatly increased. The following list includes the main pollutants in Great Britain:

Soot, tarry residues, ash, sulphur dioxide, carbon monoxide, hydrochloric, nitric and sulphuric acids, hydrogen fluoride, hydrogen sulphide, ethylene.

Sulphur dioxide is the most important pollutant. The amount deposited on the ground in England and Wales annually is thought to be in excess of 7×10^9 kg. All these pollutants produce harmful effects. Soot can clog pores of leaves as well as skins. High concentrations can reduce the amount of light reaching ground level. In either case the net result is a reduction in the rate of photosynthesis and therefore of growth. Sulphur dioxide produces its main effect on soil organisms, causing increased soil acidity. The effect on human beings is to aggravate a whole range of respiratory diseases, especially bronchitis. Very heavily polluted air is known to carry cancer-producing substances and this is reflected in the higher incidence of cancer in such regions compared with less polluted areas. A specialized kind of pollution occurs in regions where industry is linked with clear skies, bright sunshine

Concentrations of pollutants measured in central London 1954-1964

	Max Recorded Conc	Av Winter Conc
Smoke (mg m⁻³)NO_2	10	0.2
Carbon monoxide (ppm)	360*	10*
Sulphur dioxide (ppm)	2.0	0.2
Sulphuric acid (mg m⁻³)	0.7	0.01
Nitric Oxide (ppm)	1.1	0.05
Nitrogen dioxide (mgm⁻³) NO_2		0.03
*measured in traffic ppm = parts per million		

and an abundance of motor cars. In Los Angeles it reaches smog proportions in the daytime, dispersing at night.

Although **carbon dioxide** is not, strictly speaking, a pollutant, it too is added to the atmosphere in enormous quantities. Over the immense period of time represented by the thickness of the coal deposits, carbon dioxide was slowly being withdrawn and locked up in the earth's crust in the form of the fossil fuels. Over the very short period since industrialization began this carbon dioxide has been released into the atmosphere in increasing quantities by combustion of coal and oil. Carbon dioxide in the atmosphere produces a 'green-house' effect, i.e. it allows more heat to pass in than it allows to pass out, and this effect is increased by increased concentrations. Any upward shift in the average temperature, due to the increase in atmospheric concentration of carbon dioxide, could result in a significant lowering in the amount of water locked up in the polar ice caps, with a corresponding rise in sea level.

(b) Pollutants in this group are mainly:

Radio-active materials from nuclear processes, including the testing of nuclear weapons.

Chemical pesticides especially those sprayed from aircraft.

Radio-active materials

Rather ironically, a great deal of valuable ecological knowledge has been gathered from studies carried out on the effects of radio-active fall-out. Perhaps the two most important facts to emerge from such studies are that (a) airborne materials, pollen grains as well as pollutants, are dispersed on a world-wide scale with extreme rapidity, and (b) food webs are extremely efficient mechanisms for concentrating materials which are finely dispersed in the environment, whether they are essential materials or pollutants.

A radio-active isotope which has been particularly well studied is strontium 90. It is biologically indistinguishable from calcium. It has a slow rate of decay and therefore can persist as a danger to health for many years. It can be deposited in bone tissue, where its radiations can cause damage to blood-forming tissue. It falls on the ground, is absorbed by grass, is taken up by cattle, and enters human tissues in milk.

Pesticides

When applied by aerial spraying these can be dispersed in the same way as radio-active compounds. A vast number of different compounds have been used as pesticides during the past twenty years. This is partly due to the fact that pest destruction of any kind usually means being one jump *behind* the particular pest concerned. No chemical has been

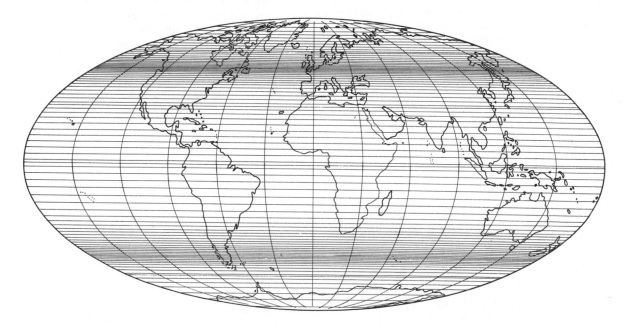

Density distribution of radioactive materials in upper atmosphere. Note: greatest concentration in belts 50°N and 50°S. (After G. M. Woodwell)

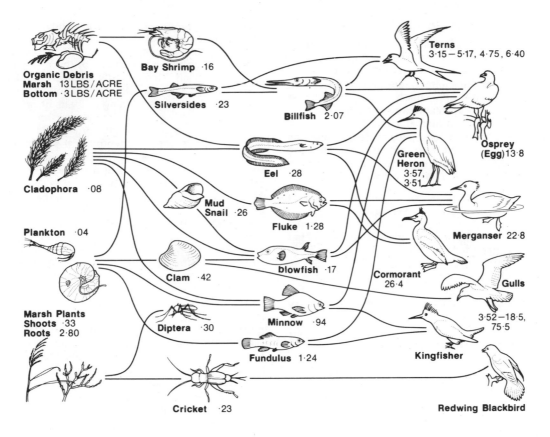

Rise in concentration of D.D.T. passing through a food web. (After G. M. Woodwell)

made, nor is it likely to be made, which has 100 per cent killing power for one specific type of organism. An attack on a particular pest with a new pesticide therefore eliminates only that proportion of the population which is not resistant to it. In the course of time the population begins to build up again by growth from the resistant survivors. The pesticide therefore becomes less effective with time. Similar limitations apply with antibiotics used to combat disease and infections in man. The more widespread and concentrated the use of a particular antibiotic the sooner it begins to lose its effectiveness. The need to use pesticides on a large scale is a result of large-scale farming of single crops, particularly in North America. There are all sorts of side effects which may take a long time to make themselves apparent. Pesticides may become inserted into food webs where they were not intended, and if they are long-lasting and not easily metabolized they become concentrated as they pass through the food web.

The side effects may be a sudden rise in the number of dead birds or an unexplained drop in soil fertility after an initial rise due to the destruction of the pest organism. This can be due to an imbalance in the soil ecosystems because of the effect of the pesticide on soil micro-organisms.

The best known pesticide is DDT, and it first came into widespread use during World War II. It is not metabolised in vertebrates, generally being deposited in the fat reserves of the body. Its toxic effects, in high concentrations, only become apparent in circumstances where the body's fat reserves are being mobilized. There seems little doubt that the distribution of DDT is now worldwide. There appear to be no regions where it is not found in human tissues in minute traces and it has even been identified in the tissues of fish-eating Antarctic birds.

A final long-term consequence of the application of pesticides against insect pests is that, although it may be unable to eliminate the insect, in the long run it may well lead to the elimination of some of its predators among the local bird population owing to the progressive concentration of the chemical as it passes along the food chain.

Land and water pollution

The story of land and water pollution is, in part, a repeat of the problems of air pollution, since air-borne pollutants eventually make their way into the soil and water. Their effects are intensified on the land by extensive application of pesticides over limited areas, and in inland waters by the concentrated discharge of industrial waste from factories. They are also complicated by the fact that pollutants in soils may be carried down to the water table and make their effects known very much later when they finally discharge into a stream or lake.

Water Pollution This poses particularly urgent problems. Man is especially sensitive to it because his needs for fresh water are continuously increasing. The need to find sources of fresh uncontaminated water is aggravated by the increasing extent to which possible sources are being contaminated. The main effects on inland waters are those due to outflow of factory waste, human sewage and the run-off of nitrates from over-fertilized land. **Factory wastes** include:

(a)   **Oil**; forming a thin, widely dispersed film on the surface, it reduces the uptake of oxygen by the water.

(b)   **Detergents**; they reduce the oxygen capacity of freshwater.

(c)   **Suspended particles**; these produce similar effects in water to those of soot in the atmosphere.

(d)   **Poisonous chemicals** such as sulphides and sulphites; acting as reducing substances, these lower the oxygen concentration in the water.

The capacity of water to absorb **sewage** depends upon the quantities involved. In the normal way sewage enters the cycle of decomposition as it does in the soil, but when the system is overloaded the rate of decomposition, caused by the rapid growth of bacteria, leads to rapid loss of oxygen.

In many areas the intensive cultivation of land, involving heavy dressing with fertilizers, has added a new pollutant in the form of **nitrates draining from the land** into inland waterways. These stimulate the growth of algae at the surface and cause depletion of oxygen. High concentrations of nitrates can also produce harmful effects on animals.

Water from some industrial processes, for example, from the cooling towers of power stations, causes **thermal pollution**. The permanent temperature rise brought about by the hot water discharged into rivers and lakes kills many native species. Also, the oxygen content is depleted, because water at higher temperatures is able to hold less oxygen in solution.

Land Pollution Mention has already been made of the way in which plant growth is impaired by industrial pollutants (p. 273), and how pesticides and radio-active substances can be transmitted through plants to animals (p. 274). Remarkably little is known about the effects of these substances on the soil itself. This is partly because of a lack of knowledge of the ecosystem of the soil. It is known that the soil supports teeming millions of organisms and that probably all the complex inter-relationships which exist between living organisms in more familiar habitats are represented in the environment of the soil. These soil organisms include not only fungi, bacteria and algae but a vast range of small arthropods, protozoa and worms of various kinds. Many occupy highly specific niches in the economy of the soil.

It is clear that the application of a particular insecticide, designed to eliminate a particular pest organism, is likely to eliminate many other species as well, an effect paralleling that of DDT on carnivorous birds (p. 275). Unless studies are conducted with the specific intention of discovering the nature of the specialized habitats of the soil, knowledge is limited to the immediate effects in terms of plant growth. Knowledge of long-term effects must otherwise wait until they actually appear. It is known that some insecticides produce adverse effects on some of the nitrifying bacteria, since peas and beans grown on affected soils fail to develop root nodules (see p. 258).

The long-term effects are likely to be less catastrophic on soil organisms than on water organisms, although it would be unwise to assume this. There is a very rapid recycling of materials in soil organisms, and probably continuous loss of materials in solution into deeper layers of the earth. This ultimately adds to the general level of water pollution.

WILD LIFE

The effects of man's activities on wild life are due to a variety of causes and it is virtually impossible to assess their extent. The fate of many of the larger and more conspicuous animals, such as birds, reptiles and mammals is well documented, and is more apparent because there are fewer of them to begin with. The seriousness of the problem is masked by the fact that in most cases destruction has had few immediate adverse effects. No species lives for ever, (see Evolution p. 245), but there can be no doubt that the extinction of a very large number of species has been accelerated as a consequence of man's actions. Although some species have been destroyed wantonly, many have disappeared because they are in competition with a more successful species.

Indirect extermination

This has occurred on a massive scale with the

progressive clearing of climax forests. The animals which live in forests are deprived of their natural habitat and therefore die out. In the tropical rainforests of the Amazon basin, which are in the process of being cleared in order to grow crops, not only are the wild animals being 'put out of business' but also tribes of South American Indians which have existed in the remoter parts of the jungle without contact with civilization are being destroyed.

Changing patterns of agriculture can also cause widespread destruction, such as the clearing of small pockets of woodland, the planting of conifers on heathland, or the drainage of marshlands. Intensified methods of farming, such as hedge removal or systematic spraying of grass verges, all contribute to the elimination not only of animal species but of plants as well. In regions where town dwellers have easy access to the countryside in large numbers, many plant species which were once common are now becoming rare due to the indiscriminate and excessive picking of their flowers, e.g. primroses and orchids.

Increasing population reflected in the spread of built-up areas, new roads and airfields also significantly diminishes the numbers of plants and animals.

Extermination by hunting

Men hunt animals for all sorts of reasons, sometimes for food, sometimes for profit, sometimes just for fun. One very well-known example is given below:

The American Bison

It is estimated that some 60 000 000 bison roamed the grasslands of the American mid-west during the early part of the nineteenth century. They provided a source of food and clothing for the hunting plains Indians, who appear to have represented the predator in balance with its prey. By the end of the century the bison was almost extinct, as a result of the great spread westward of American pioneers, whose aim was to replace the natural grazing animal, the bison, with their domesticated animals, and the natural grasslands with their own grain crops.

It may be argued that the wholesale destruction of species is a process which goes on whether man is responsible or not. There is a certain inevitability about it, as the evolutionary record shows. Nevertheless, any decrease in the richness and variety of life, whether avoidable or not, must mean a loss in every sense, quite apart from the possible long-term effects on the overall balance of nature.

47 CONSERVATION

To conserve means to maintain, to preserve, to keep the same. The adjective, conservative, means resistant to change, but in its modern usage conservation means rather more than simply keeping things as they are, as the following illustration shows.

The owner of a farm demonstrates the principles of good husbandry, or careful management, if his efforts lead to an improvement in the quality, productivity and appearance of his land. For example, the trees on his farm serve a number of purposes. They are a reserve of timber. They act as a windbreak for his crops and his cattle. They provide cover for many animals, especially birds, and a sanctuary for woodland and shade-loving plants. Trees, like all living things, are subject to disease and old age, or to overcrowding. Good husbandry requires that diseased specimens should be removed, that weaker specimens should be thinned out where there is overcrowding, and that trees taken for timber should be replaced. The farmer's policy with regard to his trees is based on the knowledge that he must plan ahead and that fifty years is a short time in the life of a tree. Before he removes a hedge he will balance any possible gain in terms of increased yield, or labour saved, against the loss due to its removal. He knows that hedges are important as windbreaks, that they help to prevent erosion. They harbour a reservoir of bird life and while he might not worry about the loss of bird song he would be concerned about the loss of a valuable ally in his fight against pest insects. Hedges also form a haven for the insect pollinators, and many plants are found in them which can be found nowhere else.

If a stream runs through the farmer's land it, too, will require his attention. A neglected river can cause serious erosion of valuable soil. Its scouring action at a bend can carry off soil on one side, replacing it with useless shingle on the other. Care is needed to see that its fish are not depleted by pollution or over-fishing, and re-stocking may be necessary from time to time.

In these and many other ways the farmer practises good husbandry. He knows that a farm is more than just a business, that other factors are involved than ensuring a high cash return for his invested money. He holds the land in trust for future generations. It will still be there when he has gone. Not only is he guided by experience of the past, but also by his concern for the future.

On a larger scale this is what conservation is all about. It is not simply a matter of preventing change. In any case this is not possible. Its chief aim is to prevent further degrading of the environment, despite all the various demands being made upon it. Conservation is, in fact, good husbandry on both a national and world-wide scale, with all the planning, foresight and cooperation that it entails.

Conservation, the sensible management of the earth and its resources, is not a particularly modern idea. Man, as a social animal, no less than as an individual, has always had to learn from his mistakes.

The story of conservation in modern Britain is a story, which, with minor differences in detail, could be applied to any country in Western Europe and North America. It is a story which begins with the attempts of a few determined individuals to arouse public interest and concern, culminating, very much later, in government action. It is also a story of piecemeal action in the many varied aspects of conservation, running together to create a flood of legislation and the appointment, in 1969, of a senior cabinet minister with overall responsibility for the whole of the environment.

THE SCOPE OF CONSERVATION

The problems of conservation can be divided for convenience into three groups. They are:

1 The remedying of mistakes.
2 Reduction in the rate of loss of non-renewable resources.
3 Population control.

The remedying of mistakes

Wildlife

Where species have become entirely extinct clearly nothing can be done. Where species may be threatened with extinction a number of actions are possible, and it is an encouraging sign of changing attitudes that more and more is being done, for example:

(a) Species may be **protected by law**. The Protection of Birds Act (1954), makes it an offence to take the eggs of, or interfere with, any wild bird.

(b) Areas in which threatened species live may be designated as a **nature reserve** and subjected to rules whose sole aim is the preservation of species within it. Examples: Brownsea Island in Poole Harbour (a bird sanctuary): Yellowstone Park in America.

(c) When a wild population falls below a certain critical level it will become extinct simply because the different members of it hardly ever meet. In such cases animals may be transferred to a **zoo** or a reserve and there encouraged to breed. There are a number of species which no longer exist outside zoos or game reserves.

(d) Any threat to the survival of the fauna of a particular area, as a direct result of man's activities, is countered by wholesale **transportation**. This was done, for example, during the building of the Kariba Dam on the Zambesi River.

A number of organizations in Britain exist for the sole purpose of conserving wild life, both plant and animal. Trees, for example, may not be felled without good reason and overall responsibility for trees in Britain is that of the **Forestry Commission** which was set up in 1919. Local authorities have the power to create by-laws, for instance, to restrict indiscriminate picking of wild flowers. (Anyone living near London cannot fail to notice how rare the primrose is.)

Pollution

More progress has been made in combating atmospheric pollution than perhaps in any other aspect of conservation. This is scarcely surprising, since the quality of the atmosphere is something that affects us all immediately and directly. The great London smog of 1952, for example, was held to be responsible for the deaths of four thousand people.

Atmospheric pollution in Britain reached its peak during the nineteenth century with the rapid development of heavy industry and the widespread use of coal as the primary fuel. The introduction of the Clean Air Act of 1954 and the gradual increase in the extent of **smokeless zones** have had a marked effect in reducing atmospheric pollution. Legislation designed to reduce the degree of pollution from industrial sources was introduced as long ago as 1893. That Act and others up to 1960 compel industries to fit collector devices in factory chimneys to trap pollutants. The main pollutant from this source is now sulphur dioxide and there appears as yet no satisfactory way of dealing with it other than by means of tall chimneys through which the flue gases are discharged at high speed. Most of current legislation is directed at containing the spread of pollution. This is being done by careful siting of new factories so that the effects of unavoidable pollutants are minimized.

Pollution originating from petrol and diesel engines rises as the number of vehicles on the roads increases. Legislation introduced in America in 1967 lays down standards for the amount of pollution allowable from exhausts of road vehicles, and there is little doubt that this will spread to other countries as their traffic densities rise, or as their car exports to America fall.

River pollution is also the target of concerted action. The Water Resources Act of 1963 set up a Water Resources Board, acting through 29 river authorities, charged with responsibility for the conservation and utilization of water. Standards are laid down, enforceable by law, governing the permitted levels of pollutants which may be discharged into rivers. Similar regulations are applied in other countries—although accidents can occasionally happen, as was the case in 1969 when the illegal dumping of chemicals caused the destruction of millions of fish along the entire length of the River Rhine.

The problems of pollution caused by the widespread use of chemicals to control pests are rather different. It is simply not possible to abolish all pesticides and chemical additives to the soil, because it would mean a drastic fall in food production. Farmers are committed to higher levels of productivity whether they like it or not. Clearly, much research into the long-term effects of pesticides is necessary, together with efforts to reduce the periods of time over which pesticides remain active. This is a continuing problem because of the adaptability of the pest organism to the particular pesticides used against it. There are a number of encouraging new techniques being developed (see p. 281).

Land Misuse

Land mismanagement at its worst calls up pictures of man-made deserts, the most ancient being the arid lands around the Mediterranean, the most recent the dustbowls of the American Mid-west. Soil erosion, however, is not a specially important problem in Britain. There are two main reasons for this: the geography and climate, and the traditional small-scale mixed farming with fields enclosed by hedges. The growth of large-scale farming, with the destruction of hedges, the abandoning of crop rotation, and the extensive use of weedkillers and fertilizers, can lead to erosion on a scale serious enough to require action, however.

The main problem in Britain has been the misuse of land. In a small, highly industrialized country, land has to carry out a number of different functions involving a number of competing interests. It must provide space for homes and factories; its fertile areas must be protected for the growing of food; areas of great natural beauty must be protected against· exploitation and destruction. Against the background of increasing population, every available space needs to be utilized in the best possible way.

The vast industrial expansion of Victorian England was accompanied by many changes among which the following should be specially noted:

(i) The rapid spread of towns, caused partly by the increase in population and partly by the continuous migration of people from the country to the towns. Most of the building was both cheap and nasty.

(ii) The increasing spread of the by-products of industrialization, factories, slag heaps, waste tips and the scars of open-cast mining and quarries.

(iii) The gradual run down of British agriculture and neglect of the countryside following the increase in the availability of cheap imported food from 1870 onwards.

The first moves to remedy this state of affairs were evident in the nineteenth century. The National Trust was founded in 1895. Various imaginative schemes for housing people under better conditions were started: Port Sunlight in 1886, Bourneville in 1879, Hampstead Garden Suburb in 1906, Welwyn Garden City in 1930. The Second World War and the need to grow more food caused great improvement in the productivity and the appearance of the countryside. The Town and Country Planning Act brought all development under the control of either local or central government. The National Parks and Access to the Countryside Act of 1949 created the National Parks Commission responsible for **national parks**, such as Snowdonia, areas of outstanding natural beauty, areas of scientific interest, local and national forest reserves.

As a result of legislation, and action by local authorities and private organizations, there appears to be a much greater sensitivity to the needs of conservation. Large stretches of the coastline are now protected from exploitation. The building of roads involves care to see that they take the route least likely to absorb good agricultural land. They are landscaped so as to disturb the appearance of the countryside as little as possible. The first step in building a housing estate used to be the removal of all the trees. They are now realized to be an asset and are preserved. Planning, especially where central government is involved, is often seen as a restriction of liberty at the personal level. Such restriction is clearly necessary in the interests of all.

Reduction in the rate of loss of non-renewable resources

Most, if not all, modern technological societies which set the pace for the so-called emergent nations, are committed to the principle of a raised standard of living. This means a policy of increasing consumption of materials, a policy of continuous economic growth. In itself this need not necessarily be harmful if it means that materials are re-cycled more rapidly. For example, it does not matter if a man changes his car for a new one every two years, or every six months, provided that its materials feed back into the cycle. It does matter, however, if the rate of accumulation of rusting cars increases four times as quickly. Part of the problem lies in the fact that the processing of raw materials into finished goods pays immediate dividends, whether it is a new suit or a new transistor radio. The incentive to possess creates the incentive to produce. There are no such incentives operating with goods that are worn out or discarded.

A serious problem is created by a modern society in the disposal of the end-products of its industry, end-products it is designed to create. Such a society generates its own special kind of pollution. Modern advertizing speaks glibly of the virtues of disposable packaging, meaning bottles, cans and plastic containers of various sorts. Disposable they may be, in the sense that they may be thrown away. In every other respect many of them are virtually indestructible. An apple core tossed into the hedge from a passing motor car, is rapidly decomposed, its materials being re-cycled. A plastic bottle, similarly disposed of, remains there for evermore.

Long-term planning must therefore embody as part of its aims the establishment of a stable system, which continuously re-cycles its materials. It must

(a) re-process its end-products. If it is possible to bring together a large number of complex materials to create, for example, a motor car, it is equally possible to create the means of reversing the process.

(b) develop other sources of power which are not exhaustible.

There are some signs that moves are being made in this direction. In some parts of the country old cars can be collected for a nominal fee and removed to a depot, compacted and then dispatched to factories which recover metal parts. The latest types of garbage disposal plant are systems for recovering useful materials together with an end-product which is germ-free and odourless and can be used to improve the quality of the soil.

There seems little doubt that electricity will be the universal energy source of the future. There are no pollutants as a result of its *use*—only, at present, in its *production* by oil- or coal-fired power stations. These are already being replaced by power stations run by nuclear reactors, although these also have waste disposal problems. The problems of pollution are already stimulating research into the development of lightweight, longlife electric batteries to replace the internal combustion engine as a portable source of power.

Much work is being done to explore the possibilities of generating electricity without using non-renewable resources. These include windmills, geothermal systems (using underground heat to produce steam to drive turbines), tidal power systems, solar radiation. Solar collectors mounted on roofs and geothermal systems can also be used as heat collectors. The heat is transferred to water which provides both the means of heat transport and a way of storing heat.

Attention is being given to the problems of heat loss. Much energy is lost unnecessarily because of poor insulation. Many of our modern glass and steel buildings consume vast quantities of gas or oil every winter which they convert into heat energy which for the most part is radiated away into the surrounding atmosphere. It is perfectly possible to design and build a school, for example, the interior of which can be kept at a comfortable temperature by the body heat of its pupils.

The cost effectiveness of various forms of transport, in energy terms, has been thoroughly studied. The closure of many railway lines in the sixties was said to be because we could not afford to run them. In energy terms we cannot really afford not to use them because, with the exception of the bicycle, rail transport is cheaper than any form of road transport.

Population control

This involves finding solutions to three quite separate problems:

(a) Control of those organisms (pests) whose numbers *increase* as a result of man's activities.

(b) Maintenance of numbers of organisms whose numbers *decrease* as a result of man's activities.

(c) Control of human population.

Control of pest organisms

Pest organisms increase in numbers because the conditions are created which favour their way of life. Wholesale, massive spraying of crops is clearly an ineffective way of re-establishing a balance because (a) it has harmful effects on other organisms and (b) it does not, except in the short term, deal satisfactorily with the pest. Three different techniques are likely in the long term to replace pesticides.

(a) Biological control by means of predators or parasites

Two spectacular examples of success in this field may be cited. The prickly pear cactus, accidentally introduced into Australia, in the early nineteenth century, rapidly became a pest. It was eventually brought under control by the introduction of a moth, native to the Argentine. A serious pest of citrus fruit trees in California, the cottony cushion scale, also an immigrant from overseas, was ultimately controlled by the importation of its natural predator, a small ladybird.

(b) Insect trapping

A great deal of research is being carried out on studies of chemical substances produced by insects. Males are normally attracted to the females by their scent. Extraction of or synthesis of the compounds which produce the scent can serve as a basis for baiting the trap. Other promising lines of research involve identification and extraction of the compounds in the plant food which attract the insect. These also can be used as the bait.

(c) Release of sterile insects

A large number of organisms are bred artificially in the laboratory, and subjected to radiation to render them sterile. They are then released in the field. When normal wild type insects mate with them no fertile eggs are produced. The end result is a reduction in fertility of the pest organism.

Maintaining population levels among wild animals

As a nomadic hunter, for example, the American Indian, man's role in a natural ecosystem is to act as a predator. He and his prey act as checks upon each other, each to some extent regulating the population of the other. With the development of a pastoral way of life and the domestication of animals, the prey, in effect, passes under complete control by man. Both the rate of reproduction and the rate of killing is regulated. Many animal species still stand in the role of the hunted prey in relation to man, particularly marine animals. Wild species need therefore to be managed. For example, the herring catch needs to be regulated in terms not of the size of the market, but of the density of herring populations. The problems here require that international agreements be made, since the seas are not subject to national control.

Human population control

Virtually all the problems of mismanagement of the earth's resources can be traced back to the biggest problem of all which is the rise in the level of human populations. It is probably not an overstatement of the case to say that unless man learns to control his own numbers he might just as well abandon any attempt to practise good husbandry as far as the environment is concerned.

Although a great deal is known about population regulation in other species there is very little positive understanding of the biological factors controlling fertility in human beings. Animal populations are controlled by external factors operating upon internal mechanisms, but the situation with human beings is complicated by the fact that we have a personal, or subjective view also to take into account. Human beings, too, appear to have a range of choice of action denied to animals. It is therefore very much more difficult to identify the biological factors operating—if, indeed, there are any at all.

In a balanced ecosystem animal populations tend to remain stable, apart from seasonal fluctuations. This they must do or the system would not be stable. In other words, fertility level (number of live offspring produced) is balanced by the death rate. Although in some species (see p. 267) fertility may vary, generally it is the death rate which acts as the main regulator of population levels. Fertility level appears to be fairly constant and is presumably the result of slow adaptation by natural selection. Rabbits, for example, have a very high death rate under normal conditions. It is not surprising that their fertility is high: they would cease to exist if it were not so. The giant tortoises of the Galapagos Islands, on the other hand, in their natural environment, are not subject to predators, and they have a long life span. Their fertility, to judge by the results of their controlled breeding in zoos, is very low.

Since fertility rates are constant, and not, in the short term, adaptable, it is not surprising that changes in death rates can produce such catastrophic changes, in either direction. Attempts have been made to assess the 'natural' level of human fertility on the assumption that man, like other mammals, is a product of slow evolution and was also at one time subject to external population control. Estimates are based on studies of primitive human communities. These are usually closely-knit societies in which there are seldom any unattached women of marriageable age. On the basis of a low

expectation of life, perhaps forty years, and a late onset of menstruation, perhaps as late as eighteen, this allows about twenty years of reproductive life. After making allowances for miscarriages, abortions and stillbirths, this suggests that an average woman might be expected to produce about six children, or, at most, eight.

The fertility rates of all human communities, including primitive ones, are, however, modified by factors which do not apply to other species. These are social factors. In primitive tribes which depend upon hunting or food-gathering for their existence, a relatively large area of land is necessary for successful survival. Their fertility level appears to be about 4-6 children per female. In agricultural communities, which need less land per person the figure is between 6 and 8. Social practices which can affect fertility level are:

Age of marriage

This is often some years after the attainment of sexual maturity.

Abortion and infanticide

Many tribes practise regular abortion or infanticide as a method of population control (see p. 270).

Perpetual widowhood

A widow in some societies is not allowed to remarry.

Restrictions on marriage behaviour

Some tribes have laws which restrict the living together of men and women during certain seasons.

Contraception

Many societies have their own particular herbal mixtures or practices which are supposed to reduce fertility. Usually these are of doubtful value. If they work, it is by faith rather than function. Population changes in advanced modern industrial countries can be illustrated by reference to Britain where conditions are in many ways similar to those in other European countries. The pattern in North America is rather different because of large-scale immigration, and in Ireland, where mass emigration has occurred.

The massive rise in population which began about 1740 was sustained for most of the nineteenth century before the pace began to slacken. It was due to the fall in death rate which, however it started, was certainly maintained by the advances in medicine and public health and sanitation resulting from the Industrial Revolution. The birth rate, or level of fertility, began to decline from about 1870 onwards, a later consequence of the Industrial Revolution. A biologist might say that here was evidence of a natural population showing an adaptive drop in its level of fertility, and he would of course be justified in saying so, whatever the social reasons that might be put forward.

At the social level the factors contributing to the lowered fertility were:

(a) the *desire* on the part of many people to limit the size of their families, and

(b) the *ability* to practise family limitation due to the spread both of knowledge about contraception and of the means to apply it.

The same sorts of reasons apply at the present time with even greater force. People living in a society such as ours are presented with choices of action which were simply not available a hundred years ago.

They have a lot to lose in terms of material possessions, leisure pursuits, and standard of living, generally, if all their efforts and their incomes are devoted to raising a large family. Such a prospect exerts very strong pressures when it exists alongside the knowledge needed to ensure small family size, and it is reinforced by the knowledge that two or three children will be better provided for than say six or seven. In the mid-nineteenth century people had no alternatives in terms of material goods. Neither did they know how to limit their families.

Fertility Control

A modern community has a wide range of methods available to it for the purposes of deliberate fertility control, all of which have drawbacks of one sort or another. It lies outside the scope of this book to deal with these in detail; nevertheless, it is important that the principles involved should be understood, if only to appreciate that the control of fertility is not the straightforward business that it might appear to be. The use of the expression 'fertility control' is deliberate since the more common term 'contraception' strictly means only what it says, preventing fusion of egg and sperm. The sequence of events involved in the production of a new individual can be broken at a number of points.

(a) The shedding of the egg from the ovary (ovulation) may be prevented. This is effected by the oral administration of pills containing progesterone, the pregnancy hormone (see p. 230). Progesterone is normally produced in the body *after* ovulation has occurred, and its purpose is to block further ovulation.

(b) The movement of eggs along the oviduct may be prevented by means of a ligature applied to the oviduct.

(c) The release of sperm may be prevented by tying off the vas deferens. This involves minor surgery, as does (b).

(d) Both eggs and sperm may be shed but prevented from meeting. This may be achieved by avoiding sexual intercourse (coitus) when an egg is actually in the oviduct, by the use of physical barriers to prevent meeting of the sex cells, or by the use of chemicals which inactivate or kill the sperms (spermicides).

(e) The fertilized egg may be prevented from implantation in the uterine wall. A foreign body in the uterus will do this. Intra-uterine devices (IUD) are generally small plastic loops.

(f) The implanted developing embryo may be removed (induced abortion).

However:

(a) can only be prescribed by a doctor and should be used only under constant medical supervision. Although the pill has been in use for a number of years and has been taken by millions of women all over the world it should be stressed that the long-term effects of its continued use are not known. Many responsible and informed people have serious reservations about the widespread use of this particular method of control.

(b) and (c) are normally only applied to people who have had as many children as they want or where there are serious health risks for the mother if she has any more children. The operation is not always successful and, where it is, its results are said to be reversible, by a further operation if necessary. All the methods listed under (d) have drawbacks, if only because they depend upon the foresight and sense of responsibility of the individuals concerned, and human beings are notably prone to error. Spermicides also probably carry a long term risk of damage to female tissues. If they destroy sperm cells it is likely that they at least damage other cells. (e) is perhaps the simplest method of fertility control. It has to be fitted under the supervision of a doctor if he considers it suitable.

(f) is the subject of much difference of opinion. It can only be carried out if the specialist doctor recommends it and it is a rather unpleasant experience for the mother. Many people feel very strongly that abortion is wrong on the same grounds that to take any life is wrong. They maintain that life begins at conception and not at birth. In any case abortion is a drastic procedure, to be regarded as neither desirable nor practicable as a regular method of control.

It can be seen therefore that the means for regulating population level in Britain exists, and there is certainly no lack of centres or services from which individual men and women can seek and obtain expert help and advice. It may be asked why it is that people do not control their numbers simply by refraining from sexual activity. The answer is that the urge to mate is a very fundamental one, and to expect human beings to exercise restraint is to expect the impossible. It is not, however, unreasonable to expect individuals to be answerable for the consequences of their actions when and where the means do exist. The problem of reducing the birth rate is not so much a matter of knowledge as a lack of a positive wish to do so on the part of individuals. In a society such as ours it is quite possible to enjoy its benefits *and* raise a large family.

At no stage so far has there been any definite attempt by government to persuade people to have smaller families, although deliberate, organized attempts to lower the fertility level can only come from such a source. If it is possible to control personal spending, by such measures as manipulating the bank rate, then it is not unreasonable to suppose that fertility level could also be varied, *in either direction*, by the thoughtful use of economic measures as well as by direct persuasion. It must be regarded as short-sighted policy in the long run not to do so.

Populations are balanced on a knife edge and the larger they become the more this balance becomes critical. If too many children are born the population outruns its supply of space, food and housing. If too few, then it becomes unbalanced because of its high proportion of old people.

One of the lessons we have learned in the last decade or so is that forecasts about population growth in the short-term are very unreliable. No one has any idea what factors are important in regulating fertility levels. We do not know what effects the forecasts of growth rates have on the growth rates themselves. The birth rate in Britain has been falling steadily since the mid-sixties. There is no acceptable explanation for this, certainly none that can be verified scientifically. Recently, for the first time, birth rate has fallen below the death rate. Whether this is the beginning of a long-term slow fall in population or a short-term fluctuation in birth rate we are in no position to know, as long as we, as a society, take no responsible collective action, and are content to leave matters to take care of themselves.

Population control in other countries

Most advanced countries are confronted with the problem of rising populations and the need to reduce their fertility levels. There seems to be no real reason why they should not do so. They all have the starting advantage of a literate population with a high standard of living and, as the example of Japan shows, quite remarkable results may be achieved if the desire is there. The Japanese, at first through the efforts of private organizations, and later by government action, succeeded in reducing the birth rate to less than half its 1948 level in about fifteen years. This was done by mounting a massive propaganda campaign, which was backed by immense technical resources

and medically trained personnel.

The situation in other parts of the world which have a population problem (South-East Asia, China, Indonesia, Africa and South America) is rather different. All these areas were subject to development, or exploitation, initially by Europeans for their own purposes. As a result they acquired some of the benefits of medical science without acquiring the industrial knowledge and machines to provide for the resulting increase in population. The results can clearly be seen in India, where the major barrier to effective fertility control is an appallingly low standard of living. The government of India is committed to a policy of population control but it is handicapped by a high level of illiteracy among its people and an acute shortage of trained medical staff. In India's case there can be no waiting for standards of living to rise in order to lower the fertility rate.

CONSERVATION—CONCLUSIONS

Many of the problems of conservation on the larger scale are problems which lie beyond the scope of action by the individual. Nevertheless, the sort of action a government may take in a democratic community is largely determined by what its people will accept, or, more positively, by what they want. The basis for constructive action must clearly be informed and educated public opinion. Even in these days biology is still too often thought of as a study of buttercups, or the dissection of dogfish, and a surprisingly large number of people have no idea how important biology is and how it affects their lives. It is still too often just another subject studied by young people who see no relevance in it other than as an examination subject. The possession of an ordinary level pass in biology is no guarantee that its owner will throw less litter about, or uproot fewer bluebells. It is an extraordinary fact that every year millions of people visit the national parks, such as Dartmoor or Snowdonia, or the coast, in order to enjoy their beauty, yet see no inconsistency in leaving such places strewn with the litter of modern living—beer bottles, cigarette cartons, ice cream wrappers and so on. Learning biology means more than just learning facts. It also means applying them with sense and understanding. It isn't only the scientist who often seems to stand apart from the object of his studies. A large number of ordinary people also forget that they are a part of the environment in which they live. The world is our home and we should not disfigure or abuse it, for we shall be the losers in the end.

1   Try to arrange to separate off a portion of the school garden, or your own back garden, if your parents are agreeable. Dig it over and remove all obvious signs of plants. Leave it undisturbed for as long as possible. Keep a careful record of all the developments that take place in it.

2 Make a continuous study of a small habitat. A rainwater butt or an aquarium with some rainwater in it, placed out of reach of larger animals, will do. Carry out regular sampling of its contents and record any changes occurring in it.

3 Visit an oakwood. Make lists of the different plants in it, under separate headings for trees, shrubs and ground cover. Notice any significant differences in the vegetation in different parts of the wood. How do you account for these?

4 Take a short country walk, armed with notebook and pencil. Make a list of all examples of land misuse you see, e.g. litter, unsightly features such as power lines and tips, or neglected fields.

INDEX

Page numbers in bold type deal with major references or with references which continue on closely following pages; numbers in italics refer to diagrams.